THE POWER OF DESIGN

THE POWER OF DESIGN
PRODUCT INNOVATION IN SUSTAINABLE ENERGY TECHNOLOGIES

Edited by

Angèle Reinders

*Delft University of Technology, Faculty of Industrial Design Engineering,
Design for Sustainability, Delft, The Netherlands
University of Twente, Faculty of Engineering Technology, Department of Design,
Production and Management, Enschede, The Netherlands*

Jan Carel Diehl

*Delft University of Technology, Faculty of Industrial Design Engineering,
Design for Sustainability, Delft, The Netherlands*

Han Brezet

*Delft University of Technology, Faculty of Industrial Design Engineering,
Design for Sustainability, Delft, The Netherlands*

A John Wiley & Sons, Ltd., Publication

Graphic design illustrations by A. Sesink and J. Sesink (2012)
Cover illustration: Virtue of Blue (2010) Design by Jeroen Verhoeven, DeMakersVan, Rotterdam, The Netherlands

Library of Congress Cataloging-in-Publication Data

The power of design : product innovation in sustainable energy technologies /
edited by Angèle Reinders, Jan Carel Diehl, Han Brezet.
   p. cm.
  Includes bibliographical references and index.
  ISBN 978-1-118-30867-7 (hardback)
  1. New products–Environmental aspects. 2. Product design. 3. Green products. I. Reinders, Angèle.
II. Diehl, Jan Carel. III. Brezet, Han.
  TS170.5.P69 2012
  658.5'752—dc23

                  2012014110

A catalogue record for this book is available from the British Library.

Print ISBN: 9781118308677

Set in 10/12pt Times-Roman by Thomson Digital, Noida, India.
Printed and bound in Singapore by Markono Print Media Pte Ltd

Contents

Case A   SolarBear: Refrigeration for the Base of the Pyramid through Adsorptive Cooling   243
Leonard Schürg, Jonas Martens, Roos van Genuchten and Marcel Crul

Case B   Environmental Impact of Photovoltaic Lighting   253
Bart Durlinger

Preface

The development of sustainable energy solutions needs an integrated view. Therefore, with *The Power of Design* we intend to fill a void in existing literature at the intersection of innovation processes, sustainable energy technologies, product development, and user behavior. As such, we aim to offer a unique publication in "product innovation in sustainable energy technologies and energy efficiency" to support recent and new developments in this growing field of industrial design engineering and research.

Many publications have been written about product design and innovation, sustainable energy technology, or user interaction. However, a book that integrates these topics and that fulfills my own need to transfer knowledge to students as an educational instructor and research supervisor in the field of industrial design engineering has not yet existed.

So far books that relate to sustainable product design usually cover the design process and the sustainability of products in relation to their potential environmental effects. Other publications on sustainable design supply information about green buildings, architecture, urban planning, and ecological landscapes, whereas our interest typically focuses on products – ranging from consumer products and vehicles to building components – but not the building as a whole.

Apart from this, existing books on product innovation tend to focus on innovation management and didn't cover our need for information and theories that could enhance creativity and innovation inside the design process. If we look at publications about sustainable energy technologies, it can be concluded that they provide information about the physical functioning of the technologies and systems' engineering, whereas the integration of these technologies in products and situations is hardly covered.

Summarizing, we could not identify a book that touches the integration of sustainable energy technologies in a product context and the related entwining of product shaping and styling, user experiences, financial aspects, environmental impacts, and the eventual manufacturing, and application of these products.

Therefore, we decided to develop and write such a book based on past experiences with courses and projects in the field of industrial design engineering at University of Twente and Delft University of Technology. For the reasons mentioned in this preface, we hope to fill a void in the existing literature on the application of sustainable energy technologies in a product context.

Due to the book's methodological approach, the reader will be able to apply its information and examples to his or her own research projects or product design processes. Therefore, you could consider this work as a reference and practical guide. The work particularly targets third- and fourth-year students and postgraduates in energy and environment and product design and engineering programs, as well as master's students studying industrial design engineering, manufacturing engineering, sustainable energy technology, environmental engineering, mechanical engineering, and electrical engineering. In addition it could be a practical guide for academic staff members involved in research, industrial design engineers, engineers in industry related to product development in sustainable energy technologies, consultants in the field of product development and sustainable energy, and architects.

We wish that the information and examples offered by *The Power of Design* will inspire the reader and initiate enthusiasm about the application of sustainable energy technologies in a product context, because we are looking forward to new and innovative products with integrated energy solutions that will reduce energy impacts in this Carbonaceous era of human history.

Angèle Reinders
Professor of Energy-Efficient Design
Design for Sustainability Program
Delft University of Technology

Department of Design, Production and Management
University of Twente

Foreword

In the second decade of this century, we are on the edge of a revolution in which all aspects of our existence will be reconceived and reformed. What will probably become known as the *Carbonaceous era* of human history is coming to an end – roughly two centuries of development based on mining millions of years' worth of long-dead animals and plants, the fossils we turned into fuels. The legacy of that carbonaceous power can be found embedded in our material world, in its physical infrastructure, and in the products, services, and systems that underpin our economies and provide for our lifestyles. But fossil fuels are a finite resource, and the demand for oil is already exhausting the easily available supply with growing economic and geopolitical repercussions. All that previously locked-up carbon we have released into the atmosphere since the Industrial Revolution is heating the biosphere and changing our climate, threatening all aspects of human life that are dependent on suitable climatic conditions.

The consequences of continuing with a carbon-based economy are unthinkable, but centuries of dependence are clearly not easy to overcome. Established systems of power have disrupted progress on global action over many decades, in spite of increasingly urgent projections from climate science. Until the turn of the twenty-first century, the consensus was that we needed action to keep the rise in global temperature below 2.0°C (compared to pre-Industrial Revolution levels). But by 2011 the International Energy Agency, in its *World Energy Outlook*, was warning that "radical" policy changes for a transition to a clean-energy future are required *before 2015* to avoid global temperature rises of 3.5°C or higher. Nevertheless the world economy is beginning to shift rapidly in unexpected ways; in that same year, more than 50% of total global investment in renewable energy took place in China.

The work presented in this handbook is from one of the leading centers in the world on design and innovation for sustainability. It is an essential guide for those designers, engineers, architects, and businesses that understand the importance – and are attracted by the challenge – of building the post-carbon future now. Whilst the range of sustainable energy technologies is generally well known, the design exploration of their contributions to the emerging future has been slow, compared to the momentum of "normal" product development. As the work in this book demonstrates, this situation is changing, but it is not simple; it requires much more innovative and creative investment than simply taking existing goods and services and replacing fossil energy inputs with renewable ones. The *systems* of resource, product, and service provision that are essential for our social existence are fossil fuel dependent, and the relatively low cost of energy compared to labor and land has left us with an energy-intensive, grossly energy-wasteful existence. Creativity and innovation must be directed to designing new products, systems, and user behaviors and lifestyles that treat energy as a precious resource and energy efficiency as an unquestioned goal. That means changes to design and innovation processes, and the authors of this book talk with real authority about the challenges and changes necessary to integrate sustainable energy technologies into product innovation.

At the core of much of the global conflict over action on climate change are the reality of global inequalities and the absolute necessity to define a trajectory for a post-carbon future that is also compatible with the millennium goals, offering a clear path for poverty alleviation and equitable development. That is an immediate economic challenge for the West (for the beneficiaries of carbonaceous development), but it adds another dimension to the challenges for innovation and design that is expertly canvassed in this book.

The *Power of Design* should have wide appeal. In the context of global doubt about the possibility of changing direction and breaking our long addiction to fossil fuels, the practice, concepts, and case studies presented by the authors are glimpses of an alternative future that is both achievable and desirable. I am reminded of a comment attributed to William Gibson: *The future is already out there, it just isn't evenly distributed.* This work is a much-needed contribution to redistributing the future – whilst we still have time.

Professor Chris Ryan
Director Eco-Innovation Lab
University of Melbourne
Australia

Acknowledgements

We greatly acknowledge the continuous assistance of Juan Manuel de Borja, who during a one-year period assisted the editors with the realization of the manuscript of *The Power of Design*.

Also we wish to thank Delft University of Technology and University of Twente for their support during the period in which the manuscript was written.

We would like to acknowledge many colleagues from industry who recognized the relevance of industrial design engineering for energy technologies and who provided access to their projects.

For their idea-formulating support in the very early stages of our methodological approach to design processes with sustainable energy technologies, we would like to thank Eric von Hippel of MIT Sloan School of Management in Boston, Arie Rip of University of Twente, and Joop Halman of University of Twente.

This work would not have existed without the inspiring contributions of many authors. Here we would like to thank with deep gratitude Chris Ryan, Stefan Kuhlmann, Joop Halman, Valeri Souchkov, Jan Buijs, Wouter Eggink, Peter Joore, Joop Schoonman, Wilfried van Sark, Frank de Bruijn, Paul Kühn, Arjan de Winter, Bram Entrop, Arjen Jansen, André de Boer, Alexandre Paternoster, Pieter de Jong, Daphne Geelen, David Keyson, Bart Durlinger, Bas Flipsen, Nils Reich, Michael Thung, Leonard Schürg, Jonas Martens, Roos van Genuchten, Marcel Crul, Satish Kumar Beella, Sacha Silvester, and Jeroen Verschelling. In the last phase of the preparation of *The Power of Design*, Anke Sesink and Joska Sesink did a great design job with regards to the infographics and several illustrations.

Also the secretaries of the Department Design, Production and Management of University of Twente are gracefully thanked for their support. Most of all we thank all the students at Technical University of Delft and University of Twente, whose work in our master's classes and graduate projects forms the basis of this book. Their enthusiastic participation resulted in scintillating (classroom) discussions, as well as many interesting product concepts partially exposed in this book and partially represented in design research papers.

About the Editors

Angèle Reinders

Since 2010 Angèle Reinders (1969) has been a part-time professor of energy-efficient design in the Design for Sustainability (DfS) program of the Faculty of Industrial Design Engineering at the Delft University of Technology, the Netherlands. Since 2002 she has been teaching and conducting research in the Department of Design, Production and Management of the University of Twente with a focus on innovative product design for integrated sustainable energy technologies. Throughout her career, Angèle Reinders has had an abiding interest in sustainable energy. Prior to joining University of Twente she worked at Delft University of Technology, where she investigated low-power energy supplies for portable applications.

In 2001, Angèle Reinders was a scientific advisor in the Asia Alternative Energy Program of the World Bank in Washington, D.C. In due course, she was a guest researcher at Fraunhofer Institute in Freiburg (1997), the Ministry of Technology in Jakarta (1998), and the Italian National Agency for New Technologies (ENEA) in Naples (2010).

From 1993 to 2000, she was associated with the Department of Science, Technology and Society of Utrecht University, where she investigated, among other things, the energy consumption of households in the European Union and life cycle aspects of energy systems for rural electrification. She received her doctoral degree from the Faculty of Chemistry at Utrecht University in 1999. Her PhD dissertation covers the analysis and simulation of the field performance of photovoltaic solar energy systems. Angèle Reinders completed a master's degree in experimental physics in 1993 at Utrecht University, specializing in material physics and energy physics. At present she is a co-founding editor of the Journal of Photovoltaics.

Jan Carel Diehl

After finishing his study in industrial design engineering, Dr. Jan Carel Diehl (1969) worked for several years as an ecodesign consultant. In his present position he is assistant professor for the DfS program at the Faculty of Industrial Design Engineering at the Delft University of Technology in the Netherlands. Within the DfS program he manages international projects on sustainable product innovation, especially in emerging markets. The main focus of his research is the knowledge transfer and implementation of sustainable product innovation into an international context. In addition his research has a special interest in cultural differences in product design and in developing products for the so-called base of the pyramid (BoP). Alongside his position at TU Delft, he is a consultant for the United Nations Industrial Development Organization (UNIDO) and United Nations Environment Programme

(UNEP) and an invited lecturer at several international universities. He is co-author of the UNEP Design for Sustainability (D4S) manual for developing economies (D4S EE).

Han Brezet

Han Brezet (1951) is research and graduate school director of the Faculty of Industrial Design Engineering (IDE) at the Delft University of Technology and professor at TU Delft's DfS program. He holds an MSc in electrical engineering from TU Delft, where he graduated in 1977 after performing research on the issue of integration of wind power in the grid. He completed his PhD in energy innovation at Erasmus University in Rotterdam. Since 1992 he has been tenured professor in sustainable product design at the IDE Faculty, where he has published more than 150 scientific and conference papers and some standard books in the DFS area after establishing the DfS program. This program was among the first in the world for life cycle–oriented applied ecodesign: the development of useful products with decarbonization, detoxification, and dematerialization as co-drivers. Among its outputs are about 30 doctoral dissertations and a few hundred MSc theses. Today, the program includes concepts and themes like product–services systems, energy-efficient and renewable energy–based products, natural materials–based products, and product design for emerging markets. Since 2012 Han Brezet has held a position as visiting Professor at the Aalborg University (Denmark).

About the Contributors

Satish Kumar Beella

Satish Kumar Beella holds a master's degree in industrial design from Indian Institute of Technology, Delhi, India. Beella's present research activities are mainly based on sustainable transportation innovations. Since 2005, he has been busy with design-based PhD research at Delft University of Technology (TU Delft) concerning alternatives for urban car transport and solutions for chain mobility options.

Jan Buijs

Jan Buijs has been for more than 25 years full professor and chair in product innovation and creativity at the Faculty of Industrial Design Engineering (IDE) of Delft University of Technology. He was educated as an industrial design engineer (MSc, Delft, 1976), and received his PhD (also at Delft) in 1984. Before working at TU Delft, he spent 10 years as a management consultant.

Marcel Crul

Marcel Crul, MSc and PhD, is research coordinator at the Design for Sustainability Programme, Industrial Design Engineering, Delft University of Technology. His research foci are on sustainable product innovation, inclusive design, and sustainable consumption in emerging economies. Currently he works on projects in Southeast Asia and North Africa. He is also involved in several research projects on sustainable living and working in the European Union.

Frank de Bruijn

Frank de Bruijn (PhD, chemical engineering) has been employed at Nedstack since 2010 as chief technology officer. In this position, Frank is responsible for the research and development of new products at NedStack.

Before joining Nedstack, he was responsible for the hydrogen, fuel cell, and CO_2 capture development of the Energy Research Center of the Netherlands (ECN). His expertise is in proton exchange membrane fuel cell (PEMFC) technology, on which he has worked since 1996; on heterogeneous catalysis; and on hydrogen technologies. Currently he is a member of the Grove Committee, which organizes the international Grove Fuel Cell Conference;

member of the Dutch Platform Nieuw Gas, a national think tank on the Gas Transition; and member of the advisory board of the scientific journal *ChemSusChem*. In 2009 he was appointed as a professor of gas conversion technology at Groningen University. He is the author of more than 30 scientific papers and co-inventor of two patents.

André de Boer

André de Boer studied mechanical engineering at the University of Twente, where he specialized in biomedical tribology. In 1979 he joined the department of Anatomy and Biomechanics of the Vrije Universiteit. In 1981 he started his PhD research at the department of Dental Physics at the State University Utrecht. In 1987 he obtained his PhD with the thesis "Mechanical Modeling and Testing of the Human Periodontal Ligament In Vivo." In 1985 he joined the National Aerospace Laboratory (NLR) to work on the dynamic behavior and stability of aerospace structures. Later he was involved in projects on thermomechanical analyses of gas turbine components and acoustics. In 1997 he became leader of the Structural Mechanics Group of University of Twente, and in that period he was also involved in the Glare certification program. In 2000 he was appointed professor of applied mechanics at the University of Twente. There he teaches students mechanical engineering and industrial design engineering structural mechanics, and is leader of the Structural Dynamics and Acoustics group at the faculty of Engineering Technology.

Arjan de Winter

Arjan de Winter has a background in mechanical engineering and finished his PhD in 1998 in a cross-over between sustainability and manufacturing. In this context, he spent a year in Japan as a post-doc at the Ministry of International Trade and Industry's (MITI) Mechanical Engineering Laboratory. Intrigued by the creation of new products, he joint Philips Applied Technologies (Apptech) as a consultant for design for manufacturing, focusing on the application of new technologies in new products. One of the early projects he participated in was the development of one of the first liquid crystal display (LCD) TVs. Lighting became a dominant factor when Arjan took up responsibility for Apptech's Lighting department's project portfolio. In 2007 Arjan joint Lighting as a group leader in product design in Lighting's front-end innovation group LightLabs, to actively drive the transition to light-emitting diodes (LEDs) and the consequent digitalization of lighting products. Focus areas so far have been to set up technology strategies, to translate these into product roadmaps, and underpinning development programs for the consumer (home) and professional retail market.

Bart Durlinger

Bart Durlinger started working at the Center for Design of RMIT in 2010. He has completed a master of science in sustainable energy technology and a bachelor of science in mechanical engineering (both from the University of Twente in the Netherlands). Recently he started working at the Center for Design of RMIT in Melbourne in the field of life cycle assessment (LCA). He has worked on LCA studies for several industry sectors, including the (bio)plastics, building, and recycling industries. He has also been involved in development work for the Packaging Impact Quick Evaluation Tool (PIQET). His interests are in the methodological aspects of LCA, its applications in industrial processes and complex systems, renewable energy technologies, and waste management.

Wouter Eggink

Wouter Eggink has been assistant professor at the University of Twente since 2007. He is especially interested in the relationships between design, technology, and society. Besides his research activities, he has taught this subject for a couple of years at the bachelor's and master's programs of Industrial Design Engineering in several courses.

Before Wouter started work at the university, he was employed as a designer and project manager for almost 7 years at D'Andrea & Evers Design, a design agency located in the Netherlands. At this design office he specialized in emotion-related product design, with special attention to the meaning of product design from a user's perspective. Before that, he worked for several years at Hollandse Signaalapparaten (now Thales Netherlands) as a mechanical system design engineer responsible for the styling of sensor systems and command centers for marine vessels. Wouter graduated in 1996 at the Delft University of Technology in industrial design engineering and obtained his PhD in design at the University of Twente in 2011.

Bram Entrop

Bram Entrop (1980) is assistant professor at the department of Construction Management and Engineering of the University of Twente, Enschede, the Netherlands. At this same university he studied civil engineering and management, and graduated in 2004 on the development of a three-step method for sustainable land use: the Trias Toponoma. After his graduation he worked as a self-employed consultant in the field of sustainable building, and he worked at the municipality of Hengelo as a sustainable building inspector. Since 2006 he has worked at the university as a PhD student and subsequently as an assistant professor. His research activities involve the process of adopting and the financial benefits of energy techniques and measures in the built environment. Furthermore, he conducts experimental research on the use of phase change materials in residential real estate in the moderate Dutch climate.

Bas Flipsen

Bas Flipsen is an assistant professor at the Delft University of Technology, faculty of Industrial Design Engineering. In 2010 he finished his PhD research on direct methanol fuel cells (DMFCs) for small portable electronic devices. He developed a design tool that generates a design of a DMFC hybrid power system for a specific device to evaluate its feasibility compared to the regularly used lithium-ion battery. With his background in aeronautical engineering, he is familiar with mathematical modeling in design. As an industrial design engineer, he recognizes the user's influence on a design's success. Both these qualities are reflected in his work as a researcher and teacher.

Daphne Geelen

Daphne Geelen (MSc, 2006) graduated on the integration of photovoltaic solar cells in consumer products at the faculty of Industrial Design Engineering of Delft University of Technology. Subsequently she continued in this area in the joint program of the Design for Sustainability program and the 3TU Cartesius Institute. She was involved in establishing a network of public and private actors for photovoltaic solar-powered product development and tutoring several product design projects involving the integration of photovoltaic solar energy in portable consumer products.

Currently she is working on her PhD research at the DUT DfS program concerning the role of social interaction in energy-related consumer behavior, more specifically focusing on the application of social interaction including gaming in product–service system design in the field of local energy production and consumption.

She organizes the Livinggreen Labs and tutors student projects that are carried out in the context of the Livinggreen.eu project.

Johannes Halman

Johannes I.M. Halman is a professor in technology innovation and risk management at the University of Twente, Enschede, the Netherlands. He also holds an endowed Schlumberger chair in Technology and Sustainable Development at the University of the Netherlands Antilles (UNA). He is the scientific director of the master of science program in Risk Management at the University of Twente.

Halman earned an MSc in construction engineering from Delft University of Technology, an MSc (cum laude) in business studies from Rotterdam School of Management at Erasmus University, and a PhD in technology management from Eindhoven University of Technology. His research interests are in the fields of innovation management, technology entrepreneurship, and risk management. He has advised international firms like Philips Electronics and Unilever on the worldwide implementation of risk management strategies within their innovation processes. Prior to his academic career, Halman worked for several years in practice. As a project manager and senior management consultant, he has initiated and conducted several construction, (re-)organization, and innovation projects. Halman has published about 150 scientific articles, mostly in international refereed scientific journals. He has received different international awards, for both his research work and his education.

He is responsible for research in management of product innovation, creativity, and multidisciplinary design and innovation teams. He teaches strategy, brand and project management, creative problem solving, innovation, and integrated new product development. He is active in both the bachelor program and the Strategic Product Design master program. He has supervised more than 230 graduates and 14 PhDs.

He has published numerous papers and articles as well as three books. His first book, *Innovation and Intervention* (1984), was awarded best management book of the Netherlands in the year 1985. His latest book, co-authored by Rianne Valkenburg, is *Integrated New Product Development*, now in its third revised and enlarged version (2005). This book has become the standard book on new product development in the Netherlands.

During the years 1992–1996, he was member of the board of the Netherlands Association of Certified Management Consultants (Ooa). From 2000 to 2007, he was chairman of the European Association for Creativity and Innovation (EACI), a nonprofit organization that stimulates debate and discussion among professionals and academics in the creativity and innovation field. And since 1991 he has been a member of the editorial board of Creativity & Innovation Management published by Blackwell in the United Kingdom.

Arjen Jansen

Arjen Jansen graduated in 1988 at Delft University. He has working experience as both a freelance industrial designer and a design consultant for various international companies. He was appointed assistant professor at Delft's Faculty of Industrial Design Engineering in 1994. Since then, his teaching activities have included design for sustainability, product engineering, mechatronics, and product design with a special focus on (outdoor) sports-related issues. Until 2011, Arjen guided some 75 MSc graduate students in various areas of industrial design engineering. In 2011 he published and defended his PhD thesis, "Human Power Empirically Explored."

Peter Joore

Peter Joore graduated as an industrial design engineer at the Delft University of Technology. After working as a product designer in industry, he worked as a senior researcher and business consultant at the Netherlands Organization for Applied Scientific Research TNO. Here he initiated and coordinated various national and international multidisciplinary innovation projects, applying state-of-the-art technology in the development of radically new products, services, and systems. In parallel he worked on his PhD research with the Design for Sustainability research group at Delft University of Technology. In 2010 he finished his thesis discussing the mutual influence between new products and societal change processes. He currently works as a professor in the area of open innovation at the NHL University of Applied Science in Leeuwarden, the Netherlands.

David Keyson

David Keyson heads the program in Sustainable Living and Work at the Faculty of Industrial Design Engineering at the Delft University of Technology, and also leads the research focus on Social Contextual Interaction Design as part of the ID Studio Lab. His educational work focuses on interactive technology design in the context of smart products and environments. Prior to joining TU Delft, he worked at Philips Research as a senior research scientist in media interaction and at Xerox in the department of Industrial Design and Human Interface. He holds a PhD from the Technical University of Eindhoven in perception and technology and an MSc in ergonomics from Loughborough University.

Professor Keyson has authored over 100 scientific papers, including papers in presence, ergonomics, applied ergonomics, displays, universal access information society and personal ubiquitous computing. He holds 16 patents relating to input devices and design principles of multimodal user–system interaction.

Stefan Kuhlmann

Stefan Kuhlmann is chair of the Department of Science, Technology, and Policy Studies (STePS), a member of the university's Institute for Innovation and Governance Studies (IGS), and leader of the Twente Graduate School program for Governance of Knowledge and Innovation.

He is a political scientist and also studied history (1978, University of Marburg, Germany); in 1986 he received a PhD in political science at University of Kassel, Germany; in 1998 he got a "habilitation" (second doctorate) in political science, also at this university. Since 1979, he has been involved in studies of research and technological innovation as social and political processes – with changing entrance points and perspectives. During the last two decades, he has analyzed science, research and innovation systems, and public policies, focusing on the dynamics of governance. Until the summer to 2006 he was managing director of the Fraunhofer Institute for Systems Innovation Research (ISI), Germany, and professor of innovation policy analysis at the Copernicus Institute, University of Utrecht, the Netherlands.

Recent research projects include study of "Knowledge Dynamics and ERA Integration" (funded by the EU PRIME Network of Excellence); "Inter-institutional Collaboration in the German Public Research System: Analysis with a Focus on Nanotechnology" and "Governance of International Research Collaboration in Nano S&T in Europe" (funded by Deutsche Forschungsgemeinschaft); and "The Societal Component of Genomics Research" (with the Department of Innovation Studies, Utrecht University; funded by the Netherlands Organization for Scientific Research).

Paul Kühn

Paul Kühn is with the Fraunhofer Institute for Wind Energy and Energy System Technology (IWES) in Kassel, Germany, and works in the Department of Energy Economy and Grid Operation. Mr. Kühn joined the institute in 2006 after completing his MSc in energy conversion and management from the University of Applied Sciences, Offenburg. In 2010 he became head of the R&D Group Reliability and Maintenance Strategies at Fraunhofer IWES.

His main research interests are in small wind turbines, and he is currently pursuing a PhD on the subject at Fraunhofer IWES/Kassel University. Other research topics include reliability analysis and maintenance strategies for wind turbines.

Jonas Martens

Jonas R.J. Martens, MSc, graduated in 2010 at the faculty of Industrial Design Engineering at Delft University of Technology. He was founder of the SolarBear initiative in 2008 and of Rikkert + Jan design and development in 2010, and he was co-founder of Better Future Factory in 2012, a design thinking collective for social innovation. He has worked with Delft University of Technology, the Lowlands Festival 2012, Studium Generale, and the Leiden municipality.

Alexandre Paternoster

Alexandre Paternoster completed the first 4 years of general science engineering in 2007 at the ENSCPB - Bordeaux, France. In 2008 he received his MSc degree with honors at the composite group of Imperial College London, United Kingdom. In February 2009 he started as a PhD student at the University of Twente, in the Structural Dynamics and Acoustics group under the supervision of Prof. Andre de Boer. He currently investigates actuation systems for morphing concepts in the scope of the Green Rotorcraft project in the framework of the Clean Sky Join Technology Initiative.

Pieter de Jong

Pieter de Jong is a PhD researcher at the University of Twente in the Netherlands, at the Structural Dynamics & Acoustics group. He acquired his MSc degree in Mechanical Engineering at the same university in 2008. He is participating in the Clean Sky project which is investigating the potential for self-powered health monitoring systems within the rotor blades of helicopters. His research concerns power harvesting within the rotor and he has explored a number of possibilities. Mechanical as well as electrical aspects are considered in optimizing the output. He will defend his PhD thesis early 2013.

Nils Reich

Nils Reich graduated from the Fachhochschule für Technik und Wirtschaft Berlin, Germany, with a thesis on a photovoltaic–wind–diesel–fuel cell hybrid system in 2003. He received his PhD in 2010 from Utrecht University, the Netherlands, with a thesis on "product-integrated photovoltaic systems." He now is employed at Fraunhofer ISE, Germany, where he is head of a team conducting PV yield assessments and certification, dealing with large-scale PV power plants.

Joop Schoonman

Joop Schoonman, applied sciences professor of inorganic chemistry at Delft University of Technology, was voted the best Dutch materials scientist of the past 80 years in the Union of Materials Science's "2006 Dutch Masters in Materials" election.

As a chemist, Schoonman's research particularly focuses on sustainable energy, an area in which he applies his expertise in the field of solid-state material chemistry. His research has led to advances in the development of batteries, solar cells, fuel cells, and hydrogen storage. Schoonman has 435 scientific publications to his name.

Schoonman studied chemistry at the University of Utrecht (1967). He obtained his PhD in 1971 and remained at the university as a researcher until 1984, when he became professor of inorganic chemistry at TU Delft. He has served as director of science for the Delft Center for Sustainable Energy since 2002.

Leonard Schürg

Leonard Schürg (MSc, Industrial Design Engineering, Delft) graduated cum laude in 2010. During his studies he founded two companies: Pointtwenty, a surfboard design office, and

Numboards, a foam-cutting company. Working on the SolarBear project shifted his interest toward renewable energies, and currently he works at a PV start-up in Germany.

Sacha Silvester

Sacha Silvester pursued his master's degree in industrial design engineering at TU Delft (1972). After that, he worked as a musician at the Theatre Company Diskus and as a research associate at the faculties of Economics and Social Sciences at the Erasmus University Rotterdam. He was co-founder of the Erasmus Centre of Environmental Studies (ESM) at the same university. In 1996 he received a PhD in social sciences at the Erasmus University Rotterdam on a series of research projects studying the effects of demonstration projects on the adoption and diffusion of energy-efficient housing in the Netherlands. In 1997 he joined the faculty of Industrial Design Engineering as an associate professor in the Design for Sustainability Program.

Currently he is associate professor at the department of Industrial Design at the same faculty, leading the applied research theme of personal mobility, and is acting chairman and member of the board of D-incert, a nationwide platform for electric mobility.

Valeri Souchkov

Valeri Souchkov has been involved in developing TRIZ and systematic innovation since co-founding the Invention Machine Lab in 1989. He was certified as a TRIZ specialist by the founder of TRIZ, Genrich Altshuller, in 1991.

Valeri started promoting TRIZ and systematic innovation in Western Europe in 1993 when he introduced public and in-house TRIZ training in the Netherlands on a regular basis. In 2000, he co-founded the European TRIZ Association (ETRIA). Currently he is a member of the Executive Board of ETRIA as well as a member of a Global R&D Council of the International TRIZ Association (MATRIZ).

Since 2003, Valeri has headed ICG Training & Consulting in Enschede, the Netherlands, where he trains and assists customers worldwide in TRIZ, systematic innovation, and technology roadmapping. He often speaks at international events and conferences, and is co-author of two books and over 70 publications. He is invited lecturer at the University of Twente and the University of Eindhoven.

Valeri holds an MSc degree in computer science and engineering from Belarus University of Informatics and Electronics.

Michael Thung

Michael Thung (MSc, 1985) studied industrial design engineering at University of Twente. He specialized in design and styling, and became interested in sustainability. Together with four students, he developed and gave a new course about sustainability in product design. Thung concluded his studies with a master's thesis considering PV technologies, innovation, and aesthetics. At present, he is a mechanical engineer in the field of product development.

Roos van Genuchten

Roos van Genuchten (MSc) earned a bachelor's degree in industrial design engineering in Delft, followed by a master's degree in strategic product design at the same faculty. After founding a company in consumer insights, The User's Advocate, during her master's studies and getting experienced in base-of-the-pyramid (BoP) projects in Medellín, Colombia, her thesis research was focused on developing the first market entry strategy for SolarBear for farmers in rural West Bengal in India. Currently she is working as a marketing manager in customer development at the Dutch Telecom company KPN.

Wilfried van Sark

Wilfried van Sark graduated from Utrecht University, the Netherlands, with an MSc in experimental physics in 1985; his MSc thesis was on the measurement and analysis of current–voltage (I–V) characteristics of c-Si cells. He received his PhD from Nijmegen University, the Netherlands, in 1989; the topic of his PhD thesis was III–V solar cell development, modeling, and processing. He then spent 7 years as a post-doc and senior researcher at Utrecht University and specialized in a-Si:H cell analysis and deposition (plasma chemical vapor deposition, both radio frequency and very-high frequency). After an assistant professor position at Nijmegen University, where he worked on III–V solar cells, he returned to Utrecht University with a focus on the (single-molecule) confocal fluorescence microscopy of nanocrystals. In 2002 he moved to his present position as assistant professor and photovoltaics (PV) group leader at the Copernicus Institute at Utrecht University, the Netherlands, where he performs and coordinates research on next-generation PV devices incorporating nanocrystals, for example luminescent solar concentrators, as well as PV performance, product design, life cycle analysis, socioeconomics, and policy development. He is member of the Editorial Board of Elsevier's scientific journal *Renewable Energy*, and member of various organizing committees of EU, IEEE, and SPIE PV conferences. He is author or co-author of over 200 peer-reviewed journal and conference paper publications and book chapters. He has co-edited three books.

Jeroen Verschelling

Jeroen Verschelling is an entrepreneur with ambitions, ideals, and a pioneering spirit. He is the co-founder and director of Kamworks Ltd. Before moving to Cambodia in 2008, Jeroen consulted renewable energy and energy conservation projects for Ecofys and other organizations like the ECN and UNDP in Europe, Africa, and the Middle East. Jeroen lived for several years with his family in a 100% solar-powered, passive house of his own design next to the Kamworks design assembly workshop in Cambodia.

1

Introduction: Challenges at the Crossroads of Energy and Design

Angèle Reinders[1,2] and Jan Carel Diehl[1]

[1]*Delft University of Technology, Faculty of Industrial Design Engineering, Design for Sustainability, Delft, The Netherlands*
[2]*University of Twente, Faculty of Engineering Technology, Department of Design, Production and Management, Enschede, The Netherlands*

1.1 Introduction

At present, many sustainable energy technologies have reached a level of maturity that allows for product integration; however, currently realized product applications are beyond the possibilities that are offered by these technologies. To better embed energy technologies in industrial product design, we believe that a better insight regarding energy technologies in relation to appropriately matched design processes will be necessary to lead to numerous new, sustainable, and energy-efficient products (and services) in the next decade. Therefore, our book provides information, theories, and examples that enable the reader to better use and apply sustainable energy technologies, matched with energy efficiency, in products in several different markets.

Product innovation in sustainable energy technologies integrates the subject areas that are relevant for the design of sustainable and energy-efficient products based on sustainable energy technologies. For this reason, it is an interdisciplinary field in its nature – it is all about energy, product design, and innovation. Therefore, in *The Power of Design* the following topics will be addressed:

- Design processes and innovation methods that can be applied towards the design of sustainable products for energy technologies, namely, platform-driven product development, the Delft innovation model, TRIZ (the theory of inventive problem solving), technology road mapping, constructive technology assessment, the innovation journey, design and styling, risk-diagnosing methodology, and the multilevel design model.

The Power of Design: Product Innovation in Sustainable Energy Technologies, First Edition.
Edited by Angèle Reinders, Jan Carel Diehl and Han Brezet.

- The functional principles of various sustainable energy technologies that are significant for product integration, namely, batteries, photovoltaic solar energy, fuel cells, small wind turbines, human power, energy-saving lighting, thermal energy technologies in buildings, and piezoelectric energy conversions.
- User behavior and energy use, from a theoretical point of view and from experiences in the practice of energy surroundings.

This chapter defines the context and scope of *The Power of Design*. As such, we will first illustrate the relevance of sustainable energy and energy efficiency. Next we will elaborate on the challenges of bringing energy-related aspects into product design. Subsequently the field of industrial design engineering will be described, as well as several approaches to design for sustainability in Western and low-income markets in emerging and developing economies.

1.2 Energy Issues: A Brief Explanation

Due to the large amount of media attention on energy issues, it is a familiar fact that world energy consumption is growing every day (Figure 1.2.1). If this trend does not change, global energy consumption will increase by 53% from 2008 to 2035 (IEO, 2011). However, large differences exist between countries; for instance, in Western countries (OECD countries), energy consumption will grow moderately by about 20%. On the other hand, in emerging economies (non-OECD countries), energy consumption will dramatically grow more than 80% over the next 25 years. A rising trend has also been observed for global household electricity consumption, which increased 3.4% per annum on average from 1990 to 2006, approximately 2% per year in Europe and 6% per year in non-OECD countries (OECD/IEA, 2009).

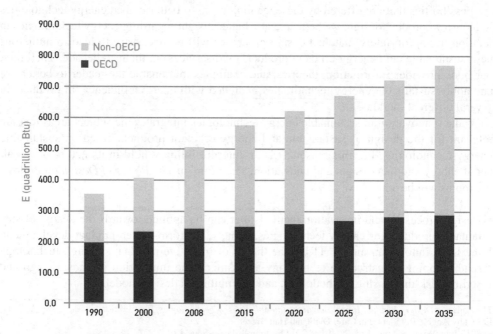

Figure 1.2.1   Global energy consumption in the period 1990–2035 (IEO, 2011)

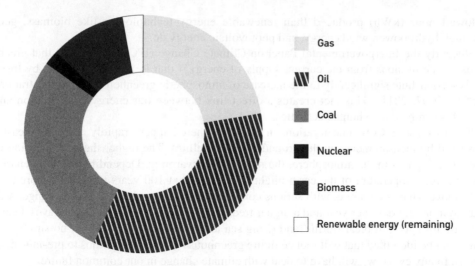

Figure 1.2.2   Share of energy sources in primary energy in 2008; data from IPCC (2011)

Currently the most commonly used fuels are fossil fuels: oil, coal, and gas. A relatively small share of our energy need is generated by renewable energy conversions. Renewable energy accounts for approximately 13% of the total energy mix, the majority of which can be attributed to biomass conversions and only 2.7% to other energy sources (IPCC, 2011; Figure 1.2.2). This distribution has a significant impact on global greenhouse gas emissions.

Figure 1.2.3 shows average greenhouse gas emissions due to the use of fossil fuels versus renewable energy technologies. Please note the logarithmic scale on the vertical axis; one can observe that fossil fuels emit approximately 10–100 times more greenhouse gases per

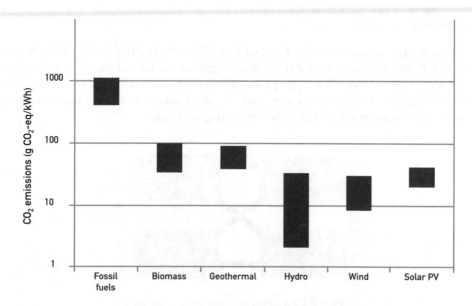

Figure 1.2.3 CO_2 emissions due to various energy conversions (Veldhuis and Reinders, 2011)

kilowatt hour (kWh) produced than renewable energy technologies like biomass, geo-thermal, hydropower, wind energy, and photovoltaic energy do.

Recently the Intergovernmental Panel on Climate Change (IPCC) concluded that green-house gas emissions from the current supply of energy – that is, energy produced by fossil fuels – contribute significantly to the increase of atmospheric greenhouse gas concentrations (IPCC, 2007, 2011). This fact creates a direct link between our energy consumption and global warming due to human activities.

The increase in CO_2 concentrations in the atmosphere happens rapidly and can be easily observed by measurements on the ground and by satellites. The higher the concentration of greenhouse gases in the atmosphere, the more global warming. Depending on the scenario, the average temperature of the earth might rise in the next 100 years to perhaps more than 5°C above what is common, with serious consequences for our climate. These changes will lead to more hot days per year and a higher frequency of extremes such as heat waves, heavy rainfall, droughts, tropical storms, and rising sea levels (IPCC, 2007). So far, no single sce-nario can be identified that will reduce future greenhouse gas concentrations to pre-industrial levels. In any event, we will have to deal with climate change in our common future.

Additionally, other important issues matter, such as our energy security. It is stated (MEZLI, 2011) that, globally, fossil fuels will be sufficiently available for the next 250 years, but the proven reserves of natural gas – at current production – will be sufficient for only 58 years. Also the availability of conventional oil reserves will decrease in the coming decades. There-fore, the price of fossil fuels is expected to remain high and will probably become higher.

To summarize, humankind emits too many greenhouse gases and the depletion of the proven reserves of fossil fuels is in sight. Moreover, significant effects of these developments are expected within this century. As such there is urgency to find sustainable energy solu-tions, because 100 years is a short period.

To find sustainable energy solutions, two possible paths are available:

- Deployment of new technologies.
- A change of lifestyle.

In various energy scenarios for the future (EPIA, 2011; IEA, 2010; IPCC, 2011; WBGU, 2003; WWF, 2011), it is assumed that 60–80% of greenhouse gas reductions in the course of this century will be drawn from energy supply and use, and industrial processes. These scenarios state that (1) renewable energy and (2) saving energy will play important roles in reducing and sustaining our energy consumption (Figure 1.2.4).

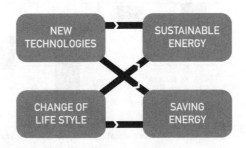

Figure 1.2.4   Pathways to energy solutions

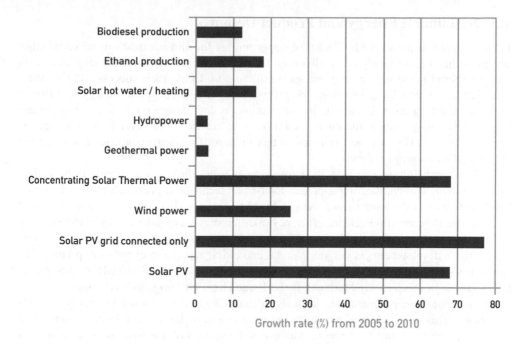

Figure 1.2.5 Global growth of renewable energy from 2005 to 2010; data from REN21 (2011)

One can think of solar energy conversions, wind energy, sustainable lighting, smart energy management, conversion of kinetic and thermal energy, new forms of energy storage, and hydrogen technology. And, indeed, in practice renewable energy is a growing market (Figure 1.2.5). In the period from 2005 to 2010, the growth rate of biofuels was around 15% and that of wind power was 25%, and various forms of solar energy conversions showed enormous growth rates between 65% and 80% (REN21, 2011).

As an example, we will shortly look at the Netherlands. The Dutch government has set targets for greenhouse gas mitigation for the short term aiming at 20% less CO_2 emissions in 2020 (compared to 1990). In the same year, 14% of the energy must be produced by renewable sources. Also the Dutch government is striving toward 20% energy savings in 2020, but this goal is not binding.

According to the Netherlands' Central Office of Statistics (CBS, 2011), industry is the largest consumer of primary energy in the Netherlands, with shares of 19% for transportation, 17% for households, and 16% for utility buildings. If this information is presented cross-sector wise, the following numbers can be derived:

- 35–40% of our energy is consumed in buildings.
- 17–19% of electricity consumption is consumed by lighting.

Adding the aforementioned 19% energy consumption due to transportation, we may conclude that at least 70% of the energy is consumed in the context of living, working, and transport. This finding applies generally to many countries, and this context is relevant for the application of sustainable energy in human livelihoods.

1.3 Sustainable Energy and Product Design

From a technical point of view, a lot of opportunities for and applications of sustainable energy technologies already exist. However, the way in which sustainable energy solutions will be offered to the market and end users will help to decide their success and thus their effectiveness in providing solutions for greenhouse gas reduction and energy security in the short and long term. Therefore, in our opinion product design could play an important role in developing sustainable energy solutions that can be incorporated into products or the environment. This will be explained in this section, which covers a brief history of the conceptual term *energy solutions*.

Until the early 2000s, as a rule, the expression *energy solution* was directly translated as *energy system*. This engineering approach focused on technical aspects, which may be translated more or less by a desire to increase a technology's energy yield. As an example, in the last 35 years due to successful R&D the efficiency of different types of solar cells was increased by a factor of approximately 4, leading to efficiencies of 40% and higher (Kazmerski, 2011). The end of these developments is not yet in sight, and efficiencies of over 50% are predicted in the nearby future. Similar upward trends in energy efficiencies can be observed for other renewable technologies such as wind turbines, fuel cells, and light-emitting diodes (LEDs).

At the same time, people aimed for a significant reduction in renewable-energy technology costs. This objective was accompanied by a vision of large-scale implementation of renewable energy. Figure 1.3.1 shows evidence of the success of this strategy in the reduction

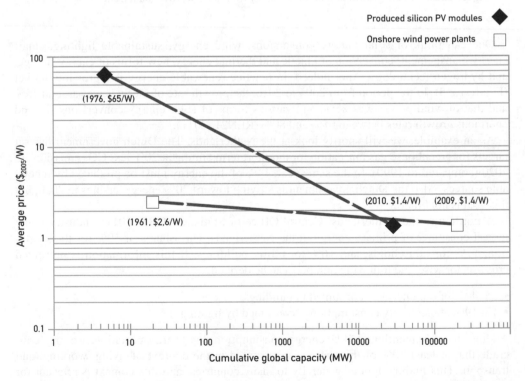

Figure 1.3.1 Price development of wind farms and PV modules versus cumulative capacity; data from IPCC (2011)

in solar panel and wind turbine prices versus their cumulative capacity; that is, in a period of 35 years the average price of photovoltaic (PV) modules went down with a factor of 50. A 50% cost reduction was achieved for wind power plants in a period of 30 years.

To stimulate the application and use of renewable energy, energy policy and legislation were developed and introduced. An example is the realization of feed-in tariffs for renewable electricity to the grid. These developments have become effective in many countries. In 2010, 96 countries had a policy for the use of renewable energy and 87 countries offered dedicated rates for renewable power fed into the grid (REN21, 2011).

Also, citizens influence energy choices. This was shown in March 2011 in Germany, where the social acceptance of nuclear energy was already under discussion. However, after the disaster of the Fukushima Daiichi nuclear power plant in Japan, the German population demanded with great protest an inherently secure and sustainable energy provision. This request was heard: The German government revised its policy to ensure the closure of 17 nuclear power plants by 2022.

Ultimately, the consumer – or end user – plays an important role in the acceptance of new forms of energy and energy efficiency, also at the level of use of energy systems. Through field research into the functioning of small-scale autonomous solar systems called *solar home systems* in Indonesia in the late 1990s (Reinders *et al.*, 1999), it was concluded that user interaction with an energy system can affect (1) the quality of the system's function, namely, electricity supply; and (2) the corresponding perception of usefulness and comfort by the end user (Figure 1.3.2).

User aspects will also become relevant with the introduction of Smart Grids. Smart Grids are power networks that intelligently integrate the behavior of all users – producers, consumers, and combinations of them – in order to efficiently provide sustainable, affordable, and reliable electricity (ETPS, 2011). The main assumption regarding smart grids is that the electricity sector will significantly become more sustainable over the next 10 years by the use of information and communications technology (ICT), renewable energy systems, and feedback on energy consumption for end users (IEA, 2011). The Smart Grids

Figure 1.3.2 Impressions of typical use circumstances of solar home systems in Java, Indonesia (pictures courtesy of Angèle Reinders) (See Plate 1 for the colour figure)

Figure 1.3.3 Different lighting products based on sustainable technologies, right: Virtue of Blue by DeMakersVan (2010) (See Plate 2 for the colour figure)

concept covers various aspects: (1) the distribution infrastructure (low and medium voltage), (2) distributed electricity generation, (3) energy management, (4) demand-side management among consumers, and possibly (5) energy storage. At the level of households and consumers, small-scale distributed energy systems in, on, and around the home – or at the district level – will become common in the nearby future. In this new situation, the end user will be confronted with various new energy technologies that will be introduced into the household, as well as with price effects due to the choice of purchasing electricity from the grid or creating self-generated power and the associated energy savings. The end user will have to learn to deal with this.

In summary, the term *energy solution* comprises energy's technical and financial aspects, societal aspects, and user aspects. Adding to this, one may assume that the visibility of small- to medium-scale energy solutions is increased by integrating them into a user environment, for which reason aesthetics will come into play. Examples of visible sustainable technologies such as LEDs, solar cells, and wind turbines are shown by different lighting products in Figure 1.3.3.

In the context of product development, the following aspects are all of equal importance for the design of a successful product: technologies, finance, society, human factors, and design and styling. Conversely stated, if these five aspects are applied in the development of an energy solution, one may speak of product design and product development. Therefore, we assume that an interdisciplinary approach to the aforementioned problems with energy consumption, durability, cost, social aspects, and users can create better solutions than using a monodisciplinary approach that focuses only on improving energy technologies. In other words, we assume that energy solutions based on these five aspects will be innovative (Figure 1.3.4).

An interdisciplinary design-based approach to energy is not just a throw in the air. It even seems necessary because the complexity, functionality, and user interaction of energy solutions are increasing as more products are integrated in an operating environment, also called *surroundings*. Figure 1.3.5 explains this statement by showing the development of energy technologies to energy systems, in this case a solar concentrator system, whose primary function is energy production. Subsequently integration takes place with products such as a fuel cell–powered car, or with surroundings like an interactive public lighting system.

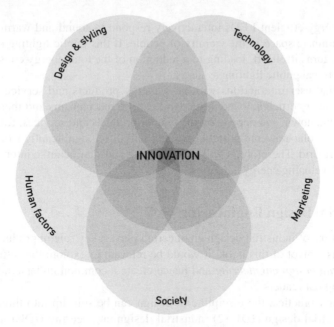

Figure 1.3.4 Innovation flower of industrial product design

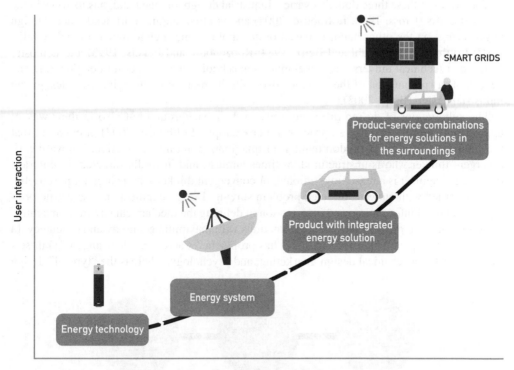

Figure 1.3.5 Energy solutions in relation to an increasing degree of integration in products or environments

In operation, energy-efficient LEDs interactively respond to sound and warmth and for this reason only illuminate space in the vicinity of people. If this public lighting system doesn't detect people, it turns itself off, leading to a reduction of the total energy consumption compared to conventional public lighting systems.

We foresee that user interactions with new energy products and services will strongly influence their energy efficiency and sustainability. The mechanisms and their effectiveness are still largely unknown, and depend on the type of products and services; for instance, user interaction with a solar-powered portable product will differ significantly from user interaction with a Smart Grid. Therefore, it will be interesting and significant to intensively consider the user when developing energy solutions.

1.4   Industrial Design Engineering

Since the discipline of industrial design engineering plays a key role in product development and hence forms a pivot of this book, it would be relevant to explore the setting and definitions of *industrial design engineering* and hence create a common understanding about this discipline among our readers.

In the scholarly practice, the discipline of design can be split up into three main subdomains: (1) industrial design (ID), (2) industrial design engineering (IDE), and (3) design engineering (DE; Figure 1.4.1).

The origin of these three domains varies. Industrial design, on one hand, has its roots in the arts and crafts (Cross, 1990; Lofthouse, 2004) and is often taught at art academies. Design engineering, on the other hand, is based on technology and technological models (Buijs, 2003; Lofthouse, 2007; Pahl and Beitz, 1984; Roozenburg and Eekels, 1998) and is mostly lectured at technical universities and engineering schools. Industrial design engineering can be seen as a combination of the previous two, with its roots in both engineering design and industrial design (Buijs, 2003).

The subdomains of design differ not only in their background but also in their way of approaching and solving design problems. For example, Lofthouse (2004) pointed out that the industrial designer is predominantly an imaginative, intuitive, creative, innovative, or divergent thinker who is unstructured, at times aimless, and inwardly directed. In contrast, the design engineer is a reasoning, rational, or convergent thinker who is logical, purposeful, and concerned with outward-directed problem solving. Design engineers use scientific principles, technical information, and imagination in defining the mechanical structure, machine, or system that can perform prespecified functions with maximum economy and efficiency. In addition, the "typical" industrial designer has proficiencies across a wide range of skill sets such as artistry, mechanical design, marketing, and psychology, whereas the "typical" design

Figure 1.4.1   The three subdomains of design

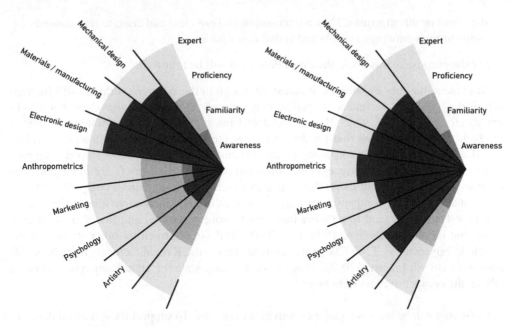

Figure 1.4.2 Differences in skills and knowledge of design engineers (left) and industrial designer (right) (Bates and Pedgley, 1998)

engineer is an expert in a restricted amount of topics, mainly related to mechanical design, materials and manufacturing, and electronic design (Bates and Pedgley, 1998; Lofthouse, 2004). The different knowledge and skill portfolios of industrial designers and design engineers are illustrated in Figure 1.4.2.

This broad variation in skills and knowledge is reflected in definitions existing within the field of IDE. For instance, the International Council of Societies of Industrial Design (ICSID, 2005) defines *industrial design* as:

> a creative activity whose aim is to establish the multi-faceted qualities of objects, processes, services and their systems in whole life-cycles. Therefore, design is the central factor of innovative humanization of technologies and the crucial factor of cultural and economic exchange.

According to Heskett (1980), industrial design is a process of creation, invention, and definition separated from the means of production, involving an eventual synthesis of contributory and often conflicting factors into a concept of three-dimensional form, and its material reality, capable of multiple reproduction by mechanical means. Finally, the Industrial Designers Society of America (IDSA, 2005) describes *industrial design* as:

> the professional service of creating and developing concepts and specifications that optimize the function, value and appearance of products and systems for the mutual benefit of both user and manufacturer. They develop these concepts and specifications through collection, analysis and synthesis of data guided by the special requirements of

the client or manufacturer. They are trained to prepare clear and concise recommendations through drawings, models and verbal descriptions.

For the purpose of this book, the IDSA definition will be followed.

At present, the role and impact of industrial design in the Western world are still increasing. Today industrial designers are called upon not only to design new products but also to manage the processes by which they are produced and get involved in the strategic aspects of production. This expanded role for design has resulted in explosive growth among professional design firms and the corporate sectors associated with design in developed countries. Even though it is not always recognized, industrial design can be a key factor in the competitiveness of a company and its products, as it improves and strengthens a company's market position and helps the company's products convey a different, innovative image. Companies that invest in design tend to be more innovative, more profitable, and faster growing than those that do not (DZDesign, 1996; EU, 2009). This is confirmed by other studies on the Dutch design sector by TNO (2005) and a recent survey of UK manufacturing firms (Confederation of British Industry, 2008). These studies clearly state the significant contribution of IDE to the competitiveness of industry.

Designing can be seen as a range of activities over time. To support the industrial designer in this process, many scholars have developed design methodologies. A common definition of design methodology is provided by Cross (1984):

> the study of the principles, practices and procedures of design in a broad and general sense. Its central concern is with how designing both is and might be conducted. This concern therefore includes the study of how designers work and think; the establishment of appropriate structures for the design process; the development and application of new design methods, techniques, and procedures; and reflection on the nature and extent of design knowledge and its application to design problems.

Much of the early methodology was compiled by engineers who applied the same "system thinking" they had used in designing their products to analyze the design process itself. The first thing that design researchers did, based on their own practices as professional product designers, was to cut the product innovation process into little pieces, which they ordered in a kind of logical way (Buijs, 2003). Within this context various design methodologies have been developed by Pahl and Beitz (1984; shown in Figure 1.4.3) and Roozenburg and Eekels (1998; see Figure 1.4.4).

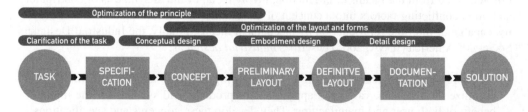

Figure 1.4.3 Flow chart representing the basic industrial design method of Pahl and Beitz (1984)

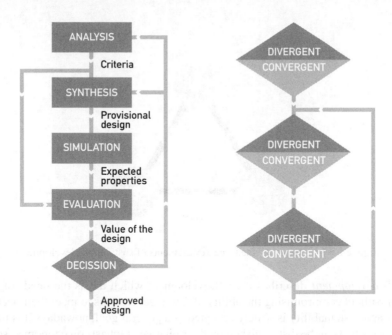

Figure 1.4.4 The basic cycle of design (Roozenburg and Eekels, 1998) characterized by divergent, convergent, and iterative activities

The basic industrial design method of Pahl and Beitz (1984), which is widely acknowledged and used in design engineering, consists of four phases: (1) clarification of the task, (2) conceptual design, (3) embodiment design, and (4) detail design. In the first three phases, product designers seek to optimize a product's working principle. The last three phases involve optimizing a product's layout and form. Thus, conceptual and embodiment design form the connecting links between technology, on one hand, and design/styling and ergonomics, on the other.

Roozenburg and Eekels (1998) developed a design methodology for industrial design based on analysis of a different model for the product design process; this resulted in the notion of a basic design cycle (see Figure 1.4.4). The *basic design cycle* encompasses all the phases a design process will go through at least once. It consists of a number of activities, each divergent, convergent, and iterative components. The cycle starts with a function that is analyzed, resulting in criteria; following that, synthesis takes place, resulting in a provisional design; that design is simulated, resulting in expected properties; these properties are evaluated, resulting in an outcome; then, finally, a decision is taken whether or not to continue (Roozenburg and Eekels, 1998). This analyze–synthesize–simulate–evaluate–decide approach is still the kernel of the present Delft innovation model, which is included in Section 2.3 of this book.

1.5 Design for Sustainability (DfS)

The main goal of *design for sustainability* is sustainable development. The concept of sustainable development was introduced and promoted by the report "Our Common Future" as a common aim for the whole world (Brundtland, 1987). The Brundtland definition of

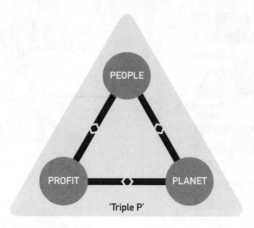

Figure 1.5.1   "Triple P," the three key elements of sustainable development

sustainable development describes it as "Development which meets the needs of a current generation without compromising the ability of future generations to meet their needs."

Even though sustainability is widely accepted as a general goal, nowadays it is interpreted in various ways that can be distinguished by their relative emphasis on economic, social, and ecological aspects.

In order to understand and manage the sustainability concept more transparently, Elkington (1998) developed the framework of the "triple bottom line" comprising the three sustainability components of *economic prosperity*, *environmental quality*, and *social justice*, which further were developed into *economy*, *environment*, and *social equity*. At present, these three key elements of sustainability are frequently referred to as the *triple P*: people, planet, and profit (see Figure 1.5.1).

Initially, sustainability in Europe largely was an environmental (planet) issue. The initial impetus was directed at what is called *design for the environment* or *ecodesign*. "Eco-design considers environmental aspects at all stages of the product development process, striving for products which make the lowest possible environment impact throughout the product life cycle" (Brezet and Hemel, 1997). *Eco* stands here for *eco*logy and *eco*nomy by looking for improvement options that decrease the environmental impact of the product lifecycle and in the meantime offering opportunities for financial benefits (so-called win–win situations). At present, according to Baumann *et al.* (2002), the UNEP "Promise" ecodesign manual (Brezet and Hemel, 1997) is most frequently used as a reference material on ecodesign.

The term *design for sustainability* has a broader and more holistic scope than *ecodesign* because, apart from environmental aspects, it also incorporates social, ethical, and equity issues into design (Figure 1.5.2). Since 1995, design for sustainability became an accepted approach (Baumann *et al.*, 2002) and a broad description of design for sustainability would be

> that industries take environmental and social concerns as a key element in their long-term product innovation strategy. This implies that companies incorporate environmental and social factors into product development throughout the lifecycle of the product, throughout the supply chain, and with respect their socio-economic

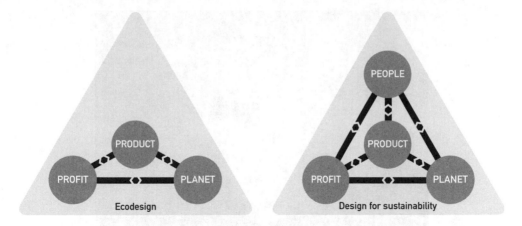

Figure 1.5.2   Ecodesign versus Design for Sutainability

surrounding, from the local community for a small company, to the global market for a trans-national company (Crul and Diehl, 2006; Diehl and Brezet, 2004).

As an example of design for sustainability, in the following we will discuss a case called the Evening Breeze project (Keskin *et al.*, 2011). Evening Breeze (Keskin *et al.*, 2011) was established in 2006. Among other products, the company developed climate systems integrated in beds (Figure 1.5.3), in agreement with their mission statement in which comfort and sustainability go hand in hand. The climate system locally cools the bed instead of the entire bedroom, and in this scenario 55% energy can be saved compared to cooling a room with air conditioning. The product was promoted as a cost-saving alternative for air conditioning in tropical resorts, since air conditioning accounts for 80% of the total energy use in hotels in the tropics.

While the initial focus of the Evening Breeze (Keskin *et al.*, 2011) bed was on saving energy, a second market study revealed that hotel guests mainly complained about noise, draft, dry air, and temperature. This stimulated a change in the proposition of the bed involving both comfort and sustainability: "creating the perfect sleeping climate by providing the

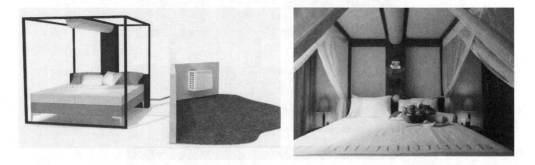

Figure 1.5.3   The original Evening Breeze, (Keskin *et al.*, 2011) targeted toward tropical resorts by Thomas van den Groenendaal and Yoeri Nagtegaal

Figure 1.5.4 The Evening Breeze Canopy, targeted toward the Western European consumer market

utmost comfort whilst having minimal environmental impact." This led to a second version of the bed, the Evening Breeze Canopy (Figure 1.5.4), targeting the Western European consumer market. This stand-alone version of the cooling unit is positioned at the head of the bed and allows customers to keep their existing bed. In terms of market success, the canopy appeared to be more successful than the original Evening Breeze bed, because for tropical resorts, the high cost of initial investment, timing, and maintenance are the most important considerations when purchasing an Evening Breeze bed, whereas for the individual user the experience of a comfortable sleeping environment prevails (Figure 1.5.5). It is interesting to note that the cooling power needed in Western European markets is lower than in the tropics, that is, the Evening Breeze Canopy will be used about 30 days a year versus 300 in a tropical location, leading to energy savings but to a lesser extent than with the original design of the Evening Breeze. Although the reasons that customers in tropical regions give for buying

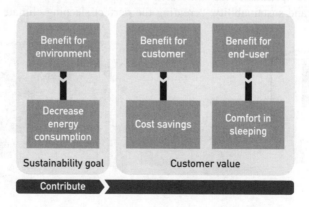

Figure 1.5.5 Translating sustainability into customer values for the Evening Breeze

the product vary (e.g., comfort, energy and cost savings, and sustainability), the value of the product is mainly their cost savings due to a reduced electricity bill. In contrast, the value for customers in the Northern Hemisphere mainly comprises luxury and comfort.

1.6 Energy Challenges at the Base of the Economic Pyramid

Whereas the main challenge in Western and emerging economies is to reduce energy and adapt to sustainable energy conversions, a great part of humankind struggles with a maybe even bigger problem, because it lacks access to modern energy services. Globally 1.5 billion people have no access to electricity; of them, 80% live in rural areas (UNDP/WHO, 2009). The majority of these people live in Southeast Asia and Sub-Saharan Africa (Figure 1.6.1), where the penetration grade of the electricity grid is low. The development of successful energy solutions for this part of world population is a challenge because of their poverty, often described by the term *base of the pyramid* (BoP) (Prahalad and Hammon, 2002; Prahalad and Hart, 2002) and identified as the 4 billion people living on an income of US $3.00 or less per day and the more than 1 billion people living on less than US $1.00 a day (World Bank, 2005), This BoP part of the world's population struggles with poverty and is characterized by a lack of access to basic needs as education, health care, and infrastructure, such as clean water supply and clean, safe, and reliable energy.

As mentioned in this chapter, 1.5 billion people don't have access to electricity but, even once connected to the grid, they often lack stable electricity supply, since rolling blackouts, ranging from 15 minutes to 6 hours a day, are common in many regions. Consequently, in these situations they also have to rely on diesel generators for productive uses and low-grade lighting sources such as candles, kerosene lanterns, and gas lamps to provide light at night. Not only is access to electricity an energy challenge for the BoP population, but also cooking and heating are causing issues. In rural regions of emerging and developing countries, 2.7 billion people cook on traditional stoves fueled by wood, dung, or other biomass resources (Aron *et al.*, 2009; Gradl and Knoblauch, 2011); this causes health problems due

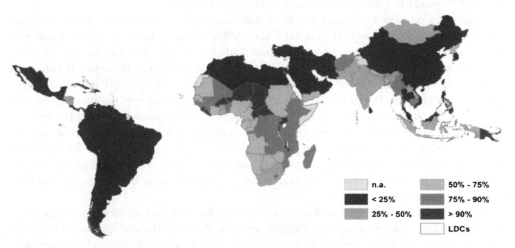

n.a.	50% - 75%
< 25%	75% - 90%
25% - 50%	> 90%
	LDCs

Figure 1.6.1 Share of the population without access to electricity (UNDP/WHO, 2009) (See Plate 3 for the colour figure)

to fumes, and accelerated deforestation due to unsustainable harvesting of biomass and wood collection, which is also known to be extremely time consuming.

As such access to clean, affordable, and reliable energy is a fundamental driver of economic growth, environmental sustainability, and social development (Gradl and Knoblauch, 2011). For example, light is one of these basic needs: It is required for education, improves the security of communities, and advances productivity (Gooijer *et al.*, 2008; Ramani and Heijndermans, 2003). Also electricity for communication (radio, television, and telephone), health care, and productive uses can be directly translated into benefits in terms of poverty reduction, development, and improved health.

In addition to lack of access to energy and the social and health problems connected to this, there is the so-called poverty–BoP penalty. This refers to the fact that poor people have to spend more than wealthier ones on the same energy product or service. BoP markets pay among the world's highest unit costs – and get some of the world's poorest quality energy (ESMAP, 2003). On average, BoP households spend about 10–15% of their income on energy (Guesalaga and Marshall, 2008; Nehme, 2007).

In general the demand for electricity in BoP markets in developing countries is growing. In addition to satisfying their basic needs of lighting, heating, and preparing and preserving food, demand grows for comfort and entertainment applications. Therefore, it is useful to develop efficient and affordable energy solutions in the form of products and services. As such, an important design challenge is to generate energy innovations targeted at the world's emerging and developing economies.

1.7 Reading This Book

Within the context of global energy issues, industrial design engineering can play an important role in the transitions from our existing carbonaceous energy systems to innovative and sustainable energy solutions that can be used in people's livelihoods. To support this process, *The Power of Design* is divided into four main sections about (1) innovation in design processes, (2) energy technologies, (3) user aspects, and (4) cases.

Depending on your experience, professional background, and interest, you could read each section independently. If you have knowledge of sustainable energy technologies and would like to become better acquainted with innovation in industrial design engineering, we recommend to start reading about innovation methods. However, if you don't have much experience with energy, you might like to start with the section on sustainable energy technologies. If you would like to get insight regarding specific product designs or considerations made during product development, you are advised to read through the section with cases. In general, the theories provided by this book will be illustrated by cases of design projects and concepts in practice. These examples will be shown in between the chapters on innovation methods and in a separate chapter that presents cases on products, design processes, and evaluation of products from the perspective of sustainability and energy efficiency.

It is our intention that the extended cases shown in this book and supplementary cases will become available through a website affiliated with *The Power of Design*, see www.wiley.com/go/reinders_design. In this digital surrounding, information and experiences with design projects in the field of sustainable energy technologies can be shared. Therefore, it would be useful to regularly visit this website and become a member of a new community of industrial design engineers, researchers, students, professional consultants, and colleagues, because changing energy use in our society isn't an issue of individuals – it matters to all of us, and we can reach results only if we learn from each other and share experiences. To quote Chris Ryan of University of Melbourne, who wrote the foreword of this book: it is a much-needed contribution to redistributing the future – whilst we still have time.

References

Aron, J.-E., Kayser, O., Liautaud, L., and Nowlan, A. (2009) *Access to Energy for the Base of the Pyramid*, HYSTRA, Paris.

Bates, D.J. and Pedgley, O.F. (1998) An industrial design team's approach to engineering design. IMC-15: Proceedings of the Fifteenth Conference of the Irish Manufacturing Committee, Jordanstown.

Baumann, H., Boons, F. *et al.* (2002) Mapping the green product development field: Engineering, policy and business perspectives. *J. Cleaner Production*, **10**, 409–425.

Brezet, J.C. and Hemel, C.G.v. (1997) *Ecodesign: A Promising Approach to Sustainable Production and Consumption*, United Nations Environment Program (UNEP), Paris.

Brundtland, G.H. (1987) *Our Common Future World*, Oxford Univeristy Press, Oxford.

Buijs, J. (2003) Modeling product innovation process, from linear logic to circular chaos. *Creativity and Innovat. Manag.*, **12**(2), 76–93.

Centraal Bureau voor de Statistiek (CBS) (2011) Industrie en Energie http://www.cbs.nl/nl-NL/menu/themas/industrie-energie/nieuws/default.htm (accessed April 14, 2012).

Confederation of British Industry (CBI) (2007) *Understanding Modern Manufacturing*. CBI, Department of Business Enterprise and Regulatory Reform, London

Cross, N. (1984) *Developments in Design Methodology*, John Wiley & Sons Ltd, Chicester.

Cross, N. (1990) The nature and nurture of design ability. *Design Studies*, **11**(3), 127–140.

Crul, M. and Diehl, J.C. (2006) *Design for Sustainability: A Practical Approach for Developing Economies Paris*, United Nations Environment Program (UNEP), Paris.

Diehl, J.C. and Brezet, H. (2004) Design for sustainability: An approach for international development, transfer and local implementation. EMSU 3, Monterrey, Mexico.

DZDesign (1996) *Diseno Industrial: Beneficio Para Las Empresas*, DZDesign, Bilbao.

Elkington, J. (1998) *Cannibals with Forks: the Triple Bottom Line of 21st Century Business*, New Society Publishers, Gabriola Island, BC.

ESMAP (2003) *Household Energy Use in Developing Countries: A Multi Country Study*, UNDP/WorldBank, Paris.

European Photovoltaic Industry Association (EPIA) (2011) *Solar Generation 6: Solar Photovoltaic Electricity Empowering the World*, EPIA, Brussels.

European Technology Platform SmartGrids (ETPS) (2011) Smart Grids European Technology Platform. http://www.smartgrids.eu/ (accessed April 14, 2012).

European Union (EU) (2009) *Design as Driver of User-Centred Innovation. C. o. t. E. Communities*, Commission of the European Communities, Brussels.

Gooijer, H.d., Reinders, A., Schreuder, D., (2008) Solar powered LED lighting – human factors of low cost lighting for developing countries. EU PV Conference, Valencia.

Gradl, C. and Knoblauch, C. (2011) *Energize the BoP: Energy Business Model Generator for Low Income Countries*, ENDEVA, Berlin.

Guesalaga, R. and Marshall, P. (2008) Purchasing power at the bottom of the pyramid: differences across geographic regions and income tiers. *J. Consum. Marketing*, **25**(7), 213–418.

Heskett, J. (1980) *Industrial Design*, Thames and Hudson Ltd., London.

ICSID (2005) Definition of design. http://www.icsid.org/about/about/articles31.htm (accessed April 12, 2012).

IDSA (2005) ID defined. http://www.idsa.org/category/seo-tags/id-defined (accessed April 12, 2012).

IEO, International Energy Outlook (2011), U.S. Energy Information Administration, Report DOE/EIA - 0484, Washington D.C.

International Energy Agency (IEA) (2010) *Technology Roadmap Solar Photovoltaic Energy*, IEA, Paris.

International Energy Agency (IEA) (2011) *Technology Roadmap Smart Grids*, IEA, Paris.

IPCC, Intergovernmental Panel on Climate Change (2011) Summary for policymakers, in *IPCC Special Report on Renewable Energy Sources and Climate Change Mitigation* (eds. O. Edenhofer, R. Pichs-Madruga, Y. Sokona, K. Seyboth, P. Matschoss, S. Kadner, T. Zwickel, P. Eickemeier, G. Hansen, S. Schlömer, and C. von Stechow), Cambridge University Press, Cambridge.

IPPC, Intergovernmental Panel on Climate Change (2007) Climate change 2007 – synthesis report, Geneva, IPCC.

Kazmerski, L. (2011) *Best Research-Cell Efficiencies*, NREL, Golden, CO.

Keskin, D., Wever, R, Brezet, H., From a sustainable product idea to a succesful business: how to translate sustainability into features that offer customer value, working paper, Design for Sustainability, Delft University of Technology, 2011

Lofthouse, V.A. (2004) Investigation into the role of core industrial designers in ecodesign. *Design Studies*, **25**(2), 215–227.

Ministerie van Economische Zaken, Landbouw & Innovatie (MEZLI) (2011) Energy report 2011, The Hague.

Nehme, F. (2007) *Lighting Africa: Catalyzing Markets for Modern Lighting*, IFCLA Program, World Bank, Washington, DC.

OECD/IEA (2009) *Gadgets and Gigawatts: Policies for Energy-Efficient Electronics*, IEA, Paris.

Pahl, G. and Beitz, W. (1984) *Engineering Design*, London Design Council, London.

Prahalad, C.K. and Hammon, A. (2002) Serving the world's poor. Profitably. *Harvard Bus. Rev.*, **80**(9), 48–57.

Prahalad, C.K. and Hart, S.L. (2002) The fortune at the bottom of the pyramid. *Strategy+Business*, **26** (first quarter), 2–14.

Ramani, K.V. and Heijndermans, E. (2003) *Energy, Poverty and Gender, A Synthesis*, World Bank, Washington, DC.

Reinders, A.H.M.E., Pramusito, Sudradjat, A. *et al.* (1999) Sukatani revisited: On the performance of nine year old solar home systems and street lighting systems in Indonesia. *Renew. Sustainable Energy Reviews*, **3**, 1–47.

REN21 (2011) Renewables 2011 Global Status Report, REN21 Secretariat, Paris.

Roozenburg, N.F.M. and Eekels, J. (1998) *Productontwerpen, Structuur en Methoden*, Lemma, Utrecht.

TNO (2005) *Design in the Creative Economy: A Summary*, Premsela Foundation, Amsterdam.

UNDP/WHO (2009) United Nations Development Programme, 2009: The energy access situation in developing countries, New York.

Veldhuis, A.J. and Reinders, A.H.M.E. (2011) Feasibility study on photovoltaic solar energy versus other energy sources in Indonesia, Project report, Department of Design, Production and Management, University of Twente, Enschede.

Wissenschaftlicher Beirat der Bundesregiering Globale Umweltveränderungen (WBGU) (2003) *Welt im Wandel: Energiewende zur Nachhaltigkeit*, Springer Verlag, Berlin.

WWF International (2011) *The Energy Report: 100% Renewable Energy by 2050*, WWF International, Gland, Switzerland.

World-Bank (2005) *World Bank Data and Statistics*, World Bank, Washington, DC.

2

Innovation Methods

2.1 Introduction to Innovation Methods in Design Processes

Angèle Reinders

*Delft University of Technology, Faculty of Industrial Design Engineering,
Design for Sustainability, Delft, The Netherlands
University of Twente, Faculty of Engineering Technology, Department of Design,
Production and Management, Enschede, The Netherlands*

2.1.1 Introduction

This section provides theory on innovation processes and innovation methods for industrial design engineering. The theory will be illustrated by results of design cases with sustainable energy technologies that have been executed with various innovation methods by students.

Here, innovation will be considered as changes of product–market–technology combinations, and the theory will refer to this framework. In this Chapter, nine innovation methods that will be discussed in this book will be briefly introduced to the reader. These innovation methods are platform-driven product development (discussed in detail in Section 2.2), the Delft innovation model (discussed in detail in Section 2.3), TRIZ (theory of inventive problem solving) and technology road mapping (in Section 2.4 and 2.5), the design and styling of future products (Section 2.6), constructive technology assessment (Section 2.7) and the innovation journey (Section 2.8), risk diagnosing methodology (Section 2.9), and the multilevel design model (Section 2.10).

First, we would like to question what innovation is and discuss the setting of innovation in product design and engineering. Innumerable definitions of *innovation* are available. Their main correspondence in meaning is related to the concept of *novelty*. For instance, *Webster's* dictionary describes "to innovate" as to make changes or to introduce new practices. Von Hippel (1988, 2005) defines an *innovation* as anything new that is actually used, whether major or minor. And Rogers (2003) describes an innovation as an idea, practice, or object that is perceived as new by an individual or other unit of adoption.

Innovation in the context of industrial design engineering is represented in Figure 2.1.1 as an innovation flower involving the following five fields: technology, design and styling,

The Power of Design: Product Innovation in Sustainable Energy Technologies, First Edition.
Edited by Angèle Reinders, Jan Carel Diehl and Han Brezet.
© 2013 John Wiley & Sons, Ltd. Published 2013 by John Wiley & Sons, Ltd.

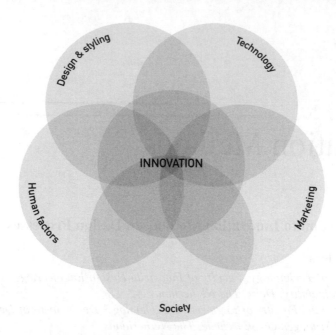

Figure 2.1.1   Innovation flower of industrial product design

human factors, marketing, and society. In Figure 2.1.1, *technology* refers to product technologies and materials, as well as manufacturing processes. *Design and styling* refers to the appearance of products and their image in the market. *Human factors* cover the user context of consumer products and the physical ergonomics. *Marketing* refers to market value costs and sales. Finally, the term *society* implies the field of policy and regulation and societal acceptance. We believe that all five components of the innovation flower are equally important to product development and the final success of a product. Therefore, in this chapter on innovation methods, we will position product development in the context of the innovation flower.

Industrial design methods (IDMs) play an important role in product development processes. IDMs convert market needs into detailed information for products that can be manufactured. IDMs vary in their approaches to problem solving. In technological design, design processes are generally viewed as heuristics. In other words, we assume that design goals can be defined by carefully describing a problem prior to formulating possible solutions to this problem. Through problem decomposition, the complexity of the problem is reduced. Next, the designer evaluates how solutions to subproblems can be merged into a solution composition that meets the design goals. The embodiment of the solution is a product.

A generally accepted approach toward this process is described in detail by Pahl and Beitz (1984) (Figure 2.1.2). It consists of four phases:

1. Clarification of the task.
2. Conceptual design.
3. Embodiment design.
4. Detail design.

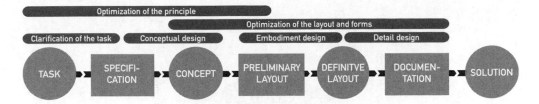

Figure 2.1.2   Flow chart representing the basic industrial design method of Pahl and Beitz

In the first three phases, product designers seek to optimize the working principle of a product. The last three phases involve optimizing the layout and form of a product. Thus, conceptual and embodiment design form the connecting links between technology, on one hand, and design and styling and ergonomics, on the other. Figure 2.1.3 shows in which phases the innovation methods that will be presented in this chapter can be applied.

Theoretically, innovations improve product performance over time. Figure 2.1.4 shows the exponential nature of such growth, which has been widely documented throughout

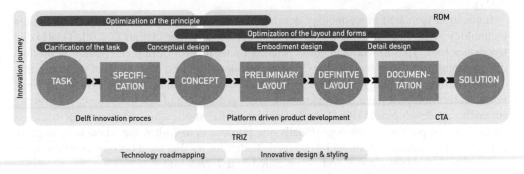

Figure 2.1.3   Flow chart representing innovation methods presented in this chapter in the framework of the basic industrial design method of Pahl and Beitz (1984). RDM: risk-diagnosing methodology; CTA: constructive technology assessment

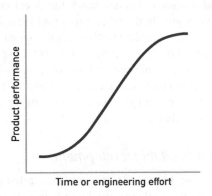

Figure 2.1.4   S curve of innovation

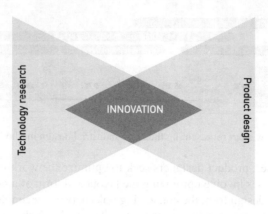

Figure 2.1.5   Innovative products result from the conjunction of technology research and product design

the history of human technology and is widely known as the *S* curve (Burgelman *et al.*, 2004).

Innovation is related to technology development. Generally it is assumed that the more a technology is developed, the less innovative its applications are. Arthur D. Little (1981) distinguishes four types of technologies depending on their level of implementation in product development. *Emerging technologies* have not yet demonstrated clear product potential. Scientists discover emerging technologies in the course of explaining working principles. Next, *packing technologies* have demonstrated market potential and are the subject of ongoing research and development (R&D) to make them fit for applications. Organic electronics is an example of packing technology. Finally, the purpose is to embed the technology in products. Such *key technologies* may be patented. Technologies that have become common to all competitors – they are a commodity – are called *base technologies*.

Technologies that are on the threshold between what Little refers to as *packing* and *key technologies* can result in innovative product design. Namely, potential applications of technology are being assessed through R&D, and they are still in the process of being integrated into products. Figure 2.1.5 shows how both technological research and product design may contribute to innovations at this stage. On one hand, the development of promising technologies may initiate product design. On the other hand, problems with the design of products based on existing technology may stimulate further technological research.

In this framework, Reinders *et al.* (2006, 2007, 2008, 2009) explored the relations between product design with existing sustainable energy technologies and the application of various innovation methods in the design process, as indicated in Figure 2.1.3. Based on these design experiences with about 70 teams of students, we can show many examples of product design with innovation methods in this chapter.

2.1.2   Platform-Driven Product Development

A product platform defines a set of related products, a so-called product family, that can be developed and produced in a time- and cost-efficient manner. Features of a product platform are modularity, connecting interfaces, and common standards (Halman, Hofer, and van

Figure 2.1.6   A solar-powered street-lighting product, exploded view by Scholder and Friso (2007) (See Plate 4 for the colour figure)

Vuuren, 2003). By using a product platform, companies can reach different markets (and customers) with less effort than by developing separate products.

An example of platform-driven product development is shown for the case of a solar-powered street-lighting product, such as in Figure 2.1.6. In order to achieve large volumes with sufficient product variations in a time-efficient manner, several modules were selected from which different product families could be composed. As a consequence, the product platforms shown in Figure 2.1.7, called City Lighting or Road Lighting, can be implemented in different markets. For a detailed explanation about platform-driven product development, see Section 2.2.

Figure 2.1.8 shows the effect of platform-driven product development in a matrix of customer value perception versus enabling technology. For instance, theoretically speaking for a manufacturer of gardening equipment, a fuel cell–powered blower would be a breakthrough product, enhancing its portfolio of conventionally powered equipment. Thus, the risk of integrating new hydrogen technology in products can be distributed over ongoing product development activities.

2.1.3   Delft Innovation Model

The Delft innovation model of Jan Buijs (2003), also known as the *innovation phase model*, aims to optimally combine the intrinsic value of technology with opportunities in the market. It consists of four phases: a strategy formulation stage, a design brief phase, a product development phase, and a product launch and use phase. For the purpose of product innovation, the strategy formulation stage, shown in Figure 2.1.9, is most important. A matrix of internal strengths of technology and external opportunities in the market results in many search ideas. Divergence, selection, and convergence of search ideas related to technology and markets

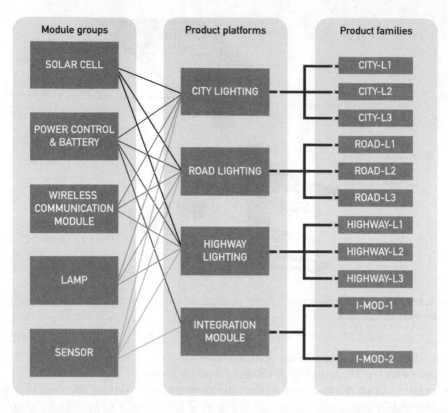

Figure 2.1.7 Product platforms of photovoltaic (PV)–powered street-lighting products consist of several modules from which different product families can be composed

lead to innovative technology–product–market combinations, the so-called *search areas*, or *search fields*.

Search field information comprises the following topics:

- Subject: the relation of the idea to markets, product (groups), and services.
- Description: a short description of the direction of the idea.
- Strength: What are the basic strengths of the idea?
- Trends: To which developments in technology and markets does the idea connect?
- Needs: Which problem(s) will be solved by the idea?
- Size: An indication of the size of the market.
- Segmentation: Which subareas and differentiation of products can be foreseen?
- Stakeholders comprise parties involved such as competitors, experts, and consumers.
- Bottlenecks: Which problems – both internal and external – can be expected?

Next, information connected to these search fields can be inputted to the design brief stage. Again divergence, selection, and convergence can be applied to select search fields to be worked out further.

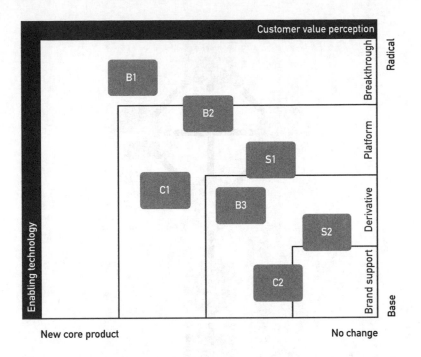

Figure 2.1.8 Product platform for gardening equipment, (B1) blower on hydrogen energy, (B2) blower with noise reduction system, (B3) blower standard, (C1) cutter with flexible drive shaft, (C2) cutter standard, (S1) chainsaw with automatic balancing system, and (S2) chainsaw standard

The creative process of the Delft innovation model is supported by visual collages and written mind maps. Figure 2.1.10 shows a communication device that has been developed by using mind maps. Mind maps of a fuel cell (Figure 2.1.11) and of lead user hikers (Figure 2.1.12) are combined to determine a suitable technology–product–market combination. For more information about this model, see Section 2.3.

2.1.4   TRIZ

TRIZ is a Russian acronym meaning the *theory of inventive problem solving* (Altshuller, 1984). TRIZ can be applied to estimate the probability of technology developments. It is a comprehensive method based on long-term patent research leading to certain basic rules governing problem solving in product development.

Each trend of technology evolution demonstrates a line of a system's structural evolution with regard to changing its physical structure, space, time, energy, supersystem relations, and other parameters. Each trend contains a number of specific patterns, which are transitions that specify how a system should be changed to move from stage A to stage B of evolution. The trends are quite generic in nature, and exploring their applicability requires a combination of modern technologies that reside outside of the industry sector in which a product belongs.

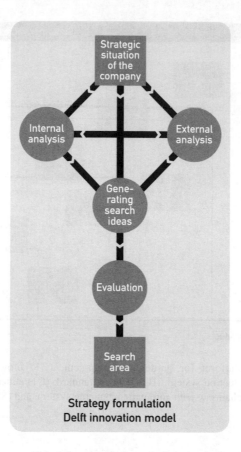

Figure 2.1.9   Scheme representing the strategy formulation stage of the Delft innovation model

Figure 2.1.10   Communication device for hikers with integrated fuel cells. Left: Front view. Right: Opened back view. Designed by Kyra Adolfsen and Benne Draijer (See Plate 5 for the colour figure)

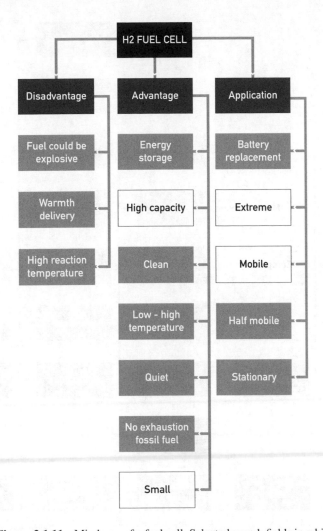

Figure 2.1.11 Mindmap of a fuel cell. Selected search fields in white

Figure 2.1.13 shows in which fields TRIZ could improve product development. Since there are many diverse TRIZ trends, the method becomes complex. Therefore, software has been developed to make it easier to apply TRIZ. Section 2.4 gives an extensive overview of TRIZ.

2.1.5 Technology Roadmapping

A technology roadmap (TRM) establishes a correlation between identified market needs and trends with existing and emerging technologies for a specific industry sector (Souchkov, 2005). Technology roadmaps usually cover 3–10 years and are used in strategic product planning, research planning, and business planning (Phaal, Farrukh, and Probert, 2001).

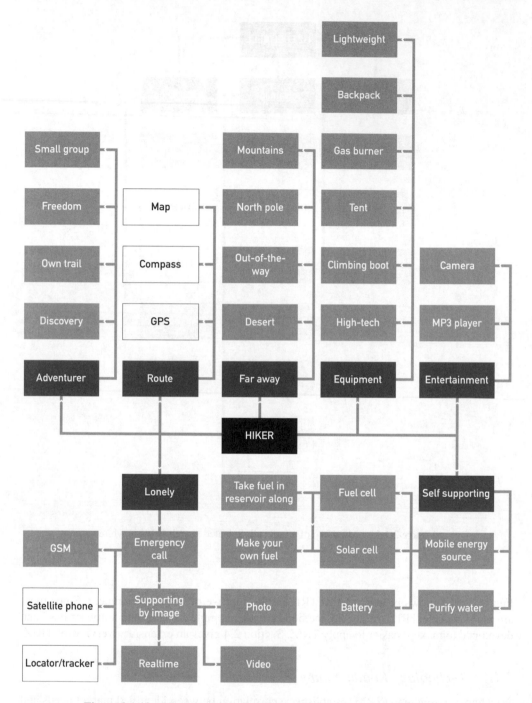

Figure 2.1.12   Mindmap of lead user hikers. Selected search fields in white

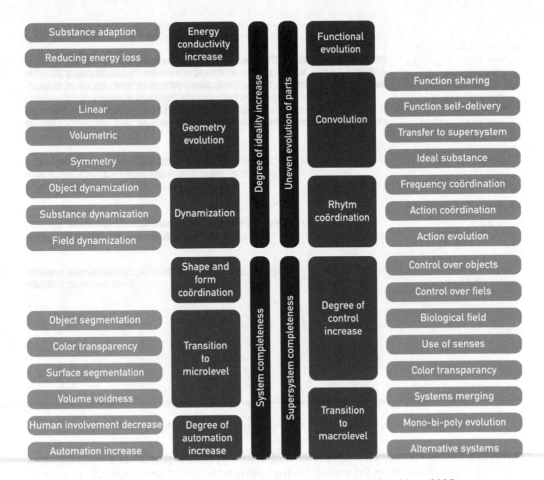

Figure 2.1.13   TRIZ trends of technology evolution, from Souchkov (2005)

Figure 2.1.14 shows an example of a TRM for the furniture industry. Theory and examples of applications of technology roadmapping will be provided by Section 2.5.

2.1.6   Design and Styling of Future Products

Design and styling are assets for technological innovation (Christensen, 1995). More specifically, cars, modern durables, and consumer products require a high degree of integration between functional and aesthetic aspects of design. Design is often regarded as purely aesthetic or superficial. Research by TNO (2005), however, has highlighted the crucial role played by design in the economy of the Netherlands and in product innovation. It was concluded that companies combining technological innovation with new product design increase their market share relative to other companies. It therefore seems likely that industrial designers will provide significant added value to products with new technologies.

A market of products with integrated technologies can be promoted by studying the image that potential users of a proposed product wish to project and designing the product so that it

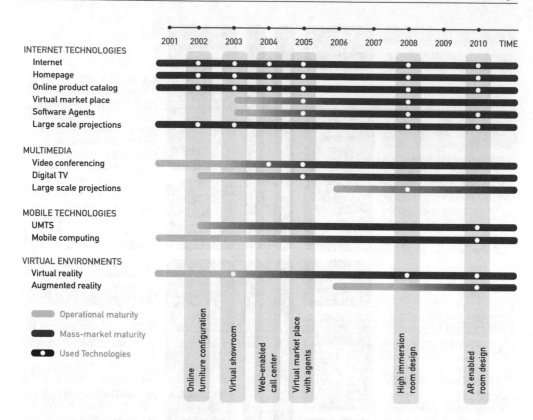

Figure 2.1.14   Technology roadmap for the furniture industry by Valeri Souchkov

meets those aesthetic requirements. A method for developing an appropriate styling for innovative products is shown in Figure 2.1.15. It comprises the use of style elements from both associative and competing products.

Figure 2.1.16 shows a design for a diving lamp with integrated fuel cells. Technology is decisive in the high-end diving market, where users require long burning hours and reliability. In the larger market of so-called family divers, however, design and styling (see Figure 2.1.17) are more important. In Section 2.6, aspects related to the design and styling of future products will be further explained.

2.1.7   Constructive Technology Assessment

In innovation journeys, many actors play a role. New products must be promoted by manufacturers and retailers, must be approved by a competent authority in accordance with industry rules and standards, and must meet with acceptance from consumers. Constructive technology assessment focuses on these processes and how to improve them (Deuten, Rip, and Jelsma, 1997). The dynamics of technological development involve a series of exchanges between technology and society. Social projects such as those shown in Figure 2.1.18 foster the acceptance of new products by teaching people about the benefits of technology for society.

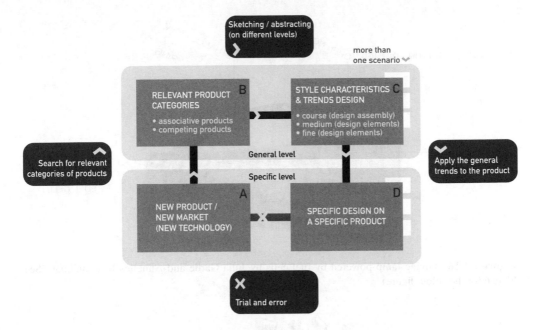

Figure 2.1.15 Method for innovative design and styling based on characteristic style elements from associative and competing product by Stevens (2005)

In Asian metropolises, transportation has always been a difficult problem. Owing to out-dated engine technologies, private taxies (referred to locally as *rickshaws*, *tuk-tuks*, *bajajs*, etc.) are a primary source of air pollution and noise. Because they require little power – about 5 kW – rickshaws could be an interesting application for proton exchange membrane (PEM) fuel cells. A positive social impact would result from the decrease of harmful emissions and increased comfort due to decreased noise. As a side benefit, PEM fuel cells could supply the rickshaws with electric power, enabling end users to operate air conditioning, stoves, and refrigerators from them. Moreover the rickshaw, as an icon of transportation cherished by many residents, might serve as a catalyst to accelerate social acceptance of fuel cell technology. In Section 2.7, the theory of CTA will be further explained.

2.1.8 Innovation Journey

In the case of technology-based product design, product development often follows certain patterns. By examining the so-called innovation journeys (Rip, 2005) of products in different fields, one may gain tools for assessing the potential innovation journeys of new products using new technologies or new situations of use; see also Section 2.8.

One pair of students used this approach to design an induction cooking device powered by fuel cells (Figure 2.1.19). They evaluated the evolution of technologies for cooking dinner, starting with plain fire and leading to the principles of electrical cooking. Comparing plain electrical elements, ceramic, halogen, infrared, and induction cooking, they found the last method most suitable in relation to customers' demands for an easy-to-use and

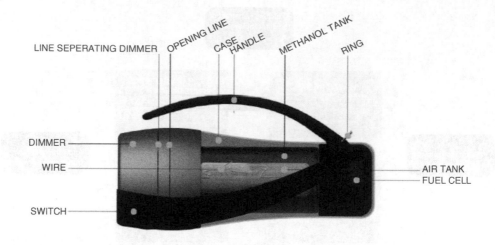

Figure 2.1.16 Diving lamp powered by fuel cells by Julia Garde and Annelies Brummelman (See Plate 6 for the colour figure)

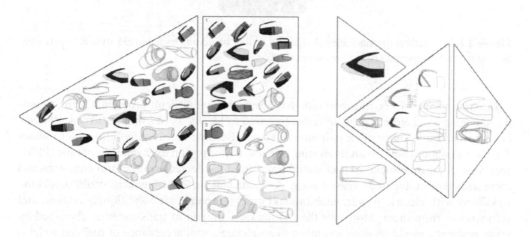

Figure 2.1.17 Design and styling of a diving lamp by diverging, categorizing, and converging associative and competitive style characteristics in two phases (Drawing by Julia Garde and Annelies Brummelman)

lightweight cooking device. The black box in Figure 2.1.19 accommodates all electrical components, including a replaceable hydrogen cartridge, a fuel cell of 500 W, and an induction coil. The black box fits snugly into the cubical case, which serves as a protective shell during transportation. The device is operated by an ergonomic user interface.

2.1.9   Risk-Diagnosing Methodology

Companies must take risks to launch new products speedily and successfully. In this scope, risk-diagnosing methodology (RDM) aims to identify and evaluate technological,

Figure 2.1.18 Fuel cell–powered rickshaw (Designed by Simon Brandenburg and Werner Helmich (2006)) (See Plate 7 for the colour figure)

Figure 2.1.19 Induction cooking device powered by fuel cells. Left: The product in use. Right: In compact carrying case (Designed by Rick Tigchelhoff and Wouter Haanstra) (See Plate 8 for the colour figure)

organizational, and business risk in product innovation (Keizer, Halman, and Song, 2002). This methodology, which will be explained in detail in Section 2.9, has been developed and applied at Philips and Unilever, and is fit for product development in diverse areas such as the automobile industry, printing equipment, landing gear systems, and fast-moving consumer goods such as shampoo and margarine. The risks evaluated by RDM are related to product family and brand positioning, technology, manufacturing, intellectual property, supply chain and sourcing, consumer acceptance, project management, public acceptance, screening and appraisal, trade customers, competitors, and commercial viability.

One pair of students applied RDM to a product platform of fuel cell–powered equipment, and drew the following conclusions:

- The main risk is acceptance of the new, unknown, fuel cell technology by users.
- Most risks related to the production, manufacturing, and distribution of the new product can be mitigated or eliminated by forming a joint venture with several stakeholders.
- A design process that involves the participation of end users can be used to create a higher degree of acceptance for new products.

References

Altshuller, G. (1984) *And Suddenly the Inventor Appeared*, Technical Innovation Center, Worcester, MA.

Burgelman, R.A., Clayton, M., Christensen, S., and Wheelwright, C. (2004) *Strategic Management of Technology and Innovation*, McGraw-Hill, New York.

Buijs, J. (2003) Modeling product innovation processes, from linear logic to circular chaos. *Creativity Innovat. Manag.*, **12**(2), 76–93.

Christensen, J.F. (1995) Asset profiles for technological innovation. *Res. Policy*, **24**(5), 727–745.

Deuten, J., Rip, A., and Jelsma, J. (1997) Societal embedding and product creation management. *Technol. Anal. Strateg.*, **9**(2), 131–148.

Halman, J.I.M., Hofer, A.P., and van Vuuren, W. (2003) Platform-driven development of product families: Linking theory with practice. *J. Prod. Innovat. Manag.*, **20**(2), 87–192.

Helmich, W. (2006) Sources of Innovation, student report, Master of Industrial Design Engineering program, University of Twente.

von Hippel, E. (1988) *The Sources of Innovation*, Oxford University Press, Oxford.

von Hippel, E. (2005) *Democratizing Innovation*, MIT Press, Cambridge, MA.

Keizer, J.A., Halman, J.I.M., and Song, M. (2002) From experience, applying the risk diagnosing methodology. *J. Prod. Innovat. Manag.*, **19**(3), 213–233.

Little, A.D. (1981) *The Strategic Management of Technology*, European Management Forum, Geneva.

Phaal, R., Farrukh, C., and Probert, D. (2001) *Technology Roadmapping: Linking Technology Resources to Business Objectives*, white paper, University of Cambridge, Cambridge.

Pahl, G. and Beitz, W. (1984) *Engineering Design*, London Design Council, London.

Reinders, A.H.M.E. and van Houten, F.J.A.M. (2006) Industrial design methods for product integrated PEM fuel cells. 2006, Proceedings of the NHA, Los Angeles.

Reinders, A., de Gooijer, H., and Diehl, J.C. (2007) How participatory product design and micro-entrepreneurship favor the dissemination of photovoltaic systems in Cambodia. Proceedings of 17th International Photovoltaic Science and Engineering Conference, Fukuoka.

Reinders, A.H.M.E. and de Boer, A. (2008) Product-integrated PV – Innovative design methods for PV-powered products. Proceedings of 23rd European Photovoltaic Solar Energy Conference, Valencia, pp. 3321–3324.

Reinders, A.H.M.E. (2008) Product-integrated PV applications: How industrial design methods yield innovative PV powered products. Proceedings of 33rd IEEE PVSC, San Diego, pp. 1–4.

Reinders, A., de Boer, A., de Winter, A., and Haverlag, M. (2009) Designing PV powered LED products – Sensing new opportunities for advanced technologies. Proceedings of 34th IEEE PVSC, Philadelphia.

Rogers, E. (2003) *Diffusion of Innovations*, Free Press, New York.

Scholder, F. and Friso, T. (2007) Sources of Innovation, student report, Master of Industrial Design Engineering program, University of Twente.

Souchkov, V. (2005) *TRIZ-Based Innovation Roadmapping*, Lecture slides, University of Twente., Enschede.

Stevens, J. (2005) *Innovative Technology: But What Should It Look Like?* Lecture slides, University of Twente., Enschede.

TNO (2005) *Design in the Creative Economy: A Summary*, Premsela Foundation, Amsterdam.

2.2 Platform-Driven Product Development

Johannes Halman

University of Twente, Faculty of Engineering Technology, Department of Construction Management and Engineering, P.O. Box 217, The Netherlands

2.2.1 Introduction

In a global, intense, and dynamic competitive environment, the development of new products and processes has become a focal point of attention for many companies. Shrinking product life cycles, increasing international competition, rapidly changing technologies, and customers demanding high variety options are some of the forces that drive new development processes. More variety will make it more likely that each consumer finds exactly the option he or she desires, and will allow each individual consumer to enjoy a diversity of options over time. In considering the implementation of product variety, companies are challenged to create this desired variety economically. In their quest to manage product variety, firms in most industries are increasingly considering product development approaches that reduce complexity and better leverage investments in product design, manufacturing, and marketing (Krishnan and Gupta, 2001). Platform thinking, the process of identifying and exploiting commonalities among a firm's offerings, its target markets, and the processes for creating and delivering offerings, appears to be a successful strategy to create variety with an efficient use of resources (Halman, Hofer, and van Vuuren, 2003). Key in this approach is the sharing of components, modules, and other assets across a family of products. Historical success stories such as the Sony Walkman, Black and Decker power tools, Hewlett Packard's Deskjet printers, Microsoft's Windows NT, and Minolta's "intelligent lens technology" have shown both the benefits and the logic behind the platform concept. Gupta and Souder (1998) even claim that thinking in terms of platforms for families of products rather than individual products is one of the five key drivers behind the success of short-cycle-time companies.

2.2.2 Definitions

The terms *product families*, *platforms*, and *individual products* are hierarchically different and cannot be used as synonyms. A *product family* is the collection of products that share the same assets (i.e., their platform) (Sawhney, 1998). A *platform* is therefore neither the same as an individual product nor the same as a product family; it is the common basis of all individual products within a product family (McGrath, 1995; Robertson and Ulrich, 1998). As a consequence, a platform is always linked to a product family, while it can serve multiple product lines in the market. The leading principle behind the platform concept is to balance the commonality potential and differentiation needs within a product family. Figure 2.2.1 shows how Skil, a developer of multiple product families (consisting of, e.g., a set of saws, drills, routers, and grinders), utilizes as much as possible the same technical components for its different brand segments even though they have their own styling and perceived worth.

One possibility to build a *platform* is to define it by means of the product architecture. This *product platform* has been defined by McGrath (1995) as a set of subsystems and

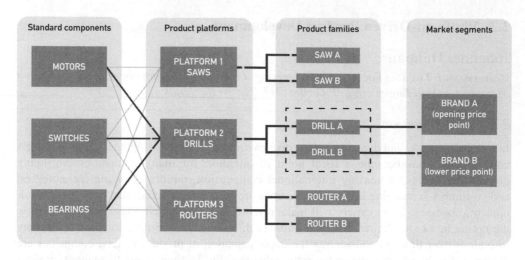

Figure 2.2.1   Platform-based development of product families within Skil

interfaces that form a common structure from which a stream of related products can be efficiently developed and produced. Baldwin and Clark (1999) define three aspects of the underlying logic of a product platform: (1) its modular architecture, (2) the interfaces (the scheme by which the modules interact and communicate), and (3) the standards (the design rules to which the modules conform). The main requirements for building a product family based on a product platform are (1) a certain degree of modularity to allow for the decoupling of elements, and (2) the standardizing of a part of the product architecture (i.e., subsystems and/or interfaces). A modular product architecture in this context is characterized by a high degree of independence between elements (modules) and their interfaces.

The typical inclination is to only think of the product architecture as the basis for a common platform of a product family. In line with discussions in literature (Meyer and Lehnerd, 1997; Robertson and Ulrich, 1998; Sawhney, 1998), we argue that a product family should ideally be built not only on elements of the product architecture (components and interfaces) but also on a multidimensional core of assets that also include processes along the whole value chain (e.g., engineering and manufacturing), customer segmentation, brand positioning, and global supply and distribution.

Process platform refers to the specific setup of the production system to easily produce the desired variety of products. A well-developed production system includes flexible equipment, for example programmable automation or robots, computerized scheduling, flexible supply chains, and carefully designed inventory systems (Kahn, 1998). Sanderson and Uzumeri (1995) refer in this respect to Sony's flexible assembly system and an advanced parts orientation system, designed specifically with flexibility, small-lot production, and ease of model change in mind. Although the costs of this multifunction machine may be twice as much as those of a comparable single-function machine, the greater flexibility possible using manufacturing equipment designed with multiple products and rapid changeover in mind offsets its initial cost. Figure 2.2.2 illustrates how process platforms can be created by

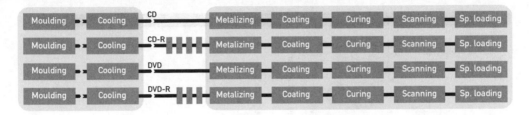

Figure 2.2.2 Process platforms in the production of CDs, CD-Rs, DVDs, and DVD-Rs

searching for potential commonalities in the production process of different product lines of CDs, CD-Rs, DVDs, and DVD-Rs.

Customer platform is the customer segment that a firm chooses as its first point of entry into a new market. This segment is expected to have the most compelling need for the firm's offerings and can serve as a base for expansion into related segments and application markets (Sawhney, 1998). Established customer relationships and knowledge of customer needs are used as a springboard to expand by providing step-up functions for higher price–performance tiers within the same segment or to add new features to appeal different segments (Meyer, 1997).

Brand platform is the core of a specific brand system. It can either be the corporate brand (e.g., Philips, Toyota, and Campbell) or a product brand (e.g., Pampers, Organics, and Nivea). From this brand platform, subbrands can be created, reflecting the same image and perceived worth (e.g., Philishave, Hugo Boss perfumes, and Organics shampoo). With a small set of brand platforms and a relatively large set of subbrands, a firm can leverage its brand equity across a diverse set of offerings (Sawhney, 1998).

Global platform is the core standardized offering of a globally rolled-out product. As an example, designing software for a global market can be a challenge. The goal is to have the application support different locales without modifying the source code. A global rollout plan details the aspects of the product that can be standardized as well as those aspects that should be adapted to country-specific conditions and customer preferences. Customization can involve physical changes in the product, and adaptation in pricing, service, positioning message, or channel (Sawhney, 1998).

2.2.3   The Creation of Platform-Based Product Families

Cost and time efficiencies, technological leverage, and market power can be achieved when companies redirect their thinking and resources from single products to families of products built upon robust platforms. Implementing the platform concept can significantly increase the speed of a new product launch. The platform approach further contributes to the reduction of two major resources (i.e., cost and time) in all stages of new product development. By using standardized and pretested components, the accumulated learning and experience in general may also result in higher product performance. Unfortunately, this is not a one-time effort. New platform development must be pursued on a regular basis, embracing technological changes as they occur and making each new generation of a product family more exciting

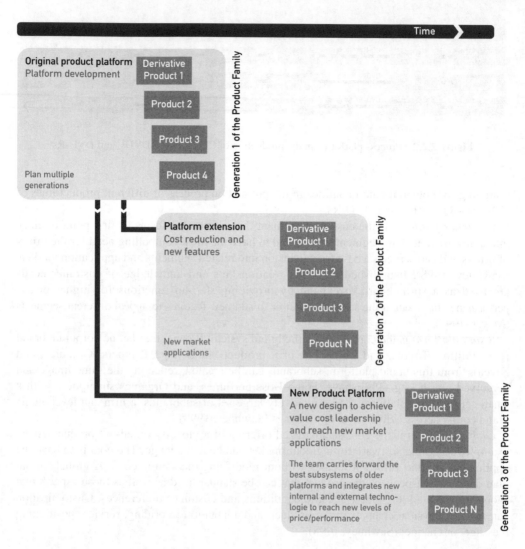

Figure 2.2.3 Product family evolution, platform renewal, and new product creation (adapted from Meyer and Lehnerd, 1997)

and value-rich than its predecessors. Meyer and Lehnerd (1997) propose a general framework for product family development (see Figure 2.2.3).

This framework represents a single product family starting with the initial development of a product platform, followed by successive major enhancements to the core product and process technology of that platform, with derivative product development within each generation. New generations of the product family can be based on either an extension of the product platform or an entirely new product platform. In case of an extension, the constellation of subsystems and interfaces remains constant, but one or more subsystems undergo major revision in order to achieve cost reduction or allow new

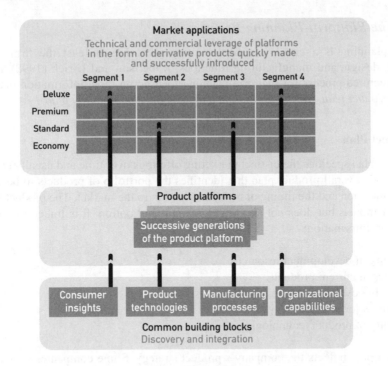

Figure 2.2.4   The evolution of product families (adapted from Meyer and Lehnerd, 1997)

features. An entirely new platform emerges only when its basic architecture changes and aims at value cost leadership and new market applications. Systems and interfaces from prior generations may be carried forward into the new design but are joined by entirely new subsystems and interfaces.

The more consistent a platform concept is defined and implemented in terms of parts, components, processes, customer segmentation, and so on, the more effective a company can operate in terms of tailoring products to the needs of different market segments or customers.

An effective evolution of a product family requires that in collective fashion three essential elements are considered (Figure 2.2.4):

- *Potential market applications*: What type of products can be offered to different market segments? Customers care whether the firm offers a product that closely meets their needs. Closely meeting the needs of different market segments requires distinctive products.
- *Potential commonalities between the distinctive products that can be achieved*: What are the design, manufacturing, and cost benefits that can be achieved by maximizing the extent to which distinctive products share, for example, common components, manufacturing processes, and distribution channels?
- *Platform strategy definition*: This is determined by making trade-off decisions between distinctiveness and commonality. In the ideal case, the platform design represents a relatively high level of commonality that is achieved without much sacrifice in distinctiveness while distinctiveness declines slowly as commonality is increased.

2.2.4   The Platform-Planning Process

Platform planning is a cross-functional activity involving at least the firm's product-marketing, design, and manufacturing functions. Robertson and Ulrich (1998) advocate a loosely structured process for platform planning focused on three information management tools: the *product plan*, the *differentiation plan*, and the *commonality plan*.

The Product Plan

The product plan specifies the distinct marketing offerings over time and usually comes from the company's overall product plan that identifies the portfolio of products to be developed by the organization and the timing of their introduction to the market. The product plan identifies major models but does not show every variant and option. It is linked to other issues and pieces of information:

- Availability of development resources.
- Life cycles of current products.
- Expected life cycles of competitive offerings.
- Timing of major production system changes.
- Availability of product technologies.

The product plan reflects the company's product strategy. Some companies choose to issue several products simultaneously; others choose to launch products in succession.

The Differentiation Plan

The differentiation plan explicitly represents the ways in which multiple versions of a product will be different from the perspective of the customer and the market. The plan consists of a matrix with rows for the differentiating attributes of the product and with columns for the different versions or models of the product, where the last one gives an approximate assessment of the relative importance to the customer of each differentiation attribute. Differentiating attributes (DAs) are those characteristics of the product that are both important to the customer and intended to be different across the products. A common pitfall in platform planning is to become bogged down in detail. The best level of abstraction results in no more than 10–20 DAs.

The team uses the differentiation plan to codify each decision about how the product will be different. On the first pass, the differentiation plan represents the ideal case of each product's differentiation for maximum appeal to customers in the target segments. On subsequent iterations, the ideal case is adjusted to respond to the need for commonality.

The Commonality Plan

The commonality plan describes the extent to which the products in the plan share physical elements. The plan is an explicit accounting of the costs associated with developing and producing each product. It consists of a matrix with rows representing the chunks of the product (listed in the first column). The term *chunk* is used to refer to the major physical elements of a product. A set of products exhibits high levels of commonality if many chunks are shared. To manage complexity, the team should limit the number of chunks to roughly the number of DAs. The remaining columns identify the products in the plan, according to the timing of their development and the metrics used in the commonality plan as number of unique parts, development costs, tooling costs, and manufacturing costs. The values of these metrics are estimated because actual values cannot be determined until the products have been designed and produced.

Managing the Trade-Off between Differentiation and Commonality

The challenge in platform planning is to resolve the tension between the desire to differentiate the products and the desire for these products to share a substantial fraction of their components. For most product contexts, an unconstrained product plan and an unconstrained differentiation plan lead to high costs. For this reason, iterative problem solving is required to balance the need for differentiation with the need for commonality. This iterative activity involves both moving along the distinctiveness–commonality curve and exploring alternate product architectures with different associated trade-off characteristics (Figure 2.2.5).

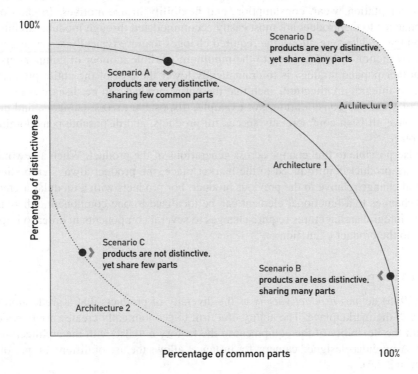

Figure 2.2.5 Trade-off between distinctiveness and commonality (adapted from Robertson and Ulrich, 1998)

2.2.5   Modular versus Integral Product Architectures

Product architecture is defined by Ulrich (1995) as the scheme by which the function of a product is allocated to physical components. The function of a product can be broken down into a number of *functional elements* (sometimes also called *functional requirements*), which are the individual operations that a product performs. The components of a product can be thought of as *physical elements*, which are the parts and subassemblies that realize the function of the product. The components are usually arranged into major building blocks, called *chunks*, made up of a number of components. The mapping between functional elements and components may be one-to-one, many-to-one, or one-to-many. Product architectures can be distinguished into *modular* and *integral* architectures. A modular architecture includes a one-to-one mapping from functional elements in the function structure to the physical components of the products, and specifies decoupled interfaces between components. An integral architecture, on the other hand, includes a complex multirelational mapping from functional elements to physical components and/or coupled interfaces between components. But which one is better?

Product Change

The architecture of a product is closely linked to the ease with which change to a product can be implemented. Products frequently undergo some change during their life due to upgrade, add-ons, adaptation, wear, consumption, and flexibility in use motives. In each of these cases, changes to the product are most easily accommodated through modular architectures. The modular architecture allows the required changes that are typically associated with the product's function to be localized to the minimum possible number of components. However, another popular strategy is to dramatically lower the cost of the entire product, often through an integral architecture, such that the entire product can be discarded or recycled in case of wear and consumption. For example, disposable razors, cameras, and cigarette lighters have all been commercially successful products, and disposable pens dominate the market place.

It is also possible to find change across generations of the product. When a new model of an existing product is introduced to the market place, the product always embodies some functional change relative to the previous product. For products with a modular architecture, desired changes to a functional element can be localized to one component, whereas products with integral architectures require changes to several components in order to implement changes to the product's function.

Product Variety

Ulrich (1995) defines *product variety* as the diversity of products that a production system provides to the marketplace. The ability of a firm to economically create variety resides not only with the flexibility of the equipment in the factory but also with the architecture of the product. A modular-designed camera, for instance, allows the use of different types of lenses (see Figure 2.2.6).

With a modular product architecture, product variety can be achieved with or without flexible component production equipment. In relative terms, in order to economically produce

Figure 2.2.6   Modularity in product design (See Plate 9 for the colour figure)

high variety with an integral architecture, the component production equipment must be flexible. This argument assumes in all cases that the final assembly process itself is somewhat flexible, that is, different combinations of components can be easily assembled to create the final product variety.

Component Standardization

Component standardization is the use of the same component in multiple products. Standardizing parts like axles, steering columns, and most importantly chassis helps the Volkswagen Auto Group to lower its production costs and cut assembly times. Shared parts also allow the Volkswagen Auto Group to produce cars from different brands such as Audi, Seat, Skoda, Scania, and Volkswagen at the same plant. A modular architecture increases the likelihood that a component will be commonly useful and also enables component interfaces to be identical across several products.

A standard interface, for instance, to connect all types of lenses (see Figure 2.2.7) allows the use of a wide variety of lenses for different types of cameras.

Component standardization has implications for the manufacturing firm in the areas of cost, product performance, and product development. Standard components can be less expensive (produced in high volume), exhibit higher performance (learning and experience), and lower the complexity, cost, and lead time of product development (known entity).

Product Performance

Ulrich (1995) defines *product performance* as how well the product implements its functional elements. Modular architectures allow for optimization of local performance characteristics by using a standard component or, when this is not available, modular architectures allow the component to be designed, tested, and refined in a focused way. Architecture also influences the improvement of global performance characteristics usually measured in

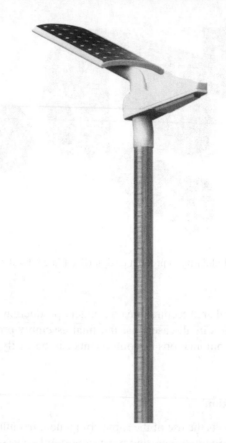

Figure 2.2.7   Outdoor lighting system (See Plate 10 for the colour figure)

such ways as efficiency, size, weight, and speed. An integral architecture facilitates the optimization of holistic performance characteristics and those that are driven by the size, shape, and mass of a product. Such characteristics include acceleration, energy consumption, aerodynamic drag, noise, and aesthetics.

The practice of implementing multiple functions using a single physical element is called *function sharing*. An integral architecture allows for redundancy to be eliminated through function sharing and allows for geometrical nesting of components to minimize the volume a product occupies. Such function sharing and nesting also allow materials to be minimized, potentially reducing the cost of manufacturing the product.

There are no deterministic approaches to choosing optimal product architecture. Ulrich (1995) concludes that in most cases the choice will not be between a completely modular or completely integral product architecture, but rather will be focused on which functional elements should be treated in a modular way (thus enabling the creation of one or more product platforms) and which should be treated in an integral way. And as pointed out by Meyer, Tertzakian, and Utterback (1997), excellence in product platform design is fundamental to the quality and success of a product family, consisting of a basic architecture composed of modules and the interfaces between these modules.

2.2.6   Measuring the Performance of Product Families

In managing product families, there are a number of frequently asked questions: *When should we renew our platform? How much will platform efforts cost, and how long should we expect them to take? What types of engineering and commercial benefits can we expect to gain from these efforts once they are completed? How can we improve our approaches and strategies for product development?*

Meyer, Tertzakian, and Utterback (1997) as well as Meyer and Lehnerd (1997) have developed measures to evaluate a platform's efficiency and effectiveness. To measure platform efficiency and effectiveness, the following data are required:

- *Engineering costs*: These costs comprise the amount of money spent on developing platforms and specific derivative products.
- *Development time*: The time spent from start to finish in platform and derivative product development. For derivative products, the development time cycle starts at the point of specific engineering work for the product itself and ends at the time of release to manufacturing.
- *Manufacturing costs*: These are the costs of upgrading a manufacturing facility to handle new products.
- *Market development costs*: These data may include expenditures on specific promotional campaigns for a new product, channel development expenses, and dealer-training programs.
- *Sales data*: Sales data for each product in a family should be aggregated across its full commercial life cycle.
- *Margins*: These data link profit to specific products.

Platform Efficiency

Platform efficiency is defined as the degree to which a platform allows the economical generation of derivative products.

$$\text{Platform efficiency} = \frac{\text{Derivative Product Engineering Costs}}{\text{Platform Engineering Costs}}$$

This metric can be used to understand development efficiency for an entire generation of derivative products, computing the average R&D costs of the derivative products for a particular platform version and then dividing that amount by that platform version's own development costs. One can also compare the platform efficiency of different platform versions across different product families.

What is a reasonable platform efficiency value? Meyer and Lehnerd (1997), in their study of firms in the electronics industry, indicated that platform efficiency values of 0.10 or less mark the presence of highly leverageable product platforms, meaning that efficient platforms allow the firm to produce derivative products at roughly 10% of the cost of developing associated base platform architectures. However, the desirable benchmark value for a particular company will be industry specific. They recommend studying a successful product family in the company, determining its level of platform efficiency, and then using that as a benchmark for similar groups inside the business.

If resources are being effectively used and learning is taking place, platform efficiency should improve with each successive platform version of a product family. Platform efficiency can be used as an indicator of platform demise. An increase in its value over successive derivative products may indicate weakness in the underlying product architecture. It can also signify a change in management or key resources, human or otherwise. Market factors can also substantially influence the efficiency with which existing platforms can be leveraged into new products, particularly when the market applications are novel to the firm.

Note that the metric, so far, has included only engineering costs. A more comprehensive understanding emerges if one considers the cost ratios for introducing new products into full-scale production.

Platform Effectiveness

Platform effectiveness is defined by Meyer, Tertzakian, and Utterback (1997) as the degree to which product platform–based products produce revenue for the firm relative to the cost of developing those products. Platform effectiveness simply compares the resources used to create products with the revenues derived from them, *over the long term*. The cost of development includes the engineering costs, manufacturing engineering costs, market development costs, and expenditures for plant and equipment. Revenues consist of net sales attributable to each derivative product within the family. Platform effectiveness for a single derivative product is represented as follows (Meyer and Lehnerd, 1997):

$$\text{Platform effectiveness for a single product} = \frac{\text{Net Sales of a Derivative Product}}{\text{Development Costs of a Derivative Product}}$$

The effectiveness measure can be aggregated for an entire generation of products, allowing one to compare the performance of successive generations:

$$\text{Platform effectiveness for a generation of products}$$
$$= \frac{\text{Net Sales of Derivative Products of a Platform Version}}{\text{Development Costs of Derivative Products of a Platform Version} + \text{the Platform Development Costs}}$$

As pointed out by Meyer *et al.*, there are many factors that can lead to declining platform effectiveness. First, the platform itself may have outlived its utility as a basis for creating specific products that are competitive in the market in their features and cost. This would lead to declining sales, affecting the numerator of the effectiveness equations. As for the denominator, R&D costs may be rising due to problems in platform efficiency. Besides, a market either in explosive growth or in free fall will greatly impact the metric.

How is *platform effectiveness* defined for your company's product lines? Meyer and Lehnerd (1997) again suggest that platform effectiveness values will vary from industry to industry and that the most pragmatic way to establish a benchmark for platform effectiveness in the company is through internal study, observing those values associated with the most successful product families.

Taken together, these measures of platform efficiency and effectiveness should be interpreted to evaluate if a specific product platforms needs renewal. When the underlying architectures of major product lines are running out of gas in their ability to (1) facilitate rapid and cost-effective development of derivative products, and (2) deliver the features required by customers at price point that would lead to continued strong sales, it will be necessary to fund and sequence new platform development projects to supersede existing platforms.

2.2.7   Managing Risk in Platform-Based Development

Platform-based development may not be appropriate for all product and market conditions. It still remains difficult to anticipate the consequences of risky platform decisions in advance.

Developing the initial platform in most cases requires more investments and development time than developing a single product, delaying the time to market of the first product and affecting the return on investment time. Also, the failure to develop new platform architectures on a continuous basis subjects the firm to substantial market risk. It stands to reason that if new platform development efforts take longer to complete than derivative products efforts, R&D aimed at platform renewal should be pursued concurrent with derivative products developments on existing platforms. That ensures a continuous stream of products embodying competitive technology (Meyer and Lehnerd, 1997; Meyer, Tertzakian, and Utterback, 1997) Long-term success and survival require continuing innovation and renewal. A potential negative implication of a modular product architecture approach is the risk of creating barriers to architectural innovation. This problem has been identified by Henderson and Clark (1990) in the photolithography industry and may in fact be a concern in many other industries as well.

On top of the fixed investments in developing platforms, platforms may also result in the over-design of low-end variants in a firm's product family to enable subsystem sharing with high-end products (Krishnan and Gupta, 2001). An additional risk concerns the balance between commonality and distinctiveness. A weak common platform will undermine the competitiveness of the entire product family, and therefore a broad array of products will lose competitiveness. Robertson and Ulrich (1998) have pointed out organizational risks related to platform development. Platform development requires multifunctional groups. Problems may arise over different time frames, jargon, goals, and assumptions. In a lot of cases, organizational forces also seem to hinder the ability to balance between commonality and distinctiveness. Engineers, for example, may prepare data showing how expensive it would be to create distinctive products, while people from marketing may argue convincingly that only completely different products will appeal to different markets. One perspective can dominate the debate in the organization.

Finally, the metrics as suggested by Meyer, Tertzakian, and Utterback (1997) can help to monitor, but they do not explicitly say when to create a new platform, and companies can fail to embark in platform renewal in a timely manner.

2.2.8   Application of Platform-Driven Product Development

The student project by Scholder and Friso (2007) comprised the conceptual development of an outdoor LED lighting system for public spaces powered by PV modules (see Figures 2.2.7 and 2.2.8).

During the design process of this project, the students applied platform-driven product development to be able to design suitable solutions for various markets. They were thinking

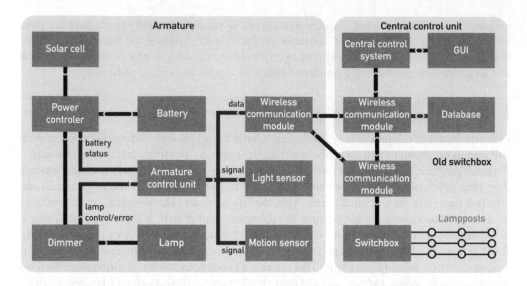

Figure 2.2.8 Scheme representing the technical functioning of this product

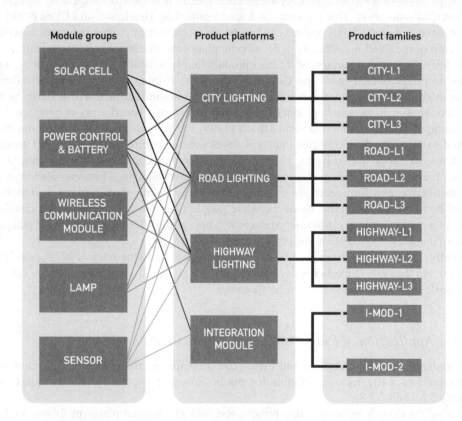

Figure 2.2.9 Product platform for the product shown in Figure 2.2.7

about different types of urban lighting such as city lighting, road lighting, and highway lighting (see Figure 2.2.9). In this way they could ensure a short time to market and guarantee a high reliability. By using a lot of standard modules, the costs for development and production are reduced, and the number of parts in stock can be reduced.

References

Baldwin, C.Y. and Clark, K.B. (1999) *Design Rules: The Power of Modularity*, MIT Press, Cambridge, MA.

Gupta, A.K. and Souder, W.E. (1998) Key drivers of reduced cycle time. *Res. Technol. Manage.*, **41**(4), 38–43.

Halman, J.I.M., Hofer, A.P., and van Vuuren, W. (2003) Platform driven development of product families: Theory versus practice. *J. Prod. Innovat. Manag.*, **20**(2), 149–162.

Henderson, R.M. and Clark, K.B. (1990) Architectural innovation: The reconfiguration of existing product technologies and the failure of established firms. *Admin. Sci. Quart.*, **35**(1), 9–30.

Kahn, B.E. (1998) Dynamic relationships with customers: High-variety strategies. *J. Acad. Market. Sci.*, **26**(1), 45–53.

Krishnan, V. and Gupta, S. (2001) Appropriateness and impact of platform-based product development. *Manage. Sci.*, **47**(1), 52–68.

McGrath, M.E. (1995) *Product Strategy for High-Technology Companies*, Irwin, Homewood, IL.

Meyer, M.H. (1997) Revitalize your product lines through continuous platform renewal. *Res. Technol. Manage.*, **40**(2), 17–28.

Meyer, M.H. and Lehnerd, A.P. (1997) *The Power of Product Platforms: Building Value and Cost Leadership*, Free Press, New York.

Meyer, M.H., Tertzakian, P., and Utterback, J.M. (1997) Metrics for managing research and development in the context of the product family. *Manage. Sci.*, **43**(1), 88–111.

Robertson, D. and Ulrich, K. (1998) Planning for product platforms. *Sloan Manage. Rev.*, **39**(4), 19–31.

Sanderson, S. and Uzumeri, M. (1995) Managing product families: The case of the Sony Walkman. *Res. Policy*, **24**(5), 761–782.

Sawhney, M.S. (1998) Leveraged high-variety strategies: from portfolio thinking to platform thinking. *J. Acad. Manag. Sci.*, **26**(1), 54–61.

Scholder, F. and Friso, T. (2007) Sources of Innovation. Student report, Master of Industrial Design Engineering program, University of Twente.

Ulrich, K. (1995) The role of product architecture in the manufacturing firm. *Res. Policy*, **24**(3), 419–440.

2.3   Delft Innovation Model in Use

Jan Buijs

Delft University of Technology, Faculty of Industrial Design Engineering, Department of Product Innovation Management, Landbergstraat 15, The Netherlands

2.3.1   Introduction

In the use of products and services, the need for new products and services will develop; namely, by using the product, users get good or bad usage experiences. In the case of good usage experiences, they will be loyal to the product and the company or brand that is offering the product; with bad usage experiences, they will look around for better product offerings and maybe change over to other products or brands. However, if a company notices that some of its clients have switched over to the products of innovating competitors, they will probably react by offering better or cheaper products, enabling their former users to switch back to the new offering.

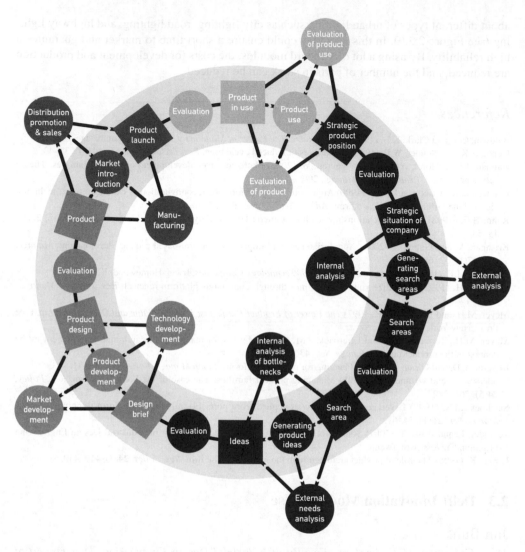

Figure 2.3.1 The Delft innovation model (Buijs and Valkenburg, 2005). Top area: Product use. Black area: Fuzzy front end. Light grey area: Development stage. Dark grey area: Market introduction (See Plate 11 for the colour figure)

To capture the product innovation process, we developed the Delft innovation model (DIM) for educating designers and engineers (Buijs and Valkenburg, 2005; see Figure 2.3.1). For the history and background of the development of this model, we refer to Buijs (2003). This model of the innovation process is one of the elements of the Delft innovation method. For more details about the overall method, we refer to Buijs (2012). In the circular representation of the Delft innovation model in Figure 2.3.1, it can be seen that the product innovation process bites itself in the tail. It shows the five stages (starting from the top and moving clockwise): (1) *product use*, then (2) *strategy formulation*, followed by (3) *design brief formulation*, then on the 8 o'clock position is the (4) *development* stage, and finally at 11 o'clock is (5) *market introduction*.

Section 2.3 is structured as follows: In Sections 2.3.2 and 2.3.3, an explication is given about the conceptual model of the innovation process within a company that was built to help practitioners in the field to structure and organize their innovation process. This is followed by Section 2.3.4, which features two practical cases in which master students of the University of Twente show their abilities to apply the model in real-life product innovation cases. Special attention is paid to their struggling with the model: They have to tweak and adjust the model for each specific case, without dumping the main message of the model. This chapter is closed by a meta-reflection by the builder of the model about their playing with his ideas (Section 2.3.5).

2.3.2   Stages of the Delft Innovation Model

The Product Use Stage

The starting point of the *product use* stage is the *product in use* (see Figure 2.3.2). This means that the consumers can buy and use the product. Now the product has to fulfill its promises. Does it do what the company, the brand image, and the nice advertisements have been promising? The new product is tested against the real-world competition.

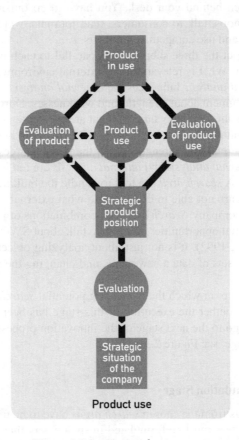

Figure 2.3.2   The *product use* stage

This *product Use* stage is the most important stage of the product innovation process, but it is also the stage the company has the least influence on. Of course they discovered the original need, they designed the product, and they are responsible for the manufacturing, distribution, and sales, but now the product has to fly for itself. Now it is up to the user to verify all the assumptions and ideas that the company has put into the new product's specifications.

Product use of a new product will lead to a change in the strategic position of the company that is offering the product. The *strategic position* is the position of the company within its competitive arena.

The *strategic position* is dependent not only on the company's strategic actions but also on the competition's actions and reactions. Evaluating this change will reveal a new *strategic situation of the company*, and must ultimately lead to the decision to start a new product innovation process. Figure 2.3.2 summarizes the *product use* stage.

The Strategy Formulation Stage

The *strategy formulation* stage should start with external investigations to check and validate the earlier recognized need for innovation in the *product use* stage. An *external analysis* cannot be executed from behind your desk. You have to go outside, into the real world. Observe clients, visit shops, talk to consumers, read journals, watch television, be at cool and hot places, and buy and use competitive products.

The reason we have put the three subprocesses parallel to each other is because they are mutually interdependent. What is relevant in the external environment depends on what the company is. The *internal analysis* influences the *external analysis*. One of the results of the *internal analysis* is a judgment about strengths and weaknesses. During the *internal analysis*, we discover what makes this company unique, what are the strategic strengths, and what are the core competences. Innovating is risky; therefore, it is better to build the innovation on the company's strategic differentiating strengths rather than on resources every company has.

The results of the *external analysis* and *internal analysis* are fed into the central process of generating search areas. A *search area* is a tool to handle the bulkiness of the relevant external world. Up front, we are not able to discriminate what external trends and developments are relevant for a given company. *Search areas* are combinations of internal company (strategic) strengths with external opportunities. We like to talk about SWOT synthesis because it is a creative activity (Buijs, 1992). It is not just about analyzing objective data; no, it is about formulating data, giving sets of data a new name, and changing the meaning of data by giving them another name.

After the evaluation step, in which the interesting potential *search areas* are selected, and in which it is checked whether the execution of this stage has been up to expectations, the company can decide to go to the next stage of the innovation process. For a summary of the *strategy formulation* stage, see Figure 2.3.3.

The Design Brief Formulation Stage

In this stage the raw ideas from the chosen *search areas* have to be transformed into concrete product ideas. Product ideas must be formulated in such a way that a new product development department can start designing the new product.

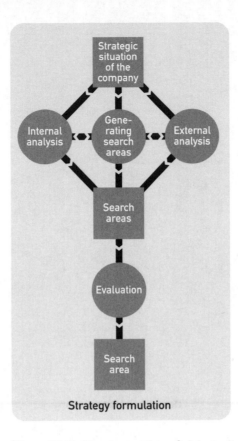

Figure 2.3.3 The *strategy formulation* stage

First we have to find out what internal bottlenecks could prohibit the company from going on with the chosen *search area*. *Internal bottlenecks* can be other strategic priorities, conflicting projects, missing knowledge, or other resources. If these bottlenecks look too big, the *search area* has to be put in the refrigerator, and probably the earlier *strategy formulation* stage has to be redone (or one of the other chosen *search areas* has to be investigated).

During the *external needs analysis*, we are really diving deep into the needs of targeted potential consumers. We are going to visit and talk to our potential clients. We have to find out what really drives people, what needs and wants they have, what they think about competing products, and why they want to spend money on substituting offerings. This is a reality check of the proposed *search area*.

Based on this external information, the third process can be started: *generating product ideas*. All kinds of creativity techniques can be used to come up with ideas for new products (see creativity gurus like De Bono, 1985; Gordon, 1961; Osborn, 1957; Rickards, 1974). You need hundreds of ideas to get one implemented. Generating ideas is the core of this stage, and all rules about creativity should be applied.

In the evaluation step, a couple of product ideas will be considered to become the future champions. They have to be formulated in the *design brief*.

For a summary of the design brief formulation stage, see Figure 2.3.4.

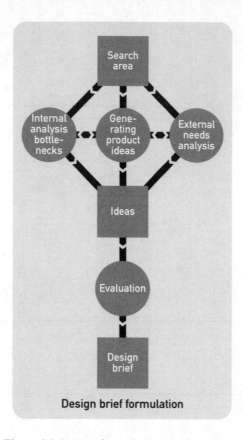

Figure 2.3.4   The *design brief formulation* stage

The Development Stage

This is the stage of all traditional product design and engineering activities. The *design brief* is a starting point, followed by three parallel processes: development process of the product, development of the market, and development of the technology.

The core is product design. Do not forget to wake up the targeted market. It is a new product, so nobody in the marketplace knows anything about it. Great emphasis on marketing and promotion activities is essential for successful innovating companies. The *development* stage ends with at least working prototypes that probably even have been tested with the potential clients.

Figure 2.3.5 represents the *development* stage of the DIM.

The Market Introduction Stage

This is the final stage of the DIM. This is the last opportunity for the innovating company to actively do something to ensure the wanted qualities of the new product. The core result is the *product launch*. We refer to this moment as "Steve Jobs in concert," because Apple has been famous for its new product introduction parties where Steve Jobs used to talk about and show the new product. In Figure 2.3.6 the *market introduction* stage is summarized.

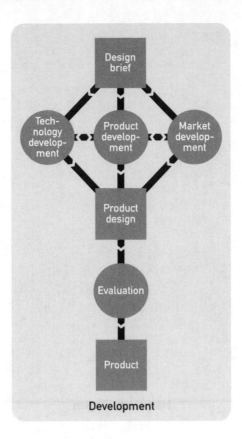

Figure 2.3.5   The *development* stage

Now the new product is for sale, and next the *product use* stage (see Figure 2.3.2) can begin, which will close the innovation cycle. This will probably cause a future change of the *strategic position* of the innovating company and its competitors. The competitive environment is also influenced and changing. The next innovation game will soon be on!

2.3.3   *Concluding Remarks on the Delft Innovation Model*

As we all know, models are just abstract representations of reality. They mimic only parts of reality, not all of it. With the introduction of the DIM, we want to help companies to improve their innovation activities, but it is not a cookbook. You have to translate our ideas about the innovation process to your own world.

In real life the five stages of the innovation process have different duration times. *Product use* can be as long as decades, *market introduction* can be as short as one day, *development* runs from a couple of weeks to 2–3 years, and *design brief formulation* and *strategy formulation* both last a couple of months, but sometimes it can take years to realize and experience the need for innovation. In our model we visualize them as five equal stages, and we suggest that all subprocesses within each stage should be executed jointly, but in reality that is not true.

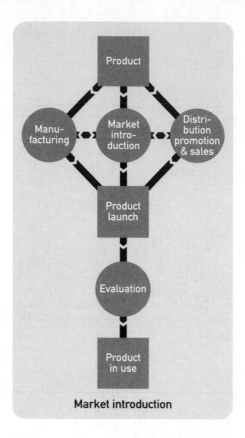

Figure 2.3.6   The *market introduction* stage

We want to conclude with some insights based on empirical studies about the product innovation process and relate them to the DIM model. About 30 years ago Cooper tried to get grip on the differences of innovation processes in the real world compared to in theory (Cooper, 1983). He discovered that not all theoretical innovation activities had been used during all innovation processes (apparently you can sometimes skip certain actions). Also he noticed different time patterns within the innovation process. From 1980 to 1987, we executed a large empirical study (the so-called *Pii study*) inside companies using an earlier version of the DIM model (Buijs, 1984, 1987). We also revealed different time patterns over the different stages of the innovation process. In 2008 we published another empirical study on different innovation patterns (Buijs, 2008). Complex projects (e.g., with new technology and/or a new client) need more activities in the first stages, and also different activities than in noncomplex projects. We also found that innovation leaders adapt an opportunistic attitude toward carrying out activities in parallel in order to gain time. It appears that project-specific circumstances lead toward skipping steps and adopting different time patterns in various product innovation projects. A possible explanation for these differences between the theoretical way innovation process should be run, and the practical way they are run, is that the theory considers innovation *processes*, while in practice companies run innovation *projects*. This could mean, for instance, that in innovation project Q the *external analysis* is skipped, because the marketing department had just finished a report about changes in the outside world. And the project leader of innovation project Q just grabs this report and uses it. So in the project there was no time spend on the *external analysis*, although the necessary information became available from another source. Another example could be a company that just acquired a small company because it developed an interesting new technology. Once again in innovation project Z there is no time spent on *technology development* (see the left-hand circle in Figure 2.3.5), but inside that acquired company other people have spent a lot of time on it. So be aware about this discrepancy between innovation projects in real life and innovation processes according to theory.

2.3.4 Applying the Delft Innovation Model in Real Life

Knowing what to do in the innovation process is one thing, but executing it is another. This section describes two applications of the method by master's-level students in the Industrial Design Engineering faculty of the University of Twente. These projects were done for the Sources of Innovation course, where students were instructed to use a variety of innovation methods in the design of a product.

UV Light for Food Preservation

Van Ettinger and Jansma (2008) applied the DIM to a student design project that required the design of a product using LED technology. This resulted in the design of an ultraviolet (UV)–LED light for food preservation in refrigerators. The DIM was used together with other design methods (innovation journey, platform-driven product development, and innovative design and styling), and the authors describe how they applied various models to their process:

> By following this path of innovation methodology, the whole industrial design process is covered. By beginning with the journey, the supporting technology will be examined and the historical aspect of the product becomes clear. *With the help of the DIM model, the clarification of the task will be sorted out and also the optimization of the elements.*

It appears that they used the DIM for task clarification, once the decision was made to explore food preservation as an outcome of the innovation journey method of the previous phase. This resulted in initial product ideas, for example the use of UV radiation to damage bacteria, and color coding to categorize food types, which were incorporated in the final design. This loosely corresponds to the strategy formulation stage of the DIM, whose starting point in the theoretical model is the strategic situation of the company. There are some differences between the DIM-in-theory and the application by these students in practice. One of the most basic is that there is no "company" for which to conduct an internal analysis in this project. Rather, the students have chosen to analyze the possibilities (strengths) of the technological possibilities and initial concept ideas: "the strength of the idea is the dual effect of controlling the e-coli bacteria (and) . . . color LEDs (to give) an indication to the customer of separating different types of food . . . minimal size . . . low energy consumption." In

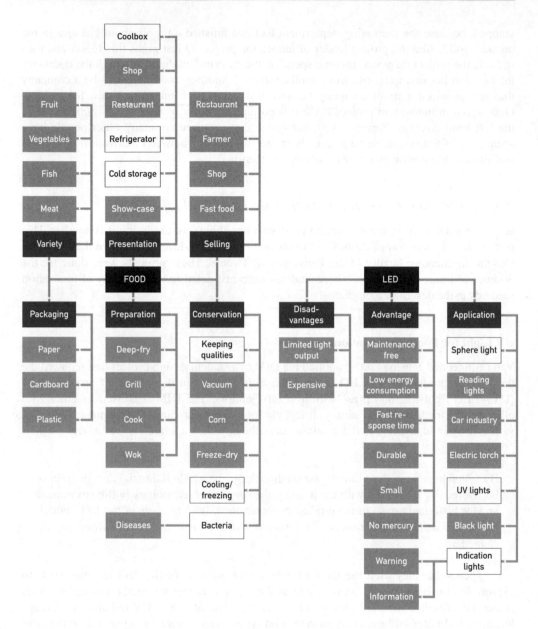

Figure 2.3.7   DIM mind maps (Van Ettinger and Jansma, 2008)

contrast to the theoretical model, challenges and weaknesses were given little consideration (although the potential psychological barriers to using UV light with food were noted). Mind maps were used to structure and visualize their DIM approach. These were in two areas, LED lighting and food, and are shown in Figure 2.3.7.

Analysis of these gave the authors the following search areas: sphere and indication light sources, ultraviolet light and radiation, specific cooling systems, and keeping the qualities of

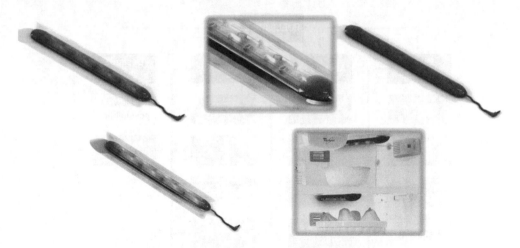

Figure 2.3.8 Final product (Van Ettinger and Jansma, 2008) (See Plate 12 for the colour figure)

food. In the theoretical model, search areas are an input to the design brief stage. In these students' work, it is an input for the platform-driven product development method, where the components needed to implement loose concepts based on the search areas are evaluated.

The resulting product after use of the DIM and other innovation methods is shown in Figure 2.3.8. Details regarding form were developed in later stages, but the fundamental decisions made while applying DIM regarding product function and use were retained.

T-Spider Fuel Cell Vehicle

Similarly, students Damkot and de Groot (2006) use the DIM early in the design process for a fuel cell–powered vehicle. The students chose to apply the method "because of its possibilities. . . . In this project, there is a new technology in a market with opportunities. The DIM Model has the capacity to combine the intrinsic values of technology with opportunities in the market." Guidelines resulting from their use of the DIM were used as inputs for platform-driven product development.

They used the innovation phase model "to optimally combine the technology with the opportunities in the city." Two mindmaps were used – one for technology, and another for cities – to visualize and structure their use of DIM.

The authors describe the mind maps shown in Figure 2.3.9 as follows: "Technology contains 4 subgroups; disadvantages, advantages, applications and Design and Styling. City contains 5 subgroups; environment, infrastructure, transportation, opportunities and problems. Transportation is split up into public transport and individual transportation. The used points for the search fields are marked green. There are six market fields and five technical fields" (see Figure 2.3.10).

After creation of the individual mind maps, a cross comparison of these technical and market-oriented search fields was made, as shown in Figure 2.3.10. A principal result of this comparison is the decision to design a small vehicle (Figure 2.3.11). As with Van Ettinger and Jansma (2008), the design here is split into different components for consideration in the next stage of platform-driven product development.

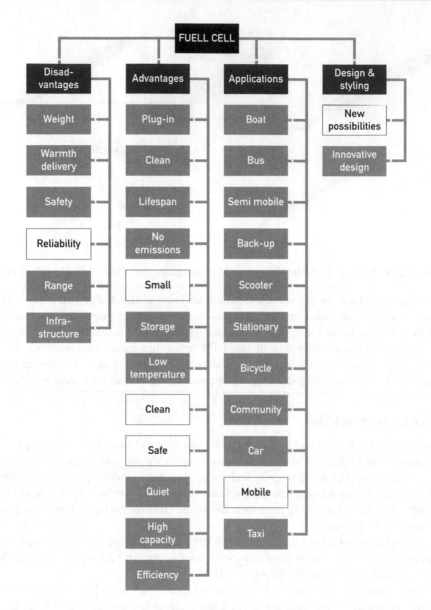

Figure 2.3.9   Search field selection using mind maps (Damkot and de Groot, 2006)

2.3.5   Reflections on the Delft Innovation Model in Practice

It should be noted that there are important differences between the DIM-in-theory and in practice. The student cases described here deal with a speculative situation where both constraints and additional resources are not present or are artificially given.

In a real-world setting, DIM is meant to provide a general approach to the design process rather than strictly prescribe a step-by-step approach. The metaphor of a recipe is given: It is

		Market fields					
		Improve mobility	Scooter	Bikeways	Roads / highways	Parking space	Router restrictions
Technical fields	Small	1	2			3	
	Clean		4				
	Safe			5			
	Mobile						
	Reliability	6					

Figure 2.3.10   Search field comparison (Damkot and de Groot, 2006)

Figure 2.3.11   Final product (See Plate 13 for the colour figure)

not so much a rigid application of the process that is intended, but rather an understanding of ingredients and their essential qualities, how they interact with each other, what can be substituted or skipped, and how to work with available resources. It is this deeper level of understanding that allows not only for improvisation, but also for greater possibilities for arriving at original outcomes.

The external environment within which innovation takes place is an important consideration. The natural inertia of organizations is to continue with the status quo. Change requires energy and is often disruptive. This places the main driving force for innovation in the external environment: such as the activities of competitors, technological developments, or access to new markets. This external location for the stimulus to innovate is another source of differentiation between applying the DIM in education and in practice. One of the educational implications is that while the real world cannot be perfectly modeled, it is important for instructors to try to approximate some of its critical elements in developing cases, and for students to be clear about assumptions they make regarding the context of their projects.

References

Buijs, J.A. (1984) *Innovatie en Interventie*, Kluwer, Deventer.

Buijs, J.A. (1987) *Innovatie en Interventie*, 2nd enlarged ed., Kluwer, Deventer.

Buijs, J. (1992) Fantasies as strategic stepping stones: From SWOT-analysis to SWOT-synthesis, in *Creativity and Innovation: Quality* (ed. T. Rickards, S. Moger, P. Colemont, and M. Tassoul), TNO, Innovation Consulting Group, Delft.

Buijs, J. (2003) Modelling product innovation processes, from linear logic to circular chaos. *Creativity Innovat. Manag.*, **12**(2), 76–93.

Buijs, J. (2008) Action planning for new product development projects. *Creativity Innovat. Manag.*, **54**(4), 319–333.

Buijs, J. (March 2012) *The Delft Innovation Method: A Design Thinker's Guide to Innovation*, Eleven Publishers, The Hague.

Buijs, J. and Valkenburg, R. (2005) *Integrale Productontwikkeling* (3rd extended version), Lemma, Utrecht.

Cooper, R.G. (1983) The new product process: An empirically-based classification scheme. *R & D Management*, **13**(1), 1–13.

Damkot, R. and de Groot, F. (2006) *T Spider*. Student report, Sources of Innovation course, University of Twente, Enschede, the Netherlands.

De Bono, E. (1985) *Six Thinking Hats*, Key Porter Books, Toronto.

Van Ettinger, P. and Jansma, S. (2008) You keep it cool . . . we keep it clean, Student report, Sources of Innovation course, University of Twente, Enschede, the Netherlands.

Gordon, J.J. (1961) *Synectics: The Development of Creative Capacity*, Harper & Row, New York.

Osborn, A. (1957) *Applied Imagination*, C. Scribner & Sons, New York.

Rickards, T. (1974) *Problem Solving through Creative Analysis*, Gower Press, Epping.

2.4   TRIZ: A Theory of Solving Inventive Problems

Valeri Souchkov

ICG Training & Consulting, Willem-Alexanderstraat 6, The Netherlands

2.4.1   Introduction

TRIZ (a Russian acronym for the "theory of solving inventive problems") was originated in 1946 by Russian military patent examiner Genrich Altshuller. His main focus of interest was to understand how inventors came up with creative solutions. To reach his goal, he studied more than 400,000 patents intentionally drawn from different areas of industries. Such massive studies helped Altshuller to capture the nature of the creative process behind producing inventions. He identified a relatively small number of high-order patterns and principles that complied with the majority of inventions and that were general for most industries (Altshuller, 1969). Another important TRIZ discovery was that technology, like any other type of human activity, does not evolve randomly. Each product in every area of technology follows a certain sequence of "inventive transformations" to meet continuously evolving market demands and requirements. Long-term studies of technology evolution revealed common patterns and trends, according to which seemingly different systems evolve. Studies of these trends and patterns form a large part of TRIZ that is known as a *theory of technical systems evolution.*

By now in total, more than 1.5 million patents and technological solutions were studied to create modern TRIZ. Although previously relatively little known outside of the former Soviet Union, TRIZ is gradually becoming a key component for supporting invention and innovation since it helps organizations and individuals to establish innovation as a scientifically based, structured, and predictable process. In short, TRIZ enables one to create new value on demand. Modern TRIZ is a result of research efforts of more than 50 years in Soviet Union, and lately in Europe and the Unites States. It offers both a theory of invention and a number of practical techniques, which help to perform the complete "idea cycle": from analysis of a given situation to generation and evaluation of new ideas.

2.4.2   Components of TRIZ

Modern TRIZ includes different methods and techniques to support different stages of the ideas generation process:

- *Analytical techniques*: A group of techniques that help to manage the complexity of problem situations, look at problems from different angles, understand real problem causes and formulate problems correctly, and extract and predict future problems.
- *Ideas generation and inventive problem-solving techniques* that define methods for solving inventive problems.
- *TRIZ knowledge bases*, which contain generic patterns and guidelines of solution strategies and patterns of "strong" solutions, which can be applied to virtually any new problem to quickly come up with new solution ideas. TRIZ also includes a unique database of scientific effects structured according to technological functions.
- *Theory of technology evolution*: Models of evolution of technical systems and techniques for forecasting future product and technology evolution; tools to explore innovative potential of systems and generate new ideas on the basis of technical systems evolution trends.
- *Psychological techniques to enhance creativity*: A group of techniques to overcome mental inertia and further develop creative imagination.
- *Evaluation and ranking techniques*: A group of techniques that help to select problems and rank generated ideas.

The use of TRIZ is organized within the systematic innovation process (Figure 2.4.1), which structures the use of different TRIZ techniques and tools according to the desired outcome. In the following subsections, we present some most important concepts and tools of TRIZ.

2.4.3   Contradiction as a Driving Force of Invention

Existing technical systems are usually improved incrementally to achieve the most optimal values of their parameters. However, when incremental improvements do not lead to a desired result, a radical improvement takes place. We usually call such radical improvements

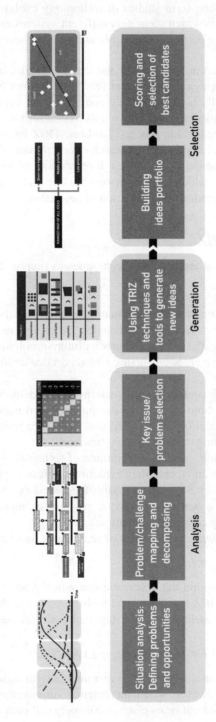

Figure 2.4.1 The systematic innovation process with TRIZ

Plate 1 Figure 1.3.2 Impressions of typical use circumstances of solar home systems in Java, Indonesia (pictures courtesy of Angèle Reinders).

Plate 2 Figure 1.3.3 Different lighting products based on sustainable technologiestechnologies, right: Virtue of Blue by DeMakersVan (2010).

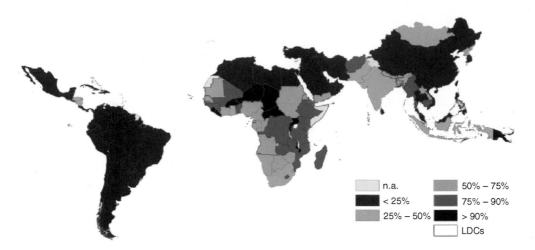

	n.a.		50% – 75%
	< 25%		75% – 90%
	25% – 50%		> 90%
			LDCs

Plate 3 Figure 1.6.1 Share of the population without access to electricity (UNDP/WHO, 2009).

Plate 4 Figure 2.1.6 A solar-powered street-lighting product, exploded view by Scholder and Friso (2007).

Plate 5   Figure 2.1.10 Communication device for hikers with integrated fuel cells. Left: Front view. Right: Opened back view. Designed by Kyra Adolfsen and Benne Draijer.

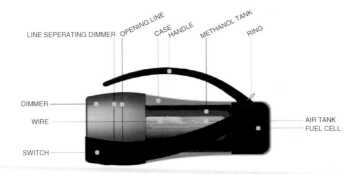

LINE SEPERATING DIMMER   OPENING LINE   CASE   HANDLE   METHANOL TANK   RING

DIMMER

WIRE

AIR TANK
FUEL CELL

SWITCH

Plate 6   Figure 2.1.16 Diving lamp powered by fuel cells by Julia Garde and Annelies Brummelman.

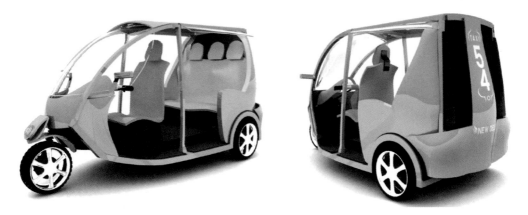

Plate 7   Figure 2.1.18 Fuel cell–powered rickshaw (Designed by Simon Brandenburg and Werner Helmich (2006)).

Plate 8   Figure 2.1.19 Induction cooking device powered by fuel cells. Left: The product in use. Right: In compact carrying case (Designed by Rick Tigchelhoff and Wouter Haanstra).

Plate 9   Figure 2.2.6 Modularity in product design.

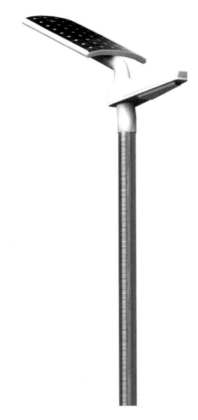

Plate 10   Figure 2.2.7 Outdoor lighting system.

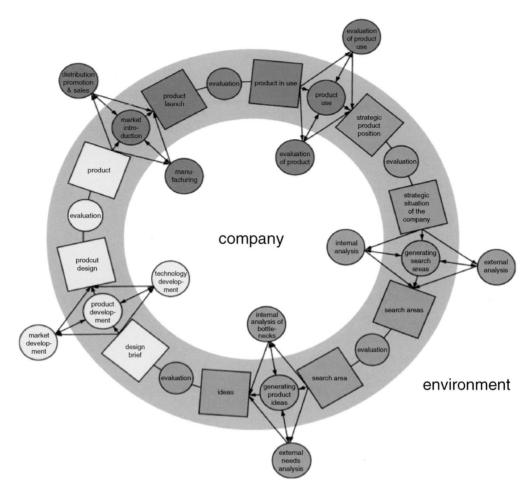

Plate 11 Figure 2.3.1 The Delft innovation model (Buijs and Valkenburg, 2005). Green area: Product use. Blue area: Fuzzy front end. Yellow area: Development stage. Orange area: Market introduction.

Plate 12   Figure 2.3.8 Final product (Van Ettinger and Jansma, 2008).

Plate 13   Figure 2.3.11 Final product.

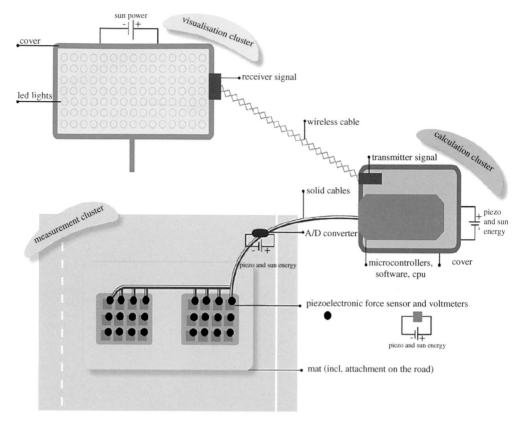

Plate 14 Figure 2.4.10 Scheme of the technical functioning of the BandOK product concept (Geurds and Van der Molen, 2011).

Plate 15 Figure 2.5.5 An impression of living cubes by Erkel and Reilink (2008).

inventions, which are new, non-obvious solutions unknown in the past. Altshuller found that in most cases, an inventive solution results from resolving a contradiction, which arises when an existing product or system is not capable of meeting new or growing market demands any longer, but known improvements of product features would cause an unacceptable change of some other product features. For instance, to increase a car's stability on a road, one of the known solutions is to increase the weight of the car. But this would require greater fuel consumption. Finding a totally novel way of improving the car's stability without increasing its weight consumption will result in a patentable invention.

TRIZ states that to obtain inventive solutions, contradictions must be eliminated without compromising. With a new solution, we should be able to achieve the desired effect in full without diminishing other advantages of a product or technology, and without causing negative side effects. For instance, coffee in a cup gets cold over time due to its contact with the much colder air. An obvious solution would be to cover a cup with a lid. But did we solve a contradiction? Not completely, because to drink coffee we will have to remove the lid, and this increases our discomfort. Therefore we have a contradiction between two parameters: the desired temperature of coffee and the degree of our comfort. According to TRIZ, the best solution will keep coffee hot without introducing any discomfort during drinking. In addition, this best solution should eliminate the contradiction without introducing new negative effects and, very importantly, without increasing costs. A good example of resolving this contradiction is cappuccino: a milk foam layer insulates coffee from the air without introducing obstacles for drinking. But, what about those who prefer other types of coffee? This contradiction is still to be solved.

With TRIZ, contradictions are solved at the abstract level by following certain generic principles rather than by numerous trials and errors. Instead of directly jumping to a solution, a problem is analyzed and formulated as a contradiction, and then a number of relevant heuristic *inventive principles* are proposed that contain the best solution strategies from previous inventors' experience (Figure 2.4.2). Each inventive principle suggests a number of

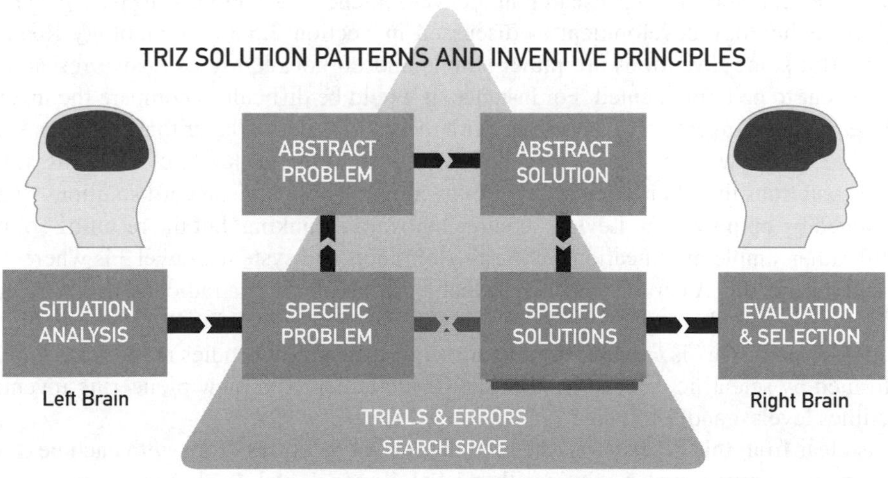

TRIZ SOLUTION PATTERNS AND INVENTIVE PRINCIPLES

ABSTRACT PROBLEM → ABSTRACT SOLUTION

SITUATION ANALYSIS → SPECIFIC PROBLEM → SPECIFIC SOLUTIONS → EVALUATION & SELECTION

Left Brain

Right Brain

TRIALS & ERRORS SEARCH SPACE

Figure 2.4.2   Solving problems with TRIZ

recommendations that can be used to solve a given problem (Altshuller and Fedoseev, 1998). Examples of such inventive principles are as follows:

Principle of Nesting

- Place one object inside another.
- Increase the number of objects nested.
- Make one object pass through a cavity of another object.
- Introduce a new process inside of an existing process.
- Increase the number of "nested" processes.
- Make process activities dynamically appear when needed and disappear when not needed.

Principle of Spheroidality

- Instead of linear parts of an object, use curved parts.
- Use rollers, balls, and spirals.
- Use rotary motion.
- Use centrifugal forces.
- If a process is nonlinear, consider increasing the degree of nonlinearity.
- Use circular flow instead of linear flow.
- Use roundabout solutions in a process.

This knowledge-based approach drastically reduces amount of time needed to find a specific solution. One of the best-known TRIZ techniques is called *40 inventive principles*. The principles are used through the so-called *contradiction matrix*, which contains references to specific inventive principles according to over 1,000 types of generic technical contradictions.

2.4.4   Five Levels of Solutions

Not all inventions are equal. Some inventions result in truly new, breakthrough technologies that require relatively a long research and development time and have to pass through all stages of technology development as discussed in Section 2.5.3, "Technology Readiness Levels." But some inventions are rather simple and do not require extensive research and development to be implemented. For instance, it would be difficult to compare the invention of a laser with the invention of adding an extra thermal insulation layer to a coffee maker.

Figure 2.4.3 shows how TRIZ distinguishes between different levels of solutions. Level 1 includes solutions that do not really require invention: These are standard solutions that can be obtained by optimization. Level 2 requires innovative thinking, but the resulting solutions are still rather simple modifications of existing products and systems. Level 3 is where "real" innovation takes off: A certain system, product, or principle finds a radically new application area. Level 4 includes so-called "pioneering" inventions: A radically new combination "function or principle" is created (usually drawn from scientific studies at level 5). And level 5 is formed by scientific discoveries that can be implemented in new pioneering inventions. Sometimes levels 4 and 5 coincide.

As is clear from this description, the total number of solutions drops with each next level. There are many more level-2 solutions than level-4 ones. Level-4 solutions represent only a fraction of all known technical solutions.

Level	Features	Examples	Example Illustration
Level 5: Discovery	Discovering a new scientific principle	X-ray discovery, (or radio waves discovery, coherent light discovery, etc.)	
Level 4: Pioneering Invention	Creating a radically new Function/Principle combination	X-Ray radiation (principle) is used to "see through" (function) a human body, thus launching a new technology area: X-Ray medical machines	
Level 3: Concept Transfer	The use of a known Function/Principle combination in a new application area (market)	X-Ray technology is brought to other areas: non -destructive testing of constructions; X-Ray security systems in airports, etc.	
Level 2: Non - linear System Change	Reconfiguring and improving an existing system (or adding new functions, etc) within the same Function/Principle/Market combination	"Pulsating" mode of an X-Ray security device to decrease energy consumption	
Level 1: Linear System Change	Solution method is known and applicable within existing Function/Principle/Market combination , only parameter value change is required	Increasing the power of X-Ray generator for testing larger objects	

Figure 2.4.3   Levels of technical solutions

2.4.5   Evolution of Technical Systems

In general, every technical system (product or technology) follows three general phases of evolution before it is replaced by a more effective system based on a new principle: birth, growth, and maturity (Figure 2.4.4).

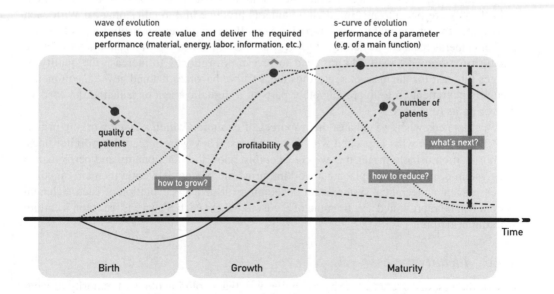

Figure 2.4.4   Evolution stages of technical systems

The curves in the figure represent the following:

- The *S curve of evolution* shows how performance of a main system parameter (of a certain function delivered by a system) changes over time provided a basic principle of function delivery is not changed. When a system is just created, the parameter's performance is usually rather low. However, the function became possible and that matters above all at this stage.
- The *wave of evolution* curve shows how a typical technical system evolves respectively its physical dimensions, energy consumptions, and costs. At the stages of birth and growth, the system tends to become more massive, energy consuming, and expensive. However, at the stage of maturity, specific values representing these parameters tend to decrease. The first electronic tape recorders that used semiconductors were bulky and costly. Later they were transformed into inexpensive small cassette recorders.
- The *quality of patents curve* shows that high-level inventions (level-3 and level-4 solutions) are usually made when a system is just created. When the system becomes mature and commercially successful, most of the patents filled protect solutions of level 2 and sometimes 1.
- The *profitability curve* indicates that during the stages of birth and grows, financial investments are usually made without any profit. Return of investments starts only when a technical system is successfully distributed at the market.
- The *number of patents curve* indicates an increase in the number of patents filled to reduce competition when a system becomes commercially successful.

Let us take, for instance, digital photography, where one of the main parameters of value is image quality. The first generation of digital cameras produced very low quality. In addition, these first cameras were bulky and costly. However, a new function was created, capturing images digitally, and it was a breakthrough invention. Then the next phase (growth) started: All efforts were put into making digital cameras produce high-quality images. With each innovation, the image quality grew rapidly. After 10 years of further evolution, even small pocket cameras reached the quality of good "old" film cameras. Now this parameter is in the "maturity" stage: We really do not need to invest extra money to increase the quality of images (at least for the mass market), thus the S Curve becomes flat and now camera manufacturers pay attention to improving other parts of a digital camera or reducing the costs of delivering its functions.

What happens when we exhaust the resources of evolution within a given working principle? The same as what happened with film cameras: They were replaced by digital ones. Such transition usually means that we considerably boost the functionality and performance of a system or product by replacing a working principle behind delivering its main function, which prevents a technical system from developing further. The evolution of each technical system can be represented as a timeline of S curves, where each S curve is based on a certain working principle.

2.4.6 Ideality

One of the first discoveries made by Altshuller was that evolution of a vast majority of technical systems follows a so-called *trend of ideality growth*. In other words, with each

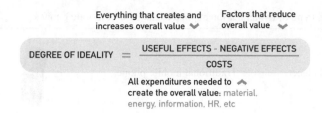

Figure 2.4.5   A system's degree of ideality expressed as a formula

successful innovative improvement, technical systems tend to become more "ideal": They increase their performance, quality, and robustness, whereas the material, energy, labor, and other types of resources needed to manufacture and provide the life cycles of these systems tend to decrease.

The trend of ideality growth plays a very important role in helping us understand how and why systems and products evolve, and define strategies for further improvements of these systems and products.

As seen in Figure 2.4.5, the overall degree of a system's ideality can be increased by increasing the overall value of the system (functionality, performance, etc.), by reducing negative effects that reduce the overall system's value (to improve quality), or by decreasing the resources needed to create and maintain the system's life cycle. Really successful innovations affect all three components in a positive way.

Increasing the degree of ideality does not always means reducing complexity. Just compare a mainframe computer 30 years ago and a modern desktop PC. The price of the desktop PC today cannot be even compared to what organizations had to pay 30 years ago for the mainframe, while the performance and functionality of a modern desktop PC are many times higher (positive effects). It is more reliable, generates less heat and noise, is easier to recycle (negative effects), and costs much less to manufacture and maintain (costs).

2.4.7   Trends of Technical Systems Evolution

During studies of the history of many technical systems, TRIZ researchers discovered that technology evolution is not a random process. Many years of TRIZ studies resulted in extracting a number of general trends governing technology evolution no matter what area technical products belong to. The practical use of these trends is possible through so-called *lines of technical systems evolution*. Each line of evolution describes patterns of transitions between old and new structures of a system.

One of the evolutionary trends of systems by transitions to more dynamic structures (*dynamization*) is shown in Table 2.4.1.

Knowing the lines and trends of technology evolution is very important to estimate what evolutionary phases a system has already passed. As a consequence, it is possible to foresee what changes the system might experience in the future and directly generate new ideas.

Table 2.4.1   TRIZ trend of dynamization applied to the case of mobile phones

1. Monolithic solid object	Traditional mobile phone.	
2. Monolithic object divided into two segments with nonflexible link	Mobile phone with a sliding part that contains a microphone.	
3. Two segments with a flexible link	Flip-flop phone of two parts.	
4. Many segments with flexible links	Phone that is made as a wristwatch: Its bracelet is made of segments that contain different parts of the phone.	
5. Completely flexible, elastic object	A completely flexible phone that can be used as a wrist band.	
6. A field replaces function of a solid object	A phone without a screen. The function of a screen is delivered by a field-visible light produced by a pico-projector.	

Currently the TRIZ trends of technical systems evolution are organized in a system (Figure 2.4.6). The trends can be used as an independent tool to reveal evolutionary potential of technical systems and produce ideas for next generations of solutions.

2.4.8   Science for Inventors

New breakthrough products often result from using the latest scientific advances. One of the TRIZ techniques, known as *databases of effects*, suggests searching for new basic principles by defining what technical function is needed and then finding a scientific principle that can deliver the function.

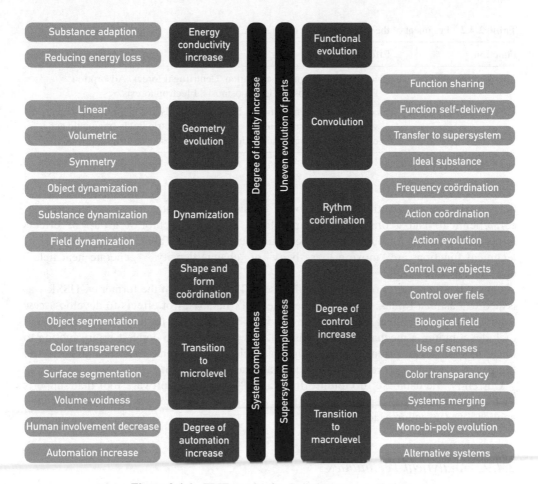

Figure 2.4.6 TRIZ trends of technical systems evolution

Studies of patent collections indicated that inventive solutions are often obtained by utilizing physical effects not used previously in a particular area of technology. Knowledge of scientific phenomena often makes it possible to avoid developing complex and unreliable solutions. For instance, instead of a mechanical design including many parts for precise displacement of an object for a short distance, it is possible to apply the effect of thermal expansion to control the displacement.

Finding a physical principle that would be capable of meeting a new requirement is one of the most important tasks in the early phases of design. However, it is nearly impossible to use handbooks on physics or chemistry to search for principles for new products. Descriptions of natural phenomena available in such texts present information on specific properties of the effects from a scientific point of view, and it is not clear how these properties can be used to deliver particular technical functions.

The TRIZ databases of effects bridge a gap between technology and science. In the databases, each general technical function is identified with a group of scientific effects

Table 2.4.2   Fragment of the TRIZ database of scientific effects

Function	Effects
To separate mixtures	Electrical and magnetic separation. Centrifugal forces. Adsorption. Diffusion. Osmosis. Electro-osmosis. Electrophoresis.
To stabilize object	Electrical and magnetic fields. Fixation in fluids that change their density or viscosity when subjected to magnetic or electric fields (magnetic and electrorheological liquids). Jet motion. Gyroscopic effect.

that can deliver the function. The databases consist of three categories of effects: physical, chemical, and geometric. A search for a needed effect or a group of effects is possible through formulation of a technical function (see Table 2.4.2). Examples of technical functions are "move a loose body" or "change density," "generate heat field," and "accumulate energy."

One of the first patents obtained with the use of TRIZ outside of the former ex-USSR was issued to Eastman Kodak. Engineers used the TRIZ databases of effects to develop a new solution for precise linear displacement of a high-end camera's flash. A traditional design includes a motor, a mechanical transmission, and an actuator to convert rotary motion to linear motion. A new patented solution uses piezoelectric effect and involves a piezoelectric linear motor, which is more reliable, cheaper, and easier to control.

Currently, the database of Goldfire Innovator^{TM}, a leading software tool that supports TRIZ, contains over 8,000 physical, chemical, and geometrical effects with examples of their applications (Invention Machine, 2011).

2.4.9   Analytical Techniques

It is well known that solving inventive problems is difficult since in most cases they are not formulated correctly. In addition, in many cases it is not clear what problem to solve. In order to deal with ill-defined initial situations, TRIZ introduces a group of tools that help us to perform the needed analysis.

One such tool is *function analysis* (see Figure 2.4.7), which decomposes systems and products to components and identifies problems in terms of undesired, insufficient, poorly controllable, or harmful functional interactions between both system components and components of a so-called *supersystem*, which is formed by everything that does not belong to the system but interacts with it. For instance, if a person takes a cup with hot coffee, the cup might be too hot and can burn his or her hand. In this case, a hand of a person is included to a function model and belongs to a supersystem of the system "cup with coffee."

TRIZ-based function analysis helps to identify a range of function-related problems within a system and rank them according to their importance. It can be used not only for finding existing problems but also for identifying resources to increase the degree of ideality of systems and products.

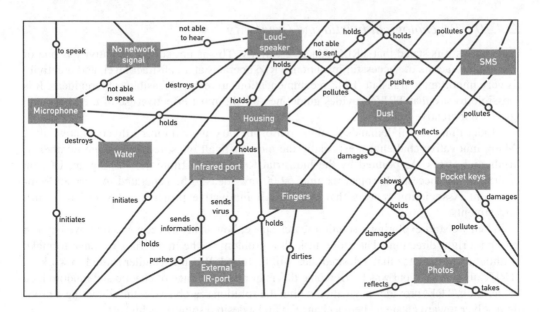

Figure 2.4.7 A fragment of the function analysis diagram

Another tool, recently introduced to TRIZ, is called root conflict analysis (RCA+) (see Figure 2.4.8). Based on a combination of classical RCA and TRIZ philosophy, RCA+ helps to "dissect" a problem formulated as an undesired effect to a tree of causally related underlying negative causes and contradictions. Such a process helps one to understand factors that contribute to the main negative effect and visualize all contradictions that create barriers preventing us from solving a problem in a straightforward way. Later, these contradictions can be directly solved with other TRIZ techniques.

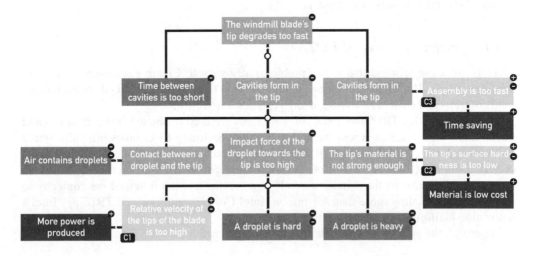

Figure 2.4.8 A fragment of the root conflict analysis (RCA+) diagram

2.4.10   Psychological Inertia and Creativity

Sometimes it is stated that TRIZ reduces creativity. This is not correct. Creativity is one of the crucial factors of successful invention. TRIZ operates at an abstract level, and creativity is very important to translate TRIZ recommendations to real and feasible solution ideas. It is possible to say that TRIZ provides guidelines for the most effective use of creativity and guides our creative search.

Modern innovation demands thinking out of the box and exploiting diverse knowledge. Many innovative challenges, especially the most difficult ones, require a huge number of trials and errors. As pointed by the Industrial Research Institute in Washington, DC, on average, one successful project requires 5,000 raw ideas to be generated. American Management Association reports that 94% of all innovative projects today fail to return investments.

When we start exploring a solution search space, how many directions should we explore to find a right direction? The more difficult a problem is, the more trials we have to make without any guarantee that a desired idea will be found. When Altshuller started to work on TRIZ, his primary goal was to overcome this major disadvantage of chaotic and random idea generation. TRIZ provides navigation within the solution search space, thus directing a problem solver toward an area where a chance to find a desired solution is highest.

Creativity is important to fight psychological inertia, which keeps us locked within existing solutions and ideas and does not let us see things differently. These barriers are difficult to overcome. TRIZ includes a section called *creative imagination development* that consists of techniques to help with developing our creative skills. Altshuller strongly believed that creative imagination can and must be developed to enable the most effective use of TRIZ. In addition, special psychological "operators" were incorporated to some TRIZ techniques to reduce our mental inertia. For instance, one TRIZ tool, the *algorithm of inventive problem solving* (ARIZ), introduced a stepwise algorithm of reformulating an initial problem by executing a number of procedures that reduce our psychological inertia and help us to recognize "hidden" resources to solve the problem.

2.4.11   Practical Value of TRIZ

TRIZ is not a magic wand that solves problems all by itself. Creative solutions are always found by people. But TRIZ eliminates a painful process of generating hundreds or even thousands of trials and errors and waiting many years for a new insight.

As reported, today TRIZ and TRIZ software tools are used in about 5,000 companies and government organizations across the globe. In 2006, Samsung Electronics officially stated that within 2004–2006, the use of TRIZ provided the company with total economic benefits of €1.5 billion. Thanks to TRIZ, Boeing Corporation designed a better refueling tanker on the basis of 767 aircraft than similar aircraft of a competitor, and it helped the company to win a contract totaling more than €1 billion. Intel Corp. recently named TRIZ as "Intel's innovation platform of the twenty first century."

In general, the use of TRIZ provides the following benefits:

- Considerably increases productivity when searching for new ideas and concepts to create new products or to solve existing inventive problems.

- Increases the ratio of useful ideas to useless ideas during the ideas generation process by providing immediate access to hundreds of unique innovative principles and thousands of scientific and technological principles stored in TRIZ knowledge bases.
- Reduces the risk of missing an important solution to a specific problem due to a broad range of generic patterns of inventive solutions offered by TRIZ.
- Uses scientifically based trends of technology evolution to identify evolutionary potential of a technology or product and select the right direction of evolution.
- Leverages the intellectual capital of organizations via increasing a number of patented solutions of high quality.
- Raises the degree of personal creativity by training individuals and groups to approach and solve inventive and innovative problems in a systematic way.
- Structures and organizes the creative phases of the innovation process.
- Supports patents strategies (circumvention, umbrellas, etc.).
- Introduces a common "innovation" language to improve communication.

TRIZ is the most powerful and effective practical methodology of creating new ideas available today. TRIZ does not replace human creativity. Instead, it amplifies it and helps to move

Figure 2.4.9 User scenario of the BandOK product by Geurds and Van der Molen (2011)

it in the right direction. As proven during long-term studies, virtually everyone can invent and solve nontrivial problems with TRIZ. However, modern TRIZ is a complex discipline; therefore, to learn and master skills with it demands time and effort. But thanks to recent advancements in TRIZ education and the development of new software and engineering support tools of systematic innovation, this process is not as difficult as in the past. Recently, TRIZ was included as a supporting tool for Six Sigma to solve problems that cannot be solved with traditional methods. TRIZ conferences and congresses are conducted annually in many countries.

2.4.12   Application of TRIZ

In 2011 Geurds and Van der Molen developed a product concept based on the use of piezo-electric sensors. The concept, called the BandOK system, measures the pressure of car tires while driving and gives feedback to car drivers on the pressure, because well-maintained tire pressure makes it safer to drive and saves fuel, money, and exhaust. During their design process, students applied TRIZ during concept selection and final design of the product, whose use is shown in Figure 2.4.9 and whose technical principles are illustrated in Figure 2.4.10.

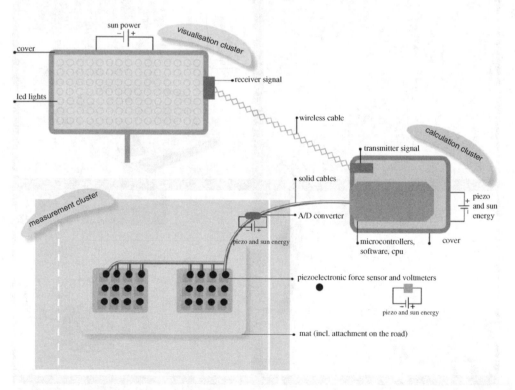

Figure 2.4.10   Scheme of the technical functioning of the BandOK product concept (Geurds and Van der Molen, 2011) (See Plate 14 for the colour figure)

Table 2.4.3 TRIZ principles applied to the BandOK concept

TRIZ principle	Ideation BandOK
Segmentation	Two measuring units on each side of the road
	Two measuring units, one for left wheels and one for right wheels
Merge	iPhone app or TomTom device instead of separate display unit
	Combine with other measuring and information devices (e.g., a check for your lights)
	Integrate speeding flash pole
Prior (anti-) action	Weather resistant
	Vandal proof
Other way around	Make display unit moveable: projection on asphalt, car screen
	Pump the tire before you see if it is flat
Dynamization	Moveable service station (e.g., ANWB Service Centrum, the Netherlands)
	Field measuring instead of physical product
Periodical	Only take a picture when a car passes
	Only give feedback when a car passes
	Only measure or be active when a car passes
High speed	Communication between camera, sensors, and display
Intermediary	Get feedback later, by e-mail or by logging into website
Replacement	Modules, platform-driven development
	Easy to repair, update
Composite	PV unit integrate in shell or pole
	Extra light source to take picture in the dark

In the embodiment design phase, the TRIZ principles shown in Table 2.4.3 were applied to their product concept with a focus on tire pressure. The ideas that were generated in this way were evaluated and partially applied in their final product concept.

References

Altshuller, G. (1969) *Algorithm of Invention*, Moskovski Rabochiy, Moscow (in Russian).
Altshuller, G. and Fedoseev, Y. (1998) *40 Principles: TRIZ Keys to Technical Innovation*, Technical Innovation Center, Worcester, MA.
Geurds, N. and Molen, P.v.d. (2011) Sources of Innovation. Master's of Industrial Design Engineering program, University of Twente.

Further Reading

Altshuller, G. (1996) *And Suddenly the Inventor Appeared: TRIZ, the Theory of Inventive Problem Solving*, 2nd edn, Technical Innovation Center, Worcester, MA.
Altshuller, G. (1999) *The Innovation Algorithm*, Technical Innovation Center, Worcester, MA.
Mann, D. (1992) *Hands-on Systematic Innovation for Technology and Engineering*, CREAX Press, Ieper.
Fey, V.R. and Rivin, E.I. (2005) *Innovation on Demand: New Product Development Using TRIZ*, Cambridge University Press, Cambridge.
Invention Machine (2011) www.invention-machine.com (accessed April 13, 2012).
Koltze, K. and Souchkov, V. (2010) *Systematische Innovation: TRIZ-Anwendung in der Produkt- und Prozessentwicklung*, Carl Hanser Verlag, München (in German).
Salamatov, Y. (1999) *TRIZ: The Right Solution at the Right Time: A Guide to Innovative Problem Solving*, Insytec B.V, Hattem.

2.5   Technology Roadmapping

Valeri Souchkov

ICG Training & Consulting, Willem-Alexanderstraat 6, The Netherlands

2.5.1   Introduction

An organization involved in product development faces a number of challenges related to its future. One of the most critical tasks is to define short-term and long-term product strategies that can be transformed into specific plans. To understand which products and technologies have to be planned for future development, especially within the long term, organizations need to forecast the evolution of their products. Any product or technology is created to meet certain needs and demands of the person or system for which it is to be developed. No technical product is developed to exist independently of the context in which it is supposed to be used: A product always delivers a function or a number of functions requested by its surrounding systems or environment. For instance, a battery is developed to provide electronic and electric devices with energy, and a flashlight's purpose is to deliver a function targeted at a higher system: a human being who wants to use the flashlight to illuminate a certain area of space.

To stay innovative, a company that produces flashlights has to therefore continuously follow the evolution of three areas: the (1) evolution of human or societal needs with respect to the functionality of devices producing light, (2) evolution of technologies that make it possible to create flashlights to meet these particular needs better, and (3) evolution of all competitive businesses that produce flashlights. In a modern global economy, ignoring just one of these areas will undoubtedly reduce the company's chances to survive and grow.

Understanding evolution means comprehending interrelations that exist between three large categories of knowledge related to a specific product: evolution of market, evolution of technology, and evolution of business. None of these areas is static; they are in continuous development.

It is also important to understand a difference between a technology and a product. A product is a technical system composed of a number of components that deliver specific functionality. A technology is a process that enables producing this functionality. For instance, a lamp can be considered a product, and a process of generating light on the basis of passing electric current through a metal wire that heats the wire and thus produces light can be considered a technology. It is obvious that a product might integrate many different subsystems that are built upon different technologies. A process of creating electric current and a process of generating light are two separate technologies. If we look at a mobile phone, we will see hundreds of enabling technologies that make it possible to create the needed functionality of its parts and components.

In turn, each subsystem of a mobile phone can be considered a separate product. However, the same functionality can be achieved on the basis of different technologies. For instance, a function of generating light can be obtained through heating metal wire, through laser radiation, or through the effect of electroluminescence in semiconductors. A particular type of technology can be described by a more general term, for example *virtual reality*. Sometimes the name of a technology can coincide with the name of a product if this product is used as a subsystem. For example, the term *photo sensor* might indicate a technology of capturing images by a light-sensitive matrix and converting captured information to digital form.

2.5.2 Technology Roadmaps

There are two major groups of factors that drive product evolution:

1. *Technology push:* Development of new technologies and, therefore, products based on scientific and technological research; and
2. *Market pull:* Market needs that should be satisfied by new- or next-generation products and technologies, and requests to create and develop these technologies.

Both technology and market are spaces with many variables. How does one successfully identify what comes next? What products will be successful? And, more importantly, what products will these be? To plan feasibly, one should be aware of the current level of market needs and business developments. However, that is not enough. To make a long-term forecast that can result in a specific product development plan, we need to know what technologies can be used to improve existing products or develop new ones. Technology roadmapping addresses this issue.

The technology roadmap (TRM) introduces connections between identified market needs and trends with existing and emerging technologies for a specific industry sector to improve existing and develop new products. Technology roadmaps provide a graphical framework for exploring and communicating strategic plans. They comprise a multilayered, time-based chart that visually links market, product, and technology information and represents their evolutions within a selected time interval, enabling market opportunities and technology gaps to be identified. Originally, technology roadmapping was introduced by Motorola in the 1980s (Willyard and McClees, 1987).

TRMs can be used in various contexts (Garcia and Bray, 1997). Primarily, they are used in strategic product planning, research planning, business planning, intellectual property creation, and protection. Typically organizations create and use TRMs in two cases:

1. *Product-planning TRMs*, where roadmaps cover the nearest future (1–3 years ahead) and are based on already available and mature technologies.
2. *Strategic emerging TRMs*, where roadmaps cover a more extended time interval (typically up to 7, sometimes up to 10 years).

A typical TRM chart for both cases is shown in Figure 2.5.1. It consists of four layers:

1. *Timeline*: This is a line that starts at the moment of creating TRM and extends to a final year selected for products planning.
2. *Business and Markets*: This box shows specific market goals and objectives that will be fulfilled by future products. For example, it can be *Mobile Internet access* or *supersonic flight*. Such goals do not define a specific product but a general objective. A product that should deliver functionality to reach the objective is defined in a lower box.
3. *Products*: This box specifies products that will meet the business and market objectives. For instance, "a camera phone" means developing a mobile phone with integrated photo or video camera. It is not necessary to define a single line of products replacing each other in time. Several products can co-exist or be launched in parallel. In addition, different products defined in TRM can utilize the same key technology. A number of different

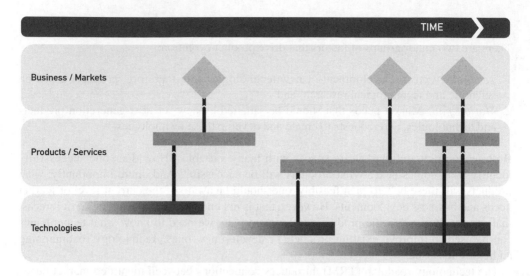

Figure 2.5.1   Time-based layered chart in a typical TRM

products can be identified by the same company that would provide mobile Internet access and that are based on the same key ultrafast wireless data communication technology.
4. *Technologies*: This box contains specific *key* technologies that are needed to build the desired products. It focuses on mature technologies that are currently not used in the existing products at all or at emerging technologies, or both. For instance, developing a camera phone requires a technology of a photo sensor. If this technology has not reached yet maturity level, we have to estimate when the technology becomes mature and thus define when a product based on that technology can be launched at the market. For this reason each emerging technology in TRM can be represented by a rectangle with a gradient that specifies the expected time of transition from immature technology to mature.

For example, if a strategic emerging TRM for mobile wireless communication were created in 1991, its simplified version could look like the one shown in Figure 2.5.2. Such a roadmap was never created at that time; the purpose of this example is to better explain the paradigm of technology roadmapping and show what the final roadmap representing a long-term plan can look like. The gradient of the bars representing technology maturity is shown in an arbitrary form for illustrative purposes only.

Note that it is extremely difficult to plan for almost 20 years ahead because certain technologies might not be available yet and new inventions that would influence the development of new products or their features are still to be created.

The diagrams shown in Figures 2.5.1 and 2.5.2 are the most simplified examples of TRM. TRM can include different layers in addition to these four layers. For instance, the following layers can be added:

• The *vision layer* represents change or updates of the vision of the organization.
• *Market niches* identify application areas for each product.

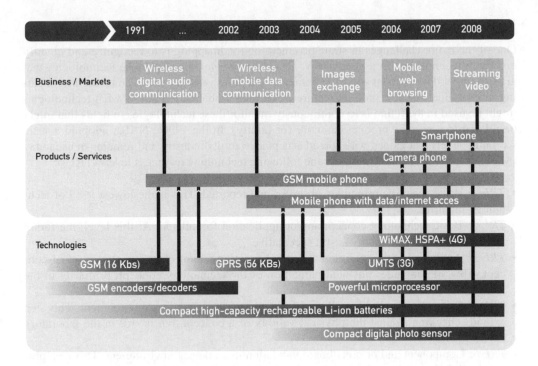

Figure 2.5.2 A TRM for a wireless mobile communication. Please note that this TRM has been made in 1991

- *Internal and external resources* are needed to achieve each step of the plan for both technology and product development. This layer might comprise financial, technical, labor, and any other type of resources.
- *R&D programs* are needed to mature the needed technologies and integrate technologies to products.
- *Integration activities* are done with partners, suppliers, and customers.

There is no unified approach to developing roadmaps. Instead, each organization that develops TRMs uses the best practices and methods available in the organization or outside and can be highly customizable (Phaal, Farrukh, and Probert, 2004).

TRMs are very flexible, and any important information that can influence the proper planning of future products, technologies, and activities might be added. In addition, each bar representing a technology or a product can be further decomposed to more specific components. For instance, a camera phone (mentioned in Figure 2.5.2) might be referred to a broader list of technologies, such as image-processing software, data storage, lens, shooting parameters control, and so forth. For this reason, to avoid the growth of complexity of a TRM, a subsystem of a more complex system can be separated into another TRM. For instance, the evolution of data storage for mobile wireless communication might require a separate TRM. However, this separate TRM must still be connected with the main TRM.

2.5.3   Technology Readiness Levels

One of the major challenges in strategic emerging roadmapping is to recognize and identify emerging technologies in a timely way. *Emerging technologies* are those technologies that are still under development and have not reached a point yet when they can be used for creating new products or new product features. The only viable way to know what technologies might be used in the future is to continuously monitor these technologies in fields that might be related to a specific product category (or family). In the 1990s, NASA adopted a scale with nine levels that gained widespread acceptance across industry and remains in use today (Mankins, 1995). The scale includes the following technology readiness levels (TRLs):

- *TRL 1:* Basic scientific principles observed and reported. This is the lowest level of technology maturity.
- *TRL 2:* Basic technology concept and/or application formulated. At this level, the future application of the technology is still speculative.
- *TRL 3:* Analytical and experimental critical function and/or characteristic proof-of-concept. This step includes physical validation of the technology concept, primarily in the laboratory environment.
- *TRL 4:* Component and/or breadboard validation in the laboratory environment. This step extends checking against the physical validity of a concept, but is still in the laboratory environment.
- *TRL 5:* Component and/or breadboard validation in a relevant environment. This step provides validation of the technology concept in a realistic environment.
- *TRL 6:* System or subsystem model or prototype demonstration in a relevant environment. This step requires the development of a prototype of a future product and proof of its feasibility in a realistic environment.
- *TRL 7:* System prototype demonstration in a space environment. This step is specific to space missions.
- *TRL 8:* Actual system completed and "flight-qualified" through test and demonstration.
- *TRL 9:* Actual system "flight proven" through successful mission operations. This step involves the elimination of all errors in a product found at the previous step.

These TRLs can be used to evaluate the readiness and maturity of any technology and its degree of applicability in forecasted products. TRLs do not depend on financial, business, or market data used in TRM; they only focus on evaluating the physical and practical validity of a technology concept.

It is obvious that for a short-term product-planning TRM, only those technologies that are already proven in a realistic environment have to be chosen (TRLs 6–9). For a long-term strategic emerging TRM, lower TRLs can be considered too.

It is also crucial to estimate properly a current technology readiness level and how much time it might take to reach TRL 9, when the technology becomes mature and can be used in a new product. Such estimations are often performed by independent science and technology experts. Large companies that have extensive R&D divisions often prefer to develop such core technologies in-house (Figure 2.5.3). Many large industrial companies invest to groups performing basic research and cooperate with universities and various academic organizations.

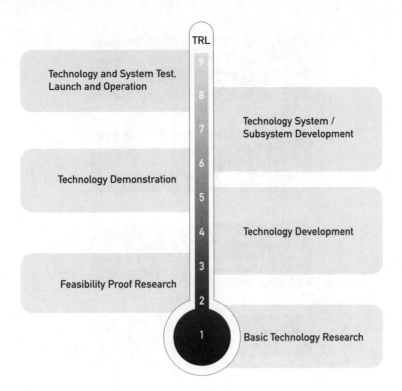

Figure 2.5.3 A "thermometer" of technology readiness levels (TRLs) mapped to technology development phases

2.5.4 TRM Process

There are three phases of the TRM process identifying the following:

1. *Market evolution:* The study of current unfulfilled and evolving market requirements and demands within a certain time frame; developing vision, and identifying the marketing objectives and time intervals needed to reach these objectives.
2. *Product evolution:* Studying what changes will be necessary in the existing products that would meet the identified market objectives, and identifying what new products can be created on the basis of the mature and emerging technologies within the defined time frame.
3. *Technology evolution:* Examining key and critical technologies (both existing and emerging) that would support the development of products identified at phase 2.

The process of creating TRM is highly iterative. It is performed in both top-down and bottom-up manners. A TRM process integrates both technology push and technology pull; therefore, after each phase of the TRM process is completed as a "first draft," information gathered during the next phase can be used to introduce changes and modifications to the information gathered during the previous phase. Often, completely new information can be

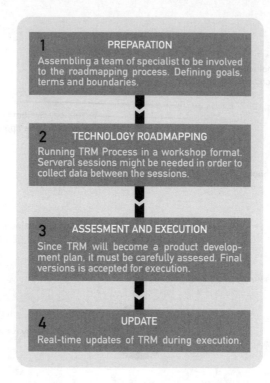

Figure 2.5.4   A general process including TRM

introduced; for instance, finding a new technology can lead to identifying a new market opportunity by creating a new product.

To properly prepare and finalize TRM, preparation and assessment activities are required. Once created and approved, TRM must be reassessed and updated regularly to avoid wrong investments, or add new technologies or the results of market studies that were omitted in the original and follow-up versions. Figure 2.5.4 represents all these activities, including a TRM process (stage 2).

Preparation and analytical studies (phases 1 and 2 of the process described in Figure 2.5.3) can take time before one starts the first workshop on visualizing TRM. All findings during these stages are recorded in a document or in a number of documents that will accompany a created visual TRM. This document usually contains detailed descriptions of the follows:

- Results of market studies and vision.
- Future expected product features.
- Emerging technology readiness levels.
- Argumentations of decisions.
- Risks.
- Future collaborations.
- Estimated investments.

Today a number of software tools exist to support TRM. However, large organizations often prefer to develop their own tools adapted to their needs and goals. These tools can be integrated with other documents created by an organization for different purposes.

2.5.5   Benefits from TRM

In summary, TRM:

- Provides a visual framework for integrated product and technology planning.
- Helps with establishing the monitoring of key and critical technologies.
- Establishes a plan of key activities within a relatively long term.
- Visualizes critical information along a timeline.
- Identifies technology gaps.
- Helps with defining innovation strategies.
- Improves communication among different parties involved in business, technology, and product development.
- Improves decision making by bringing together critical issues along different activities.
- Identifies needed collaboration and cooperation activities organization-wide and with third parties that will be needed to realize technology and product plans.
- Helps with identifying resources, cost, and time targets.
- Helps to reduce short- and long-term risks.

TRM can be considered a part of technology foresight activities. It is a useful tool to identify business opportunities and to plan diverse joint activities across different units and groups in an organization.

2.5.6   Application of TRM

In 2008, master's-level students Erkel and Reilink developed the living cubes concept shown in Figure 2.5.5. *Living cubes* is a modular lighting system with interactive tiles that react to

Figure 2.5.5   An impression of living cubes by Erkel and Reilink (2008) (See Plate 15 for the colour figure)

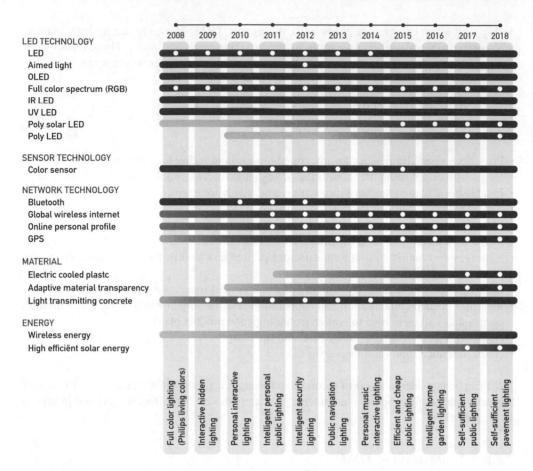

Figure 2.5.6 A technology roadmap made in the framework of living cubes by Erkel and Reilink (2008)

the environment in public spaces like, for instance, parks or shopping malls. In this project a TRM was developed for various technologies that could be applied in lighting products. This TRM is shown in Figure 2.5.6.

The interactive tiles with LED lights are all connected with each other to one central central processing unit (CPU) tile that processes all data to decide which tiles emit a certain color of light with a certain intensity. The living cubes concept has different functionalities during day and night. During day, the only functionality is decoration: making a public space interactive and personal. During night, the decorative functions remain and additionally lighting and security functions become active.

All light-emitting tiles have one or more light sensors (depending on the size of a tile) that can detect an object or person standing on top of the tile. When a tile detects a person, it sends a signal to the CPU tile that decides the appropriate reaction. All kinds of configurations can be created with different interactions. The following interactions are integrated in this concept:

Personal color: As soon as an individual person is standing on a light-emitting tile, the tile
   will emit a certain color of light. This light is chosen based on the person's preference.

This preference is detected by the information on the person's mobile phone using a Bluetooth connection. This technology will be integrated into the concept in the near future. When this technology is not yet integrated or if the person doesn't has a mobile profile, a random color is chosen.

Fading colors: The light-emitting tiles on which a person is standing have a special individual color. When a person walks over these tiles and is no longer standing on a certain tile, this tile still emits the person's individual color of light. After some time, the color fades (lower intensity) until it does not emit that color of light anymore. This system shows when a person has been sitting on a bench and it also shows from where person is coming. The purpose of this feature is to create an interactive personal and dynamic atmosphere in a public space like a park.

Blending colors: When two persons are standing close to each other, they both have their individual colors of light surrounding them. The tiles between those two persons blend the two colors. When, for example, a couple is sitting on a bench the bench blends the two colors. This creates a special atmosphere.

References

Erkel, H. and Reilink, D. (2008) Sources of innovation. Master of Industrial Design Engineering program, University of Twente.

Garcia, M.L. and Bray, O.H. (1997) *Fundamentals of Technology Roadmapping*, Sandia National Laboratories, Strategic Business Development Department, Albuquerque, NM.

Mankins, J.C. (1995) *Technology Readiness Levels: A White Paper*, NASA Office of Space Access and Technology, Advanced Concepts Office, Huntsville, AL.

Phaal, R., Farrukh, C., and Probert, D. (2004) Customizing roadmapping. *Res. Technol. Manage.*, **47**, 26–37.

Willyard, C.H. and McClees, C. (1987) Motorola's Technology Roadmapping Process. *Res. Manage.*, **30**, 13–19.

Further Reading

Laube, T., Abele, T., and Ein Leitfaden (2005) *Technologie-Roadmap: Strategisches und taktisches Technologiemanagement*, Fraunhofer-Institut Produktionstechnik und Automatisierung (IPA), Stuttgart (in German).

Phaal, R. (2003) Strategic roadmapping: Linking technology resources to business objectives. *Int. J. Technol. Manage.*, **26**, 1–18.

Phaal, R., Farrukh, C., and Probert, D. (2010) *Roadmapping for Strategy and Innovation: Aligning Technology and Markets in a Dynamic World*, University of Cambridge, Institute for Manufacturing, Cambridge.

Tolfree, D. and Smith, A. (2009) *Roadmapping Emergent Technologies*, Matador, Leicester, UK.

2.6 The Design and Styling of Future Things

Wouter Eggink

University of Twente, Faculty of Engineering Technology, Department Design, Production and Management, The Netherlands

2.6.1 Introduction

Design and styling of new products used to be easy. Modernism depicted that the new product should be designed to fit the utility function of the object efficiently, and that the shape should be geometric and clean. Postmodernism of the 1980s, however, re-introduced the idea of cultural reference and emotion in product design. From then on, products had to communicate not only a utility function but also cultural context and moral values (Krippendorff, 2007). Increased interest in the perception of these products by their intended users added a role for psychology and emotion in product design in the late 1990s (Desmet, 2002). Finally, the idea of product experience, introduced in the beginning of the twenty-first century, added an important role for the use of products (Green, 2002). From then on, using products should also be pleasurable and provide a memorable experience (Schifferstein and Hekkert, 2007). A lot of work has been done on mastering this increasing complexity, mostly relying on step-by-step methodology.

The design and styling of innovative products are even more complicated, depending on their innovativeness. When incremental innovation is applied, the characteristics of relating products can provide a reference for the design of new ones. One can think of the classic example of the first generation of cars, which looked like "carriages without horses." However, in the case of breakthrough innovations, the intended new objects have no real "predecessor" in their existence, in other words, there is no reference to determine how they should look. To match the example of the introduction of the car, one can think of the first successful airplane by the Wright brothers, which showed little resemblance to birds (or anything else).

This newness is best suited with an umbrella methodology based on insights from creativity. Our design educational practice shows that the search for suitable product concepts beyond the obvious can be facilitated by bringing the design challenge to a higher level of abstraction (Eggink, 2009b), as for instance depicted by TRIZ, and more recently also adopted by the Vision in Product Design (ViP) method (Hekkert and Dijk, 2001). To stay with our example of the car as a carriage without a horse, one should not try to design the next generation of carriages, but rather "a means of individual people transport with the use of combustion engine technology."

With regard to the development of innovative aesthetics in particular, in our postmodern twenty-first-century society, the image of products has sometimes become more important than the product itself (Baudrillard, 1994; Jameson, 1991). Then it is not the physical product, but the perception of the product by the user, that determines its existence (Crilly, Moultrie, and Clarkson, 2004). Bernhard Bürdek argued as early as 1991 that the functionalist idea of "form follows function" cannot play its central role in the development of aesthetics anymore, because of the increased importance of "visually anonymous" electronics and information technology (Bürdek, 1996). Electronics and interfaces simply do not have a particular shape, like steam engines or bicycles used to have. Naylor and Ball (2005) illustrated this ambiguity effectively with the supposedly functional shape of a neutral box that could be "anything" (Figure 2.6.1). This abstract nature of technology even increases when considering developments in bio- and nanotechnology (Drukker, 2009). Interestingly, in contrast, developments in materials and computerized production techniques enable us to make nearly any shape we want (Figure 2.6.2). In this context, the communication function of aesthetics – telling the user what the product is and how to use it – is therefore very important. It is this combination of circumstances that provides the designer with a powerful influence on the recognition and acceptance of innovative product concepts by their intended users.

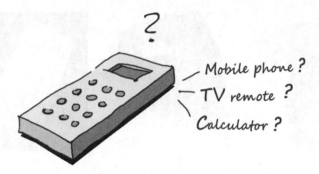

Figure 2.6.1   The Generic Keypad by Naylor and Ball (2005, p. 62) illustrates the problem of product ambiguity

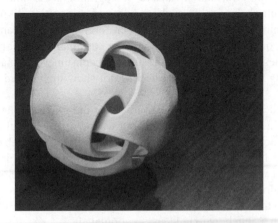

Figure 2.6.2   Example of a complex shape that is produced with the use of rapid prototyping techniques

2.6.2   Communication

The communication function of design is easily understood, following the communication model of Crilly, Moultrie, and Clarkson (2004) (Figure 2.6.3). The consumer response at the right side of Figure 2.6.3 is steered by the information that is received through the senses.

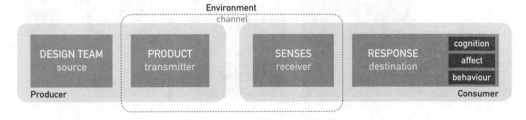

Figure 2.6.3   Basic model of product communication by Crilly, Moultrie, and Clarkson (2004, p. 551)

Figure 2.6.4 Products that communicate how to hold and use them (left and middle), and an example that does not (right)

The interpretation of that information leads to (re)cognition, a certain level of affect (like or dislike), and a particular behavior. To evoke the desired recognition, liking, and behavior, the product has to transmit the right information. Therefore, the design team has to put the right idea into the styling of the product. For example, a sustainable car concept has to transmit the idea of environmental friendliness. If the intended consumer associates the styling of the concept with spoiling fuel instead, he or she will not show the desired behavior (using and liking the car).

The communicated idea can support the usability of the product. The do-it-yourself cordless drill and jigsaw in Figure 2.6.4 are clear about where the user should put his or her hands to hold and manipulate the objects. The rather smoothly curved handlebar sections reflect the form-language of the human body. Especially with the jigsaw, this is in sharp contrast with the rather technical expression of the 'dangerous' part of the device, surrounding the saw blade. The cordless drill also has a shape that literally points towards the hole that is to be drilled. In contrast to that, the styling of the "white box" design of the little iPod on the right, does not provide such a clue how to hold and manipulate it.

The idea that the product transmits does not necessarily have to do with the primary functionality of the object. The iPod Nano in Figure 2.6.5 expresses the idea of "simplicity,"

Figure 2.6.5 The iPod Nano from 2005 successfully communicates "simplicity" (left), whereas the music device on the right communicates the opposite

which is a metaphor for the interface's ease of use. The rather chaotic styling of the ghetto blaster on the right is expressing "complexity" as a metaphor for *value for money*; the consumer gets a lot of options when purchasing the device. Especially with the design of innovative products, where the perception of the consumer is already challenged by a lack of reference from predecessors, the central message to communicate should not be too complex. When the central idea of the product is clear, the shape and functionality of the product can be designed in a focused way, with the result that the consumer will recognize the benefits easier (Eggink, 2009a).

The iPod, for example, was one of the greatest recent successes in handheld devices. But its shape was not very pleasant to hold in your hand, and successors with better functionality. However, the central idea of "simplicity" was very dominantly implemented, stressing the problem that often occurred when introducing new consumer electronics: people being afraid that they could not cope with the new product's complexity.

This communication of the design idea should be regarded as a cultural phenomenon; the response of the consumer is determined by his or her cultural context or so-called *cultural capital*. In fact, the message is best understood when the sender and receiver share a common frame of reference. A red car, for instance, connotes "fastness" and "danger," because people learn that red depicts danger as in traffic lights. They can also associate the color with the fast cars of the Ferrari brand.

The designer has to constantly search for the right cultural references that contain the desired message, then transform the characteristics of these cultural references into the styling of the new object. This transformation of cultural meanings can be executed on different levels of abstraction (Nijkamp and Garde, 2010), from the simple use of the color red for "danger" to the implementation of form complexity as a metaphor for advanced technology that can be seen in the object of Figure 2.6.2, which was issued to promote the possibilities of rapid prototyping.

2.6.3 Acceptance

New products have to build an "audience." Early industrial designer Raymond Loewy said in his autobiography, "A lot of people are open to new things, as long as they look like the old ones" (Loewy, 1951). Loewy translated this in the famous MAYA (most advanced yet acceptable) principle, which suggests that things should look new, but not too much. This is unfortunately a relative and therefore very subjective guideline, because it is never clear what is "too much." This MAYA principle has a background in social studies on people's aesthetic preferences, where it is argued that people long for both ease of recognition and the excitement of newness (Armstrong and Detweiler-Bedell, 2008). In other words, it is the ever-existing paradox of comfort versus thrill. However, in their illustrative research on typicality, novelty, and aesthetic preference, Hekkert, Snelders, and van Wieringen (2003) conclude that people have aesthetic preference for objects that are both typical *and* novel:

> In sum, it seems that our results provide an empirical basis for the industrial design principle coined MAYA by Raymond Loewy. . . . In order to create a successful design, the designer should strike a balance between novelty and typicality in trying to

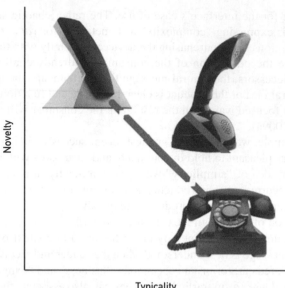

Novelty

Typicality

Figure 2.6.6 Example of the relationship between novelty and typicality in telephone designs. The gray arrow depicts the (normal) highly negative correlation

be as innovative as possible while preserving, as much as possible, the typicality of the design. The fact that this is feasible is due to the fact that the correlation between novelty and typicality, although highly negative, falls short from being perfect. (p. 122)

In fact, the interpretation by Hekkert, Snelders, and van Wieringen (2003) turns the MAYA principle into a very useful tool. The transformation of terms puts an end to the end-less discussions on whether an advanced design is still acceptable or too advanced, because now a new design should be both advanced (novel) *and* acceptable (typical). This is, of course, only possible when typicality and novelty are considered two different variables, and not each other's opposites. This is best understood when the opposite of typical is seen as *different*, and the opposite of novel is seen as *expected* (Eggink, 2010). The historic example of the red Ericophone in Figure 2.6.6 is a telephone design where both typicality and novelty score high. The Ericophone is novel because of the upright position of the handset (with the dial at the bottom of the base) and typical because of the familiar shape and form language of the handset, which is copied from the traditional black model.

This principle can be implemented in many different ways. In the Gina concept car by German car manufacturer BMW, for example, the bodywork is innovatively executed in fabric, covering a tubular frame. Because the fabric was made in one continuous piece covering the whole body, this solution resulted in a form of wrinkling of the bodywork when the doors were opened. This wrinkling of body coverings is very new within the styling of cars, but at the same time very recognizable for the user because it associates with skin. Master student Industrial Design Engineering Jan Willem Peters explored the principle in his master gradua-tion project and came up with a range of other examples. Amongst others he designed a

Figure 2.6.7 *Lampshade-Fireplace* and *Tent* concept by Jan Willem Peters, combining familiar shapes in a new context (tent) and combining a new functionality with a familiar context (fireplace)

lampshade-fireplace, and a tent with the shape of a leaf (Figure 2.6.7). The lampshade-slash-fireplace looks very familiar in a living room context, but is new because of the gel burner that is providing a real fire, situated somewhere one does not expect. The tent concept transforms the shape of a leaf into a small shelter. The familiar structure of the veins of the leaf is combined with the structural support for the tent.

So the combination of newness and recognition is typically achieved when something widely known is presented or applied in a new, unfamiliar context (Eggink, 2011b). This principle can be traced back to the practice of "displacement" by the early twentieth-century Surrealists, and is still a powerful mechanism for cultural meaning making (Figure 2.6.8).

2.6.4 Method

When the novelty–typicality principle is combined with the outline for stimulating creative solutions on a higher level of abstraction as was derived from TRIZ, we can put this in a

Figure 2.6.8 *Aphrodisiac Telephone* by Salvador Dalí, from 1936 (1936, courtesy Museum Boymans-van Beuningen, Rotterdam; photo Tom Haartsen) and *Prada Value Meal* by Tom Sachs (1997) share the same principle of meaning making

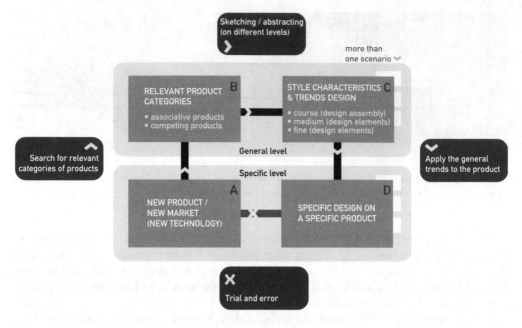

Figure 2.6.9 Basic model of innovative design and styling, by Stevens (2008–2010)

scheme as depicted in Figure 2.6.9. The design challenge of the new product is elevated to a general level on the left. Here the central idea to be communicated should be established from relevant product categories and related cultural references. From there, the implementation of newness and typicality on an abstract level can be facilitated by associative and competing products or imagery.

On the right side of the model in Figure 2.6.9, the insights should be applied to the specific product design, resulting in a concrete object that incorporates the abstract product idea. Within this process, an exploration of different design possibilities, based on future trends and scenario development, could ensure that the specific design fits the desired future context.

2.6.5 Examples

Two examples will be given of the actual development of the styling of future concepts. Both originate from the teaching practice at the Industrial Design Engineering curriculum of the University of Twente in the Netherlands.

The first example is from a design course in cooperation with Dutch sports car manufacturer Donkervoort. The assignment was to develop future concepts for a genuine sports car experience in a society context that has more and more traffic restrictions. Leendert Verduijn, Bernd Worm, and Frank Scholder approached the problem by first visualizing the intrinsic contradiction of the assignment (Figure 2.6.10).

The visualized argument is that developments in electronic driver assistance, in combination with road safety and increased density, will put the driver in a stringent harness

Figure 2.6.10 Visuals depicting the central problem statement and the desired solution for the design: combining the racing car with the intrinsic freedom of city running

(Figure 2.6.10, left), leading to illusionary freedom (Figure 2.6.10, middle). The solution for a true sports car experience has to lie in substantially less car instead of more car (Figure 2.6.10, right). Presented in this way, the visualizations stimulate the designer to look for creative solutions that lie behind the obvious, and they also support the evaluation of design ideas (Eggink, 2011a).

The implementation of this abstract idea is shown in Figure 2.6.11. Instead of a sports car, the students designed a minimalistic construction that stays as close to the body as possible, reminiscent of the inspiration source from popular youth culture, city running or Parkour. The eventual concept is a clear combination of novelty and typicality (Figure 2.6.11, right). The back half looks like a lightweight motorcycle, and the front half like a skateboard (the associative and competing products as mentioned in the upper-left corner of the basic model in Figure 2.6.9). The posture of the driver resembles that on a reclining bicycle, but the absence of the driving train renders a different effect. The sense of freedom for the user is established by minimizing the parts in the front. Because no part of the construction is in the sight of the user, the shape reflects an urge forward and there is no visible protection that blocks the user's idea of freedom.

The second example is from a master course named Create the Future, where student groups first have to create a future context for their designs through systematic scenario development (Eggink, Reinders, and van der Meulen, 2009). Within this future context, the groups have to develop a product concept, and in this particular example the group created a future electric mobility concept named *Amplified Walking*. The central idea is that the

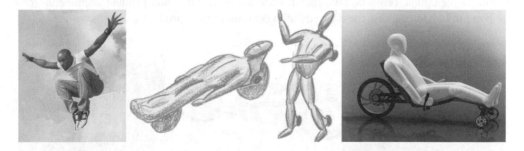

Figure 2.6.11 From central idea to design concept: The electric driven tricycle can be folded into the shape of a wheeled suitcase so that the fun can be combined with public transport before or after use

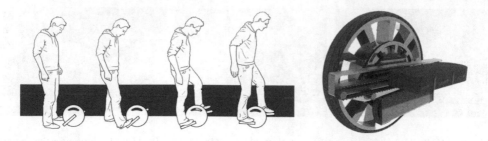

Figure 2.6.12 The amplified walking concept, by Alfred Doppenberg, Ronald van Galen, Mark Grob, and Elias van Hoek

electrically driven device amplifies the movements that you normally make when you are walking, stepping, or running (Figure 2.6.12). Ingenious electronic feedback will keep the user stabilized in the same manner as the two-wheeled Segway does.

The amplified walking concept was developed for a future scenario context where consumer involvement with new developments was very high. This resulted in a product architecture based on building blocks that should be possible to construct in endless combinations, like a form of mass customization. The group presented three different product architectures, likely to be built by different user groups. All three also combined a form of newness with recognition of the typical features of the user group's main inspiration sources (Figure 2.6.13). The first one combines the appearance of a unicycle with the styling of a white iPad. The second one is a mix of skateboard and mountain bike. The third concept resembles a typical fitness device, targeted on elderly people.

2.6.6 Conclusions

Design and styling have major influences on perceptions of innovative product concepts. The right styling depends on a profound mix of both novel and typical styling items. This combination of items has to communicate the desired attributes of the product, by means of association with the cultural background of the intended user. For clear acceptance of the communicated message, it should be not too complicated. One central idea is easier to communicate and can at the same time perform a guiding role in the design's detailed development. Two example projects have shown how styling features and product architecture can go hand in hand to express the desired feelings around future product concepts.

Figure 2.6.13 Three possible implementations of the amplified walking concept

References

Armstrong, T. and Detweiler-Bedell, B. (2008) Beauty as an emotion: The exhilarating prospect of mastering a challenging world. *Rev. Gen. Psych.*, **12**(4), 305–329.

Baudrillard, J. (1994) *Simulacra and Simulation*, University of Michigan Press, Ann Arbor.

Bürdek, B.E. (1996) *Design; Geschiedenis, Theorie en Praktijk van de Produktontwikkeling*, Ten Hagen & Stam, The Hague.

Crilly, N., Moultrie, J., and Clarkson, P.J. (2004) Seeing things: Consumer response to the visual domain in product design. *Design Studies*, **25**(6), 547–577.

Desmet, P.M.A. (2002) *Designing Emotions*. PhD dissertation, Delft University of Technology, Delft.

Drukker, J. (2009) *Things to Come; een economisch-historische visie op de toekomst van het industrieel ontwerpen*, University of Twente, Enschede.

Eggink, W. (2009a) A chair to look to the moon: What we can learn from irrational design history for contemporary design practice. *Design Principles and Practices: An International J.*, **3**(1), 103–114.

Eggink, W. (2009b) A practical approach to teaching abstract product design issues. *J. Eng. Design*, **20**(5), 511–521.

Eggink, W. (2010) The reinvention of the ready made. Proceedings of the 7th Design and Emotion Conference 2010, Spertus Institute, Chicago, pp. 1–12.

Eggink, W. (2011a) Disruptive images: Stimulating creative solutions by visualizing the design vision. Proceedings of the 13th International Conference on Engineering and Product Design Education, City University London, pp. 97–102.

Eggink, W. (2011b) The rules of unruly design. Proceedings of the 4th World Conference on Design Research, October 31–November 4, Delft University of Technology, pp. 1–11.

Eggink, W., Reinders, A., and Meulen, B.v.d. (2009) A practical approach to product design for future worlds using scenario-development. Proceedings of the 11th International conference on Engineering and Product Design Education, University of Brighton, UK, pp. 538–543.

Green, B. (2002) Pleasure with products: Beyond usability – introduction, in *Pleasure with Products; Beyond Usability* (ed. W.S. Green and P.W. Jordan), Taylor & Francis, London, pp. 1–5.

Hekkert, P. and Dijk, M.v. (2001) Designing from context: Foundations and applications of the ViP approach. Designing in Context: Proceedings of the Design Thinking research Symposium (5), DUP Science, Delft, pp. 1–11.

Hekkert, P., Snelders, H.M.J.J., and van Wieringen, P.C.W. (2003) "Most advanced, yet acceptable": Typicality and novelty as joint predictors of aesthetic preference in industrial design. *Brit. J. Psychol.*, **94**(1), 111–124.

Jameson, F. (1991) *Postmodernism, or, The Cultural Logic of Late Capitalism*, Duke University Press, Durham, NC.

Krippendorff, K. (2007) On the essential contexts of artifacts, or on the proposition that "Design Is Making Sense (of Things)." *Design Issues*, **5**(2), 9–39.

Loewy, R. (1951) *Never Leave Well Enough Alone*, Simon & Schuster, New York.

Naylor, M. and Ball, R. (2005) *Form Follows Idea: An Introduction to Design Poetics*, Black Dog, London.

Nijkamp, M. and Garde, J.A. (2010) A practical approach to translate social cultural patterns into new design. Proceedings of the 12th International Conference on Engineering and Product Design Education, Norwegian University of Science and Technology, Trondheim, Norway.

Schifferstein, H.N.J. and Hekkert, P. (eds.) (2007) *Product Experience*, Elsevier, Amsterdam.

Stevens, J. (2008–2010) *Basic Model of Innovative Design & Styling*. Unpublished PowerPoint presentations, Sources of Innovation course, University of Twente/Stevens IDE Partners.

Further Reading

Hekkert, P. and Dijk, M.v. (2011) *Vision in Design: A Guidebook for Innovators*, BIS, Amsterdam. This book is about the Vision in Product Design (ViP) approach; a rather chaotic but intriguing book that argues for the application of intuition in design methodology.

Naylor, M. and Ball, R. (2005) *Form Follows Idea: An Introduction to Design Poetics*, Black Dog, London. An inspiring book on the possible role of intuition in design activity.

Heskett, J. (2002) *Toothpicks & Logos: Design in Everyday Life*, Oxford University Press, Oxford. A comprehensive work addressing the relation between design activity and culture.

Verbeek, P.-P. (2000) *What Things Do: Philosophical Reflections on Technology, Agency, and Design*, Penn State University Press, Altoona, PA. On the relation between people and technology, from a philosophical perspective.
Schifferstein, H.N.J. and Hekkert, P. (eds) (2007) *Product Experience*, Elsevier, Amsterdam. On the relation between people and technology, from a psychological perspective.

2.7 Constructive Technology Assessment

Stefan Kuhlmann

University of Twente, Department of Science, Technology, and Policy Studies,
P.O. Box 217, The Netherlands

2.7.1 Introduction

Technology assessment (TA) aims to support the designing and shaping of technology in society. TA is supposed "to reduce the human costs of trial and error learning in society's handling of new technologies, and to do so by anticipating potential impacts and feeding these insights back into decision making, and into actors' strategies" (Schot and Rip, 1997, p. 251). TA practices are intended to create, communicate, and apply knowledge about and reflection on the potential and actual interaction of (new) technology and societal actors and forces.[1] An overarching definition is offered by the TAMI project, a joint effort of European TA agents, to summarize the state of the art of TA: "Technology assessment is a scientific, interactive and communicative process which aims to contribute to the formation of public and political opinion on societal aspects of science and technology" (TAMI project, 2003, p. 4).

The first concepts of TA were developed in the United States at the end of the 1960s. At that time a need to assess the potential, normally unexpected negative effects of new technologies was perceived that led in 1972 to the creation of the Office of Technology Assessment (OTA) as a research-based service and early warning mechanism of the US Congress; OTA worked for more than 20 years with quite some success, but in it was closed down by a Republican majority in Congress (Smits *et al.*, 2010; Smits and Leyten, 1991). The OTA served as a model for various forms of parliamentary TA services established in European countries, directly or indirectly linked to parliamentarian decision making (such as in the United Kingdom, Germany, Denmark, the Netherlands, and Switzerland).

Since the late 1990s, TA concepts have increasingly adopted a *dedicated design ambition*. In the meantime, there is a considerable variety of TA approaches such as *expert TA*, *participatory TA*, *interactive TA*, *rational TA*, *real-time TA*, and constructive TA (CTA).

In particular, CTA aims to understand social issues arising from new technologies and to influence design practices. This development has been driven by three independent but interrelated forces:

1. After the nuclear and computer technologies of the twentieth century, today *new and emerging technologies* are again attracting our attention, such as nanoscience, nanotechnology, and life sciences (e.g., genomics). These new technologies are characterized

---

[1] Since the early days of TA, experts have uttered a dedicated design ambition; see, for example, Coates (1975); Bimber and Guston (1997); Rip, Misa, and Schot (1995); and Smits, Leyten, and den Hertog (1995). This holds even more for most recent TA works such as Robinson (2010) or Te Kulve (2011).

by a considerable heterogeneity of related knowledge bases, new forms of inter-disciplinary exchange (e.g., *translational research* in biomedicine), as well as generic fields of application with potentially far-reaching effects on economies and societies.

2. More than in earlier days, TA experts feel confident about being able to cope with the *Collingridge dilemma* (Collingridge, 1980), that is, they see a good chance to effectively *shape technological development* at early stages, before design and societal embedding have become irreversible. This hope draws not a small amount on the insights of international interdisciplinary science, technology, and innovation studies (STIS): Today we have at our disposal a good socioeconomic understanding of innovation processes (e.g., Dosi, 1982), framed by long-term specific technological, economic, social, political, and cultural "regimes" (Nelson and Winter, (1977, 1982); Rip and Kemp, 1998); stretching across multilevel systems (e.g., Geels and Schot, 2007); and shaped by specific forms of grown *de facto* governance. The better one understands these interrelations, the more likely and robust a prospective "modulation" of technological developments will become (Rip, 2006).

3. As a third force we have seen since the late 1990s, at least in Europe, a growing interest in TA concepts drawing on *participatory* elements in the design of new technologies, now conceptualized as a process of innovation. "Users" are interfering in innovation (e.g., Oudshoorn and Pinch, 2003; van Oost, Verhaegh, and Oudshoorn, 2008; Von Hippel, 2005) – think of the enthusiasm of Linux-based open-domain software communities, or of the dedication of voluntary contributors to the various Wiki databases.

On such grounds, a "realistic," (i.e., modest though dedicated) design ambition toward technologies can emerge: If in today's polyvalent society literally "everything goes," why shouldn't it be possible to design and shape technology in an explicit and at the same time reflexive way?

The present chapter[2] will sketch a dynamic concept of TA – with a focus on CTA – and its contribution to the governance of technological innovation. A constructivist and reflexive TA concept will be suggested: Informed by heuristics-based analyses, CTA will be presented as modulating ferment in the social process of technological innovation and as a building brick of its emerging *de facto* governance. Finally we will suggest the metaphor of TA as a dance of three elements:[3] The "practice" of technological innovation; the "theory" of science, technology, and innovation studies; and the "policy," that is, public and private governance ambitions.

2.7.2   New Attention for the Design and Governance of Science, Technology, and Innovation

Increasingly politicians, industrial actors, societal groups, and technology experts are concerned about inappropriate attempts at steering technological development and thus hampering the realization of desired innovation effects. In the past, all too often

---

[2] The text draws partially on Kuhlmann (2007, 2010) and Rip (2008).
[3] See also Kuhlmann (2007) and Smits and Kuhlmann (2004). The dancing metaphor was used earlier by Arie Rip (1992) with respect to the relation of science and technology, inspired by Derek de Solla Price's discussion of this relation (1963).

development followed the model of the "economics of technoscientific promise" (Felt *et al.*, 2007): Promises to industry and society, often far reaching, are a general feature of technological change and innovation, and are particularly visible in the mode of governance of emerging technosciences (biotechnologies and genomics, nanotechnologies, neurosciences, or ambient intelligence), all with typical characteristics. They require the creation of a fictitious, uncertain future in order to attract resources: financial, human, political, and so on. They come along with a diagnosis that we are in a world competition and that we (Europe, the United States, etc.) will not be able to afford our social model if we don't participate in the race and become leaders in understanding, fueling, and exploiting the potential of technosciences. The model

> works with a specific governance assumption: a division of labor between technology promoters and enactors, and civil society. Let us (= promoters) work on the promises without too much interference from civil society, so that you can be happy customers as well as citizens profiting from the European social model. (Felt *et al.*, 2007, p. 25)

Under this model of techno-economic promises politics, science and industry take the lead, while the innovation needs and expectations represented in the society appear to remain in a rather passive consumer role.

Felt *et al.* (2007) suggest as an alternative model the "economics and socio-politics of collective experimentation," which are characterized by emerging or created situations that allow people to try out things and learn from them. The main difference with the other model is that "experimentation does not derive from promoting a particular technological promise, but from goals constructed around matters of concerns and that may be achieved at the collective level. Such goals will often be further articulated in the course of the experimentation" (Felt *et al.*, 2007, p. 26f). This model requires a specific division of labor in terms of participation of a variety of actors, who are investing because they are concerned about a specific issue (see also Callon, 2005). "Users matter" in innovation – that has been shown not in the least by our UT colleague Nelly Oudshoorn and her team (Oudshoorn and Pinch, 2003). Examples of such demand- and user-driven innovation regimes include the information and communication sector (where the distinction between developers and users is not sharp), sports (e.g., von Lüthje, Herstatt, and von Hippel, 2005), or the involvement of patient associations in health research (e.g., Rabeharisoa and Callon, 2004) and pharmacogenomics (e.g., Boon *et al.*, 2008). The concept of "open innovation" debated around the user-driven development of nonpatented open-source software, and more generally in Hank Chesbrough's influential book (2003) is largely overlapping with the collective experimentation concept. The governance of such regimes is precarious since they require the long-term commitment of actors who are not always equipped with strong organizational and other relevant means, and there is always some room for opportunistic behavior. Nevertheless, the promise is innovation with sustainable effects.

2.7.3   Constructive Technology Assessment

So we are in need of a governance of technological innovation that builds on exchange, debate, negotiations, and cooperation between companies, science, civil society, and political systems. Which role can TA adopt in this setting?

Which TA? Smits *et al.* (2010; see also Smits and Leyten, 1991) differentiate "Watchdog TA" and "Tracker TA." Watchdog TA is supposed to fulfill an early-warning function for political decision makers. Related projects are often conducted by centralized (parliamentary) TA agencies. Actors in the innovation process do not play an active role in Watchdog TA. In contrast, Tracker TA, or CTA, aims to proactively and constructively interfere in the process of technology development and design.

TA Methods (Source: TAMI project, 2003, pp. 48–49)

Scientific methods include the "Delphi method, expert interviews for collecting expert knowledge, modeling and simulation, cost/benefit-analysis, systems analysis, risk analysis, material flow analysis, trend extrapolation, scenario technique for creating knowledge to think about the future . . . discourse analysis, value research, ethical analyzes, value tree analysis."

Interactive methods include "consensus conferences, co-operative discourses, public expert hearings, focus groups, [and] citizens' juries are belonging, currently in part supported by using electronic media."

Communication methods include "newsletters, opinion articles, science theater, (interactive) websites and various types of networking."

CTA as a constructive design process starts from the assumption that actors involved in technology development can find themselves in two basically different positions: as insiders or outsiders to the development process. Garud and Ahlstrom (1997) suggested differentiating the positions of enactors and comparative selectors: *Enactors* are technology developers and promoters aiming to enact new technology; they "construct scenarios of progress, and identify obstacles to be overcome. They thus work and think in 'enactment cycles' which emphasize positive aspects" (Te Kulve, 2011, p. 31). *Comparative selectors*, in contrast, observe technological development process from the outside and may be in a situation to compare the enactors' offers and performance with other, parallel developments. There are professional comparative selectors (such as regulatory bodies like the US Food and Drug Administration) using professional tools for their assessments, and amateur comparative selectors (such as critical consumers and nongovernmental organizations [NGOs], the latter increasingly turning professional) (Te Kulve, 2011, p. 32). Enactment cycles and comparative selection cycles come together in *bridging events* (Garud and Ahlstrom, 1997), with insiders and outsiders interacting. Such interaction may lead to variations in the technological design process and can have an impact on the direction of selection decisions. Such bridging events "can be constructed on purpose, by actors from enactor or selector positions, and by more disinterested actors such as Constructive Technology Assessment (CTA) agents" (Te Kulve, 2011, p. 33).

CTA agents start from the assumption that enactors and selectors of new technology make assessments all the time, so rather than making assessments (by TA agents) CTA aims to create and orchestrate bridging events as dedicated "spaces" for interaction, learning, and reflection (Rip and te Kulve, 2008). Both enactors and selectors can undergo *first-order* or *second-order learning processes*: According to Argyris and Schön (1978), first-order

learning links outcomes of action to organizational strategies and assumptions that are modified so as to keep organizational performance within the range set by accepted organizational
norms. The norms themselves remain unchanged. Second-order learning concerns inquiries
that resolve incompatible organizational norms by setting new priorities and relevance of
norms, or by restructuring the norms themselves together with associated strategies and
assumptions, hence escaping tunnel vision and crossing borders.

CTA agents, in order to create dedicated spaces for first- and second-order learning, organize *interactive workshops*, often enriched by the elaboration of alternative *sociotechnical
scenarios*, stimulating the debate of participants (see the "TA Methods" in this section).
Such scenarios "capture ongoing dynamics and develop assessments of future developments.
They show the effects of interactions between enactors and selectors which provides more
substance to interactions in workshops as actors can draw upon the scenarios for inspiration"
(Te Kulve, 2011, p. 34).

The interfering character of CTA requires an explicit understanding of the room for
maneuver of the involved actors – up to now, still a weak point of CTA concepts. To
which extent can the participants effectively manage or steer the development, variation,
and selection processes? The *governance context* of CTA-supported design processes
needs to be explored and understood. *Governance* means the coordination and control of
autonomous but interdependent actors either by external authority or by internal mechanisms of self-regulation or self-control (Benz, 2007, p. 3). This is of particular relevance
when it comes to the development and design of new and emerging technology: The
potential areas of application, markets, concerned actors and audiences, and dimensions
of potential effects are still in flux; and political arenas, decision criteria, and policy
means are not yet determined.

2.7.4   Governance: CTA and Design in an Institutional Context

CTA and technological design are taking place in "inherited" economic and institutional
environments. Any realistic attempt to shape technological development effectively has to
understand the driving forces and hampering factors of governing this institutional context.

A useful heuristic for this purpose is offered by the school of evolutionary-economic
analyses of technology dynamics. It is built on the findings of "innovation studies," in particular on the seminal work of Nelson and Winter (1977, 1982): In their "search of a useful
theory of innovation" and convinced of the stochastic, evolutionary, and organizationally
complex and diverse character of innovation, the authors observed different technological
"regimes" characterized by longstanding specific "search strategies" of engineers, determining to some extent the development trajectory of a given regime. Drawing on this basic
observation, other authors have defined *technological regimes* as "the complex of scientific
knowledge, engineering practices, production process technologies, product characteristics,
user practices, skills and procedures, and institutions and infrastructures that make up the
totality of a technology" (Van den Ende and Kemp, 1999, 835). Rip and Kemp (1998) explicitly added to the "grammar" of a regime the public and private strategies and *policies* of
relevant actors: Technology is conceptually and as an artifact socially constructed, including
the governance of a regime.

Finally, these conceptual elements were combined in the heuristic of a *multilevel perspective* on sociotechnical transitions (e.g., Geels and Schot, 2007), characterized by niche

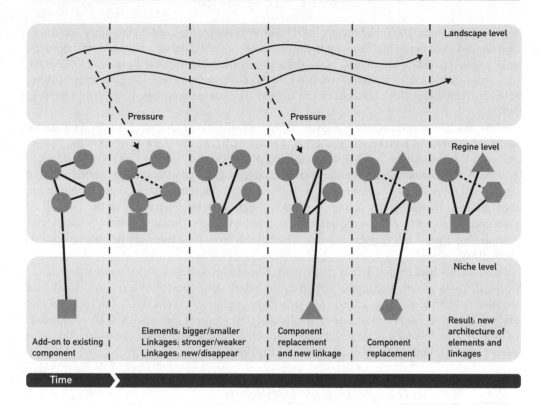

Figure 2.7.1 Schematic presentation of reconfiguration of a technological regime (Geels and Schot, 2007, p. 412)

innovations on a micro level (developing in emerging or created and protected incubation spaces), sociotechnical regimes on the meso level, and wider sociotechnical landscapes on the macro level (macroeconomic, cultural, and macropolitical developments). Studied with the help of this heuristic, one can see *regime transitions*, sometimes incremental and sometimes radical, sometimes driven by basic changes in the overarching landscape or stimulated through niche innovations undermining dominant regimes (see Figure 2.7.1).

Such transition processes are fueled by promises and expectations about technological options and innovation (Van Lente, 1993). Actors anticipate and assess their options *vis-à-vis* changing regimes and create *de facto* new patterns (Rip, 2001) that can trigger "irreversibilities" (Callon, 1991) resulting in "endogeneous futures" (Rip, 2001). In other words, the analysis of the transition of sociotechnical regimes offers a heuristic help to open up the allegedly fatal transition point of the Collingridge dilemma for empirical research. This holds also for the governance of technological developments in a regime context: We can better understand the options and limitations for dedicated shaping.

Here it is useful to clarify the underlying concept of *governance*: The concept is used here as a heuristic, borrowed from political science, denoting the dynamic interrelation of involved (mostly organized) actors; their resources, interests, and power; the fora for debate and arenas for negotiation between actors; the rules of the game; and the policy instruments applied (e.g., Benz, 2007; Kuhlmann, 2001). Governance profiles and their quality and direction are reflected not a small amount in the character of public debates between stakeholders, policy makers, and experts. Think of the debates on genetically modified organisms (GMOs), or – still more in *status nascendi* – debates on the governance of an emerging, cross-cutting science, technology, and innovation (STI) field like nanotechnology (e.g., Joly and Rip, 2007).

How much leeway do actors in a given regime actually have? One has to understand the *de facto governance* of a given social context. Conceptually we can draw here upon the "actor-centered institutionalism" of Mayntz and Scharpf (1995) and Scharpf (2000). The *de facto* governance of sociotechnical regimes can be analyzed as a web of cognitive, normative, and regulatory rules. Actors have to cope with these rules; while inevitably reproducing them, they are also incrementally changing them through deviating behavior. In the context of emerging sociotechnical regimes, actors cannot achieve more (but also cannot achieve less) than to shape what will happen anyway, while at the same time rules are being transformed (Rip, 2008).

Against this background TA, in particular CTA, can be understood as a *modulating factor* of the *de facto* governance and of the co-evolutionary development of a regime. The better we understand the *de facto* governance in a given regime, the more CTA can in a realistic and constructive manner modulate technological development and design. CTA as a means of *reflexive governance* is aware of the limits of dedicated governance; this awareness is even a strategic underpinning of its ambition (Rip, 2006; Voss *et al.*, 2006).

2.7.5   CTA as a Dance: Strategic Intelligence

CTA as a means of reflexive governance aims to make the diverging perspectives and interests of relevant actors visible and debatable aiming to increase the learning capacity. Here we suggest the metaphor of a *dance* of "practice," "policy," and "theory" (see Figure 2.7.2); they can be seen as partners on a dancing floor, moving to varying music and exposing different configurations.

"Theory" is represented by the arsenal of dedicated and methodologically rich STIS.[4] The multilevel analysis of regime transitions sketched out in this section is of particular value for the application of CTA as a means of reflexive governance (e.g., Konrad, Truffer, and Voss, 2008; Markard and Truffer, 2008). As a dancing partner, STIS (theory) would move about in the worlds of technology designers in a constructivist and reflexive manner (Robinson, 2010), analyzing the perspectives and interests of the other dance partners and reflecting on its own position (sometimes even changing its own beliefs): in other words, CTA becomes a means of reflexive governance.

The "dance floor" for CTA can be conceptualized as a "forum," defined as institutionalized space specifically designed for deliberation or other interaction between heterogeneous actors with the purpose of informing and conditioning the form and direction of strategic social choices in the design and governance of science and technology (see Figure 2.7.3; Edler *et al.*, 2006). The debates on a forum can be supported with insights from strategic

---

[4] For an overview, see Silbey (2006), Hackett *et al.* (2007), and Fagerberg *et al.* (2006). Scientific journals such as *Research Policy* (rather economics oriented) and *Science, Technology, and Human Values* (rather sociologically oriented) enjoy a high reputation.

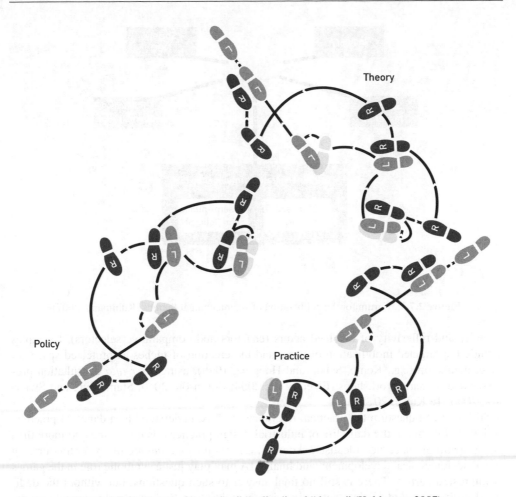

Figure 2.7.2   Dance of "practice," "policy," and "theory" (Kuhlmann, 2007)

intelligence (SI). SI has been defined as a set of sources of information and explorative as well as analytical (theoretical, heuristic, and methodological) tools – often distributed across organizations and countries – employed to produce useful insight in the actual or potential costs and effects of public or private policy and management (Kuhlmann *et al.*, 1999). Strategic intelligence is "injected" and "digested" in fora, with the potential of enlightening the debate. SI can draw on semipublic intelligence services (such as TA agencies or statistical offices), on "folk" intelligence provided by practitioners or NGOs, and in particular on research-based statistical STIS.

2.7.6   Limits to CTA and Reflexive Governance of Technology Design

Today constructive TA is using a broad spectrum of scientific, interactive, and communicative methodologies and SI instruments in order to modulate on various fora, the *de facto* governance of technological development and design, aiming to increase the learning

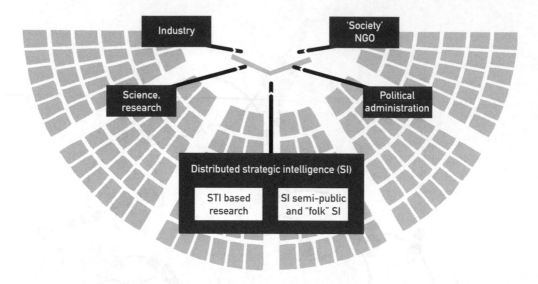

Figure 2.7.3   Forum for the deliberation of sociotechnical themes (Kuhlmann, 2007)

capacity and reflexivity of involved actors (enactors and comparative selectors). Examples include the targeted mobilization of users and the creation of niches as protected space for experimentation (e.g., Kemp, Schot, and Hoogma, 1998), using *inter alia* consultation processes and scenario workshops (Elzen *et al.*, 2004; Robinson, 2010; Stemerding and Swierstra, 2006; Te Kulve, 2011).

Still, one can question the practical relevance of CTA, understood as a dance of practice, policy, and theory in the daily life of industrial design practices. Is this in the end more than just a dream of concerned idealists? After all, CTA is just *one among many factors* driving actual sociotechnical development, and finally TA may play just a symbolic role in the games of interested parties. There is still no final answer to such questions; but without the dedicated *political will* of leading actors in science, technology, industry, politics, and society, there will be no reflexive TA as constructive design practice and governance.

2.7.7   Application of CTA

Researchers executed a constructive TA for the introduction and use of a new type of asphalt comprising phase change materials that would reduce the freezing of road surfaces and hence the use of salt in winter. Goals of the CTA were to obtain information about the acceptance of the new product by society, which will be done by looking into political, social, environmental, economic, and cultural effects, starting with mapping the actors that play a role. These actors are shown in Figure 2.7.4. In the rest of this subsection, the roles of the actors will be explained by the students.

Road-Building Companies

Road-building companies play a big role in the implementation of new road surface material, because they have to build the new roads with it. If production requires new techniques and

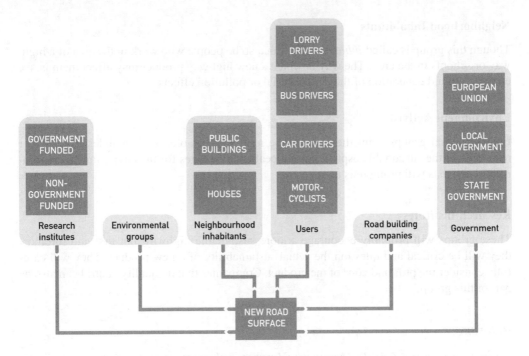

Figure 2.7.4   Scheme representing the interfaces between various actors and a new type of asphalt

perhaps also higher costs, the companies might not be pleased. However, the use of advanced and sustainable techniques could also work to their advantage if it fits with an image the company would like to depict. New supplier contracting is probably necessary, because the needed phase change materials are produced by specialized companies.

Government

The government has an important voice in the implementation of a new highway product, because a lot of planning is involved. When can the new roads be built? How much will it cost? How long will it take, and what kind of impact will it have on traffic flow? How much maintenance will be needed in the future? What is the product lifetime of the new material? These are just a few of the questions that can be asked about the long- and short-term influence of new highways.

Vehicle Drivers

This group consists of car owners as well as motorcyclists, bus drivers, and lorry drivers. Their interests have to do with costs and safety. If the new highways come with higher road taxes, they will not be happy about it. But perhaps if safety can be improved and is in balance with the higher costs, they might be willing to concede.

Neighborhood Inhabitants

Though this group is called *inhabitants*, it can also be people who work in the area of a highway or schools in the area. The way in which a new highway product may affect them is, for example, sound generation of the new material or pollution effects.

Environment Activists

Environmental groups want the world to be more sustainable. This includes using fewer resources. If the sustainable asphalt has noticeable advantages for the environment, environmental activists will promote is use.

Research Institutes

These groups will probably encourage a more sustainable highway, but first and foremost they will be critical and question the actual sustainability of a new product. They will carefully consider the pros and cons of the product. Companies that do quality control can also be part of this group.

References

Argyris, C. and Schön, D.A. (1978) *Organizational Learning: A Theory of Action Perspective*, Addison-Wesley Publishing, Reading, MA.

Benz, A. (2007) Governance in connected arenas – political science analysis of coordination and control in complex control systems, in *New Forms of Governance in Research Organizations: From Disciplinary Theories towards Interfaces and Integration* (ed. D. Jansen), Springer, Heidelberg, pp. 3–22.

Bimber, B. and Guston, D. (1997) The end of OTA and the future of technology assessment. *Technol. Forecast. Soc.*, **54**(2), 125–130.

Boon, W.P.C., Moors, E.H.M., Kuhlmann, S., and Smits, R.E.H.M. (2008) Demand articulation in intermediary organisations: The case of orphan drugs in the Netherlands. *Technol. Forecast. Soc.*, **75**, 644–671.

Callon, M. (1991) Techno-economic networks and irreversibility, in *A Sociology of Monsters: Essays on Power, Technology and Domination* (ed. J. Law), Routledge, London, pp. 132–165.

Callon, M. (2005) Disabled persons of all countries, unite, in *Making Things Public: Atmospheres of Democracy* (ed. B. Latour and P. Weibel), ZKM/MIT, Karlsruhe, pp. 308–313.

Chesbrough, H.W. (2003) *Open Innovation: The New Imperative for Creating and Profiting from Technology*, Harvard Business School, Boston.

Coates, V. (1975) *Readings in Technology Assessment*, George Washington University, Washington, DC.

Collingridge, D. (1980) *The Social Control of Technology*, Pinter, London.

De Solla Price, D.J. (1963) *Little Science, Big Science*, Columbia University Press, New York.

Dosi, G. (1982) Technological paradigms and technological trajectories: A suggested interpretation of the determinants and directions of technical change. *Res. Policy*, **11**, 147–162.

Edler, J., Joly, P.-B., Kuhlmann, S. *et al.* (2006) Understanding *"Fora of Strategic Intelligence for Research and Innovation."* The PRIME Forum Research Project, Karlsruhe.

Elzen, B., Geels, F.W., Hofman, P.S., and Green, K. (2004) Socio-technical scenarios as a tool for transition policy: An example from the traffic and transport domain, in *System Innovation and the Transition to Sustainability: Theory, Evidence and Policy* (ed. B. Elzen, F. Geels, and K. Green), Edward Elgar Publishing, Cheltenham, UK, pp. 251–328.

Fagerberg, J., Mowery, D.C., and Nelson, R.R. (2006) *The Oxford Handbook of Innovation*, Oxford University Press, Oxford.

Felt, U., Wynne, B., Callon, M. *et al.* (2007) *Taking European Knowledge Society Seriously*, European Commission, Brussels.

Garud, R. and Ahlstrom, D. (1997) Technology assessment: A socio-cognitive perspective. *J. Eng. Technol. Manage.*, **14**, 25–48.

Geels, F.W. and Schot, J. (2007) Typology of sociotechnical transition pathways. *Res. Policy*, **36**(3), 399–417.

Hackett, E.J., Amsterdamska, O., Lynch, M., and Wajcman, J. (eds) (2007) *The Handbook of Science and Technology Studies*, 3rd edn, MIT Press, Cambridge, MA.

Joly, P.B. and Rip, A. (2007) A timely harvest. *Nature*, **450**(8), 1–11.

Kemp, R., Schot, J., and Hoogma, R. (1998) Regime shifts to sustainability through processes of niche formation: The approach of strategic niche management. *Technol. Anal. Strateg.*, **10**(2), 175–198.

Konrad, K., Truffer, B., and Voss, J.-P. (2008) Multi-regime dynamics in the analysis of sectoral transformation potentials: Evidence from German utility sectors. *J. Cleaner Production*, **16**, 1190–1202.

Kuhlmann, S., Boekholt, P., Georghiou, L. *et al.* (1999) Improving Distributed Intelligence in Complex Innovation Systems. Final report of the Advanced Science & Technology Policy Planning Network (ASTPP), a Thematic Network of the European Targeted Socio-Economic Research Programme (TSER), Brussels/Luxembourg 1999, Office for Official Publications of the European Communities, http://publica.fraunhofer.de/eprints/urn:nbn: de:0011-n-555104.pdf (accessed April 14, 2012).

Kuhlmann, S. (2001) Governance of innovation policy in Europe: Three scenarios. *Res. Policy*, **30**, 953–976.

Kuhlmann, S. (2007) Governance of innovation: Practice, policy, and theory as dancing partners. Inaugural Lecture, University of Twente.

Kuhlmann, S. (2010) TA als Tanz: Zur Governance technologischer Innovation. Neue Aufgaben des Technology Assessment, in *Technology Governance: Der Beitrag der Technikfolgenabschätzung* (ed. G. Aichholzer, A. Bora, S. Bröchler, M. Decker, and M. Latzer), Edition Sigma, Berlin, pp. 41–60.

Lüthje, C., Herstatt, C., and von Hippel, E. (2005) User-innovators and "local" information: The case of mountain biking. *Res. Policy*, **34**(6), 951–965.

Markard, J. and Truffer, B. (2008) Technological innovation systems and the multi-level perspective: Towards an integrated framework. *Res. Policy*, **37**(4), 596–615.

Mayntz, R. and Scharpf, F.W. (1995) Der Ansatz des akteurzentrierten Institutionalismus, in *Gesellschaftliche Selbstregelung und politische Steuerung* (ed. H.G. Dieselben), Campus, Frankfurt, pp. 39–72.

Nelson, R. and Winter, S. (1977) In search of a useful theory of innovation. *Res. Policy*, **6**, 36–76.

Nelson, R.R. and Winter, S.G. (1982) *An Evolutionary Theory of Economic Change*, Belknap Press of Harvard University Press, Cambridge, MA.

van Oost, E.C.J., Verhaegh, S.J.S., and Oudshoorn, N.E.J. (2008) From innovation community to community innovation user-initiated innovation in wireless Leiden. *Sci. Technol. Hum. Val.*, **34**(2), 182–205.

Oudshoorn, N. and Pinch, T. (eds) (2003) *How Users Matter: The Co-construction of Users and Technologies*, MIT Press, Cambridge, MA.

Rabeharisoa, V. and Callon, M. (2004) Patients and scientists in French muscular dystrophy research, in *States of Knowledge: The Co-Production of Science and Social Order* (ed. S. Jasanoff), Routledge, London, pp. 142–160.

Rip, A. (1992) Science and technology as dancing partners, in *Technological Development and Sciences in the Industrial Age* (ed. P. Kroes and M. Bakker), Kluwer Academic, Dordrecht, pp. 231–270.

Rip, A. (2001) Assessing the impacts of innovation: New developments in technology assessment. In OECD Proceedings, Social Sciences and Innovation, Paris (OECD), pp. 197–213.

Rip, A. (2006) A coevolutionary approach to reflexive governance – and its ironies, in *Reflexive Governance for Sustainable Development* (ed. J.-P. Voss, D. Bauknecht, and R. Kemp), Edward Elgar Publishing, Cheltenham UK, pp. 82–100.

Rip, A. (2008) "A Theory of TA?" Lecture delivered at ITAS, Karlsruhe, February 11.

Rip, A. and te Kulve, H. (2008) Constructive technology assessment and socio-technical scenarios, in *The Yearbook of Nanotechnology in Society, Vol. 1: Presenting Futures* (ed. E. Fisher, C. Selin, and J.M. Wetmore), Springer, Berlin, pp. 49–70.

Rip, A. and Kemp, R. (1998) Technological change, in *Human Choice and Climate Change, Vol. 2: Resources and Technology* (ed. S. Rayner and L. Malone), Batelle Press, Washington, DC, pp. 327–400.

Rip, A., Misa, T., and Schot, J. (eds) (1995) *Managing Technology in Society: The Approach of Constructive Technology Assessment*, Pinter, London.

Robinson, D.K.R. (2010) *Constructive Technology Assessment of Newly Emerging Nanotechnologies: Experiments in Interactions.* PhD dissertation, University of Twente.

Scharpf, F.W. (2000) *Interaktionsformen: akteurzentrierter Institutionalismus in der Politikforschung*, Leske + Budrich, Opladen.

Schot, J. and Rip, A. (1997) The past and the future of constructive technology assessment. *Technol. Forecast. Soc.*, **54**, 251–268.

Silbey, S. (2006) Science and technology studies, in *The Cambridge Dictionary of Sociology* (ed. B. Turner), Cambridge University Press, Cambridge, pp. 536–540.

Smits, R. and Kuhlmann, S. (2004) The rise of systemic instruments in innovation policy. *Int. J. Foresight and Innovation Policy (IJFIP)*, **1**(1–2), 4–32.

Smits, R. and Leyten, A. (1991) *Technology Assessment: Waakhond of Speurhond (Technology Assessment: Watchdog or Tracker? Towards an Integral Technology Policy)*, Vrije Universiteit Amsterdam, Amsterdam.

Smits, R., Leyten, A., and den Hertog, P. (1995) Technology assessment and technology policy in Europe: New concepts, new goals, new infrastructures. *Policy Sci.*, **28**(3), 272–299.

Smits, R., van Merkerk, R., Guston, D., and Sarewitz, D. (2010) Strategic intelligence: The role of TA in systemic innovation policy, in *The Theory and Practice of Innovation Policy: An International Research Handbook* (ed. R. Smits, S. Kuhlmann, and P. Shapira), Edward Elgar, Cheltenham, pp. 387–416.

Stemerding, D. and Swierstra, T.E. (2006) How might scenario studies help us to think about the normative implications of genomics and predictive medicine? In *Questions éthiques en médecine prédictive* (ed. A. de Bouvet, P. Boitte, and G. Aiguier), John Libbey Eurotext, Paris, pp. 81–88.

TAMI project (2003) Europäische Akademie *et al.*, Technology Assessment in Europe: Between Method and Impact, European Commission, STRATA project, http://www.ta-swiss.ch/?redirect=getfile.php&cmd[getfile] [uid]=944 (accessed April 14, 2012).

Te Kulve, H. (2011) Anticipatory Interventions and the Co-Evolution of Nanotechnology and Society. PhD dissertation, University of Twente.

Van den Ende, J. and Kemp, R. (1999) Technological transformations in history: How the computer regime grew out of existing computing regimes. *Res. Policy*, **28**, 833–851.

Van Lente, H. (1993) *Promising Technology: The Dynamics of Expectations in Technological Developments*, WMW-Publikatie 17, Universiteit Twente, Enschede, the Netherlands.

Von Hippel, E. (2005) *Democratizing Innovation*, MIT Press, Cambridge, MA.

Voss, J-.P., Bauknecht, D. and Kemp, R. (eds) (2006) *Reflexive Governance for Sustainable Development*, Edward Elgar, Cheltenham.

2.8 Innovation Journey: Navigating Unknown Waters

Stefan Kuhlmann

University of Twente, Department of Science, Technology, and Policy Studies, P.O. Box 217, The Netherlands

2.8.1 Introduction

Product design and creation can been regarded as comprising an "innovation journey," biased by unforeseen setbacks along the road. In fact, real and sustainable innovation success should rather be viewed as "by-products along the journey" than as end result. If one takes a closer look to the contingencies during such a journey, the retrospective attributions of success to certain approaches or persons will prove to be misleading; such rash "attributions reinforced top managers' belief that managing innovation is fundamentally a control problem when it should be viewed as one of orchestrating a highly complex, uncertain and probabilistic process" (Van de Ven *et al.*, 1999, p. 59).

An innovation journey should be imagined as a journey into unknown waters or an uncharted river (Van de Ven *et al.*, 1999, p, 212). This metaphor helps "to develop an empirically grounded model of the innovation journey that captures the messy and complex progressions" while travelling (pp. 212–213). Consequently, according to van de Ven and colleagues, "innovation managers are to go with the flow – although we can learn to maneuver the innovation journey, we cannot control it" (p. 213).

2.8.2   Method

Still, while conceiving innovation processes as uncertain open ended processes of more or less organized social action, one can identify certain patterns and a number of typical key components characterizing the journey and helping the actors to navigate along uncharted rivers. Van de Ven *et al.* (1999, pp. 23–25) suggest the following components (see Figure 2.8.1): During an initiation period, (1) an innovation project is (often quite slowly) put in motion, sometimes (2) triggered by a "shock," and (3) project plans are developed, less as a journey map but rather to legitimate the project *vis-à-vis* corporate management. In the developmental period (4) the initial innovation idea proliferates into numerous variations, but soon (5) setbacks and mistakes are encountered "because plans go awry or unanticipated environmental events significantly alter the ground assumptions of the innovation." Projects often end in vicious cycles, or (6) actors decide, often after power struggles, to change the criteria of success and failure. (7) Various innovation personnel join the project and leave it, experiencing euphoria and frustration, while (8) investors and top management accompany the process, serving as checks and balances on one another. (9) Interaction with other organizations can have a supportive or negative impact on the innovation project, while (10) wider sectorial infrastructures are being developed with competitors, government agencies, and others. During the implementation or termination period (11), innovation adoption occurs "by linking and integrating the 'new' with the 'old' or by reinventing the innovation to fit the local situation." After implementation (12), investors and top management "make

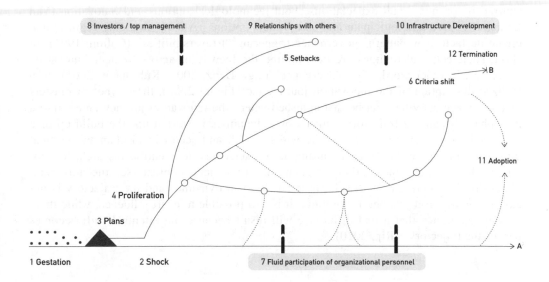

Figure 2.8.1   Key components of the innovation journey (Van de Ven *et al.*, 1999, p. 25)

attributions about innovation success or failure. These attributions are often misdirected but significantly influence the fate of innovations" (Van de Ven *et al.*, 1999, p. 24).

Regarding navigating innovation journeys by learning "to go with the flow," this advice sounds somewhat humble and unambitious. Do innovation actors (managers, users, and policy makers) really have no serious chance to guide innovation projects toward desired targets? Rip (2010) suggests that actors interested in strategic interventions will be more successful if they understand the co-evolutionary nature of the overall process and its institutional environment: "At least, they can avoid being [un]productive, as would happen when using a command-and-control approach while technological innovations are following their own dynamics."

2.8.3   Discussion about Innovation Journeys

In order to better understand the options and limitations for interventions in an innovation journey, it is useful to apply a heuristic offered by the school of evolutionary-economic analyses of technology dynamics.[5] This was discussed in Section 2.7.4 but is worth repeating here. This heuristic is built on the findings of "innovation studies," in particular on the seminal work of Nelson and Winter (1977, 1982), who observed different technological "regimes" characterized by longstanding specific "search strategies" of engineers, determining to some extent the development trajectory of a given regime. Rip and Kemp (1998) added to this the public and private strategies and policies of relevant actors: Technological innovation is socially constructed, including the governance of a regime.

As discussed in Section 2.7.4, scholars combined these conceptual elements into the heuristic of a "multilevel perspective" on sociotechnical transitions (e.g., Geels and Schot, 2007), characterized by niche innovations on a micro level, sociotechnical regimes on the meso level, and wider sociotechnical landscapes on the macro level. Studied with the help of this heuristic, one can see regime transitions (see Figure 2.8.1) that are spurred by promises and expectations about technological options and innovation (Van Lente, 1993). Actors anticipate and assess their options *vis-à-vis* changing regimes and create de facto new patterns governance triggering "irreversibilities" (Callon, 1991). In short, sociotechnical regimes are determining the leeway of actors to steer innovation processes, in other words the "governance" (e.g., Benz, 2007; Kuhlmann, 2001). In a stylized description of the innovation journey (see Figure 2.8.3), three types or clusters of governance activities can be distinguished where the innovation journey enters into a new phase because a trajectory with its own dynamics is started up: the build-up of a protected space, stepping out into the wider world, and sector-level changes (vertical dimension in Figure 2.8.3), each cutting across activities in scientific research, technological development and markets, regulation, and societal context (see the horizontal dimension in Figure 2.8.3). "Each phase has its own dynamics and the trajectory is not easy to modify. But just before 'gelling,' it is still possible to exert influence, while there is some assurance that a real difference will result because the intended shift becomes part of the trajectory" (Rip, 2010).

---

[5] See, for a more detailed discussion, Section 2.7, "Constructive Technology Assessment."

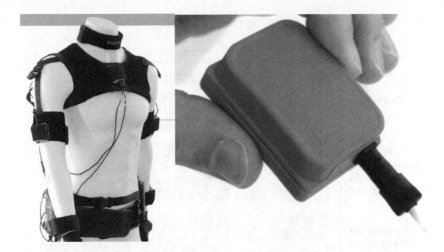

Figure 2.8.2 Xsens' MVN Biotech motion measurement system (left) and MTx motion tracker (right) (See Plate 16 for the colour figure)

2.8.4   Example from Practice

An example of a successful innovation journey (actually several journeys) is provided by Xsens.[6] Today Xsens is a leading developer and global supplier of 3D motion-tracking products based upon miniature inertial sensor technology. The company was founded in 2000 by two graduates of the University of Twente, the Netherlands. Inspired by the possibilities of tiny motion sensors for measurement of the performance of athletes, they specialized in sensor technologies and sensor fusion algorithms. In 2000 Xsens launched its first measurement unit, which was used for human motion measurement and industrial applications.

After more than 10 years of experience and several trips and setbacks along "uncharted rivers," Xsens today is recognized for its motion-tracking and motion capture products with best-in-class performance, outstanding quality, and high ease of use. Clients of Xsens include Electronic Arts, NBC Universal, INAIL Prosthesis Center, Daimler, Saab, Kongsberg Defence Systems, and many other companies and institutes throughout the world. Xsens is working with many industry partners, including Autodesk, Sagem, and Siemens.

As an innovation actor, Xsens navigated through all three stylized phases of a journey: building up a protected space, stepping out into the wider world, and deling with sector-level changes:

1. The company started with a promising technological option (inertial tracking) with a potential to be developed in numerous directions. In 2000 the founders depended on other partners and explored various market niches – in this case, there was almost no "protected space," facilitated by a parent organization, for example a large company, or by public innovation policy (in other cases, say the windmill energy technology in Denmark and Germany, facilitated by electricity feed-in law; see Hendry and Harborne, 2011). But

---

[6] The text of this paragraph draws on http://www.xsens.com/ and http://en.wikipedia.org/wiki/Xsens (accessed January 9, 2012).

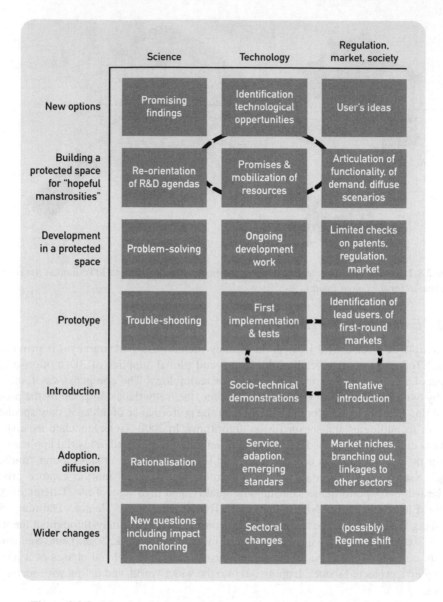

Figure 2.8.3 Mapping the innovation journey in context (Rip and Schot, 2002)

through exploration and learning, Xsens experienced quite a lot of cumulative development. Initially the company tried to launch the human motion measurement devices as speedometer for joggers, assuming that joggers are always ready to spend money for trendy gadgets; this failed. Shortly afterward, the innovators explored the application of the tracking sensors with disabled people.

2. This helped to make steps toward the integration of several "motion trackers," which facilitated applications with ergonomic research in industry, creating again new

application potential – the scope of achieved and accepted innovations made it possible "to step out into the wider world."

3. One of the most successful products – developed after several setbacks and learning loops – was the MVN Inertial Motion Capture suit,[7] a cost-efficient system for full-body human motion capture (see Figure 2.8.2). The MVN is based on unique, state-of-the-art miniature inertial sensors, biomechanical models, and sensor fusion algorithms. Meanwhile it has found applications in film and commercials, game development, training and simulation, live entertainment, biomechanics research, sports science, rehabilitation, and ergonomics – and some of the applications even helped to induce sector-level changes.

References

Benz, A. (2007) Governance in connected arenas – political science analysis of coordination and control in complex control systems, in *New Forms of Governance in Research Organizations. From Disciplinary Theories towards Interfaces and Integration* (ed. D. Jansen), Springer, Heidelberg, pp. 3–22.

Callon, M. (1991) Techno-economic networks and irreversibility, in *A Sociology of Monsters: Essays on Power, Technology and Domination* (ed. J. Law), Routledge, London, pp. 132–165.

Geels, F.W. and Schot, J. (2007) Typology of sociotechnical transition pathways. *Res. Policy*, **36**(3), 399–417.

Hendry, C. and Harborne, P. (2011) Changing the view of wind power development: More than "bricolage." *Res. Policy*, **40**, 778–789.

Kuhlmann, S. (2001) Governance of innovation policy in Europe – three scenarios. *Res. Policy*, **30**, 953–976.

Nelson, R. and Winter, S. (1977) In search of a useful theory of innovation. *Res. Policy*, **6**, 36–76.

Nelson, R.R. and Winter, S.G. (1982) *An Evolutionary Theory of Economic Change*, Belknap Press of Harvard University Press, Cambridge, MA.

Rip, A. (2010) Processes of technological innovation in context – and their modulation, in *Relational Practices, Participative Organizing (Advanced Series in Management, Volume 7)* (ed. C. Steyaert and B. Van Looy), Emerald Group Publishing Limited, Bingley, UK, pp. 199–217.

Rip, A. and Kemp, R. (1998) Technological change, in *Human Choice and Climate Change, Vol. 2, Resources and Technology* (ed. S. Rayner and L. Malone), Batelle Press, Washington, DC, pp. 327–400.

Rip, A. and Schot, J.W. (2002) Identifying loci for influencing the dynamics of technological development, in *Shaping Technology, Guiding Policy* (ed. R. Williams and K. Sørensen), Edward Elgar, Cheltenham, UK, pp. 158–176.

Van de Ven, A.H. *et al.* (1999) *Research on the Management of Innovation: The Minnesota Studies*, Oxford University Press, Oxford.

Van Lente, H. (1993) *Promising Technology: The Dynamics of Expectations in Technological Developments*, WMW-Publikatie 17, Universiteit Twente, Enschede, the Netherlands.

2.9 Risk-Diagnosing Methodology

Johannes Halman

University of Twente, Faculty of Engineering Technology, Department of Construction Management and Engineering, P.O. Box 217, The Netherlands

2.9.1 Introduction

This chapter introduces risk-diagnosing methodology (RDM), which aims to identify and evaluate technological, organizational, and business risks in new product development.

---

[7] See http://www.xsens.com/en/general/mvn (accessed January 9, 2012).

RDM was initiated, developed, and tested within a division of Philips Electronics, a multi-national company in the audio, video, and lighting industry. Also Unilever, one of the world's leading companies in fast-moving consumer goods, decided to adopt and implement RDM on a worldwide basis. Since its initiation, RDM has been applied to new product development projects in areas as diverse as the development of automobile tires, ship propellers, printing equipment, landing gear systems, lighting and electronic systems, and fast-moving consumer goods in various industries in Germany, Italy, Belgium, the Netherlands, Brazil, China, and the United States. RDM proved very useful in diagnosing project risks, promoting creative solutions for diagnosed risks, and strengthening team ownership of the project as a whole (Halman and Keizer, 1994; Keizer, Halman, and Song, 2002).

2.9.2   Requirements for an Effective Risk Assessment

The true nature of project risk is determined not only by its likelihood and its effects, but also by a project team's ability to influence the risk factors (see e.g., Keil *et al.*, 1998; Sitkin and Pablo, 1992; Smith, 1999). Thus a project activity should be labeled "risky" if the following criteria apply:

- The likelihood of a bad result is great.
- The ability to influence it within the time and resource limits of the project is small.
- Its potential consequences are severe.

To be effective, a risk assessment needs to help identify potential risks in the following domains:

- *Technology:* Product design and platform development, manufacturing technology, and intellectual property.
- *Market:* Consumer and trade acceptance, public acceptance, and the potential actions of competitors.
- *Finance:* Commercial viability.
- *Operations:* Internal organization, project team, co-development with external parties, and supply and distribution.

The most powerful contribution of risk assessment comes at the end of the feasibility phase of the innovation process, at the contract gate (see also Figure 2.9.1). At this stage, the transition to the actual product development and engineering of a particular product or product range takes place; uncertainty has to be managed, taking into account the potential risks relating to all the aspects of manufacturability, marketability, finance, human resources, and so on. In this phase of the project, management still has the ability to substantially influence the course of events and make a considerable impact on the eventual outcome (Cooper, 1993; Wheelwright and Clark, 1992). However, a periodical reassessment of potential risks in subsequent phases is still required.

In sum, it would appear that a comprehensive risk assessment approach would do the following:

- Evaluate each potential risk on its likelihood, its controllability, and its relative importance to project performance.

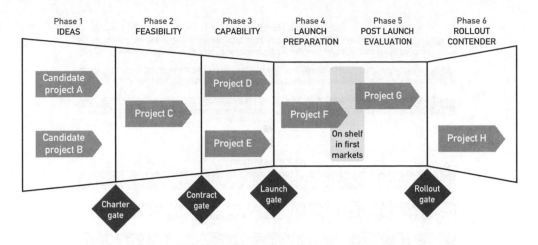

Figure 2.9.1   A typical innovation funnel

- Take a cross-functional perspective by identifying and evaluating technological, market, and financial as well as operational risks.
- Conduct the risk assessment at the end of the feasibility phase, and periodically reassess the project for unforeseen risks and deviations from the risk management plan.
- Identify and evaluate the product innovation risks individually, and generate, evaluate, and select alternative solutions in subgroups and plenary sessions.

2.9.3   The Risk-Diagnosing Methodology (RDM)

The purpose of RDM is to provide strategies that will improve the chance of a project's success by identifying and managing its potential risks. RDM is designed to be applied at the end of the feasibility phase, and should thus address such issues as consumer and trade acceptance, commercial viability, competitive reactions, external influential responses, human resource implications, and manufacturability. In this subsection, we will describe the successive steps (see also Figure 2.9.2) of RDM.

Initial Briefing

The first step of RDM is meant to build a full understanding of the conditions to be met at the start of the RDM process and to make the necessary appointments. The initial briefing takes place between the risk facilitator and project manager. This initial briefing should cover both general and project-specific topics. Project-specific topics include its objectives and unique characteristics, its stakeholders, the nature of its current phase, and the commitments required from its participants. More general topics include how information about the project will be made available to the risk facilitator, how this information will be kept confidential, who will participate in the RDM process (e.g., stakeholder, a project manager, a project team, and experts), how participants will be informed of their involvement, when and where the RDM process will take place, and what may be expected from it. Special care must be taken to include in the team technological, business, and marketing expertise. The output of this initial

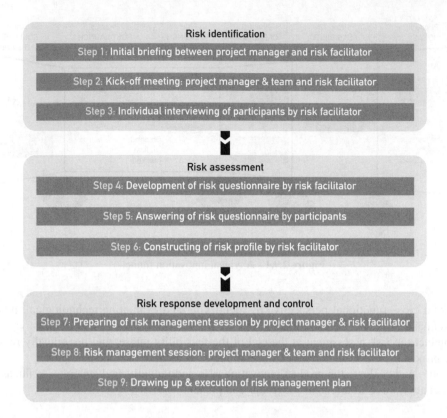

Figure 2.9.2   Outline of risk-diagnosing methodology (RDM)

briefing is twofold: agreements between the project manager and risk facilitator on actions to be taken, and invitations to a "kick-off" meeting for participants in the RDM process.

Kick-off Meeting

The objective of this meeting is to make sure that all participants know what to expect during the RDM process and are willing to cooperate. During the kick-off meeting, the following topics are addressed: objectives of and steps in the RDM process; the expected input, level of involvement, and amount of time from participants; the confidentiality of the interviews and other information provided by participants; and the expected output of the RDM process. After the kick-off meeting, agreements are made on the date, time, and location of the interviews and on the date, time, and location of a plenary risk management session.

Individual Interviewing of Participants

The objective of this step is to develop a comprehensive overview of all critical aspects in the innovation project. To enable participants to describe freely what they see as the riskiest aspects of the project, the risk facilitator interviews all participants individually. Each interview takes about 1.5 hours, during which the participant is led to think carefully on the project and its risks, and on his or her contribution specifically. Every participant is asked to study the project plan and the reference list of potential risk issues (see Appendix 2.9.A). In every new interview, the preceding interviews are taken into account (without mentioning the respondent's name) to test the completeness and correctness of the already gathered data. The protocol for the interviews is as follows:

- A short introduction of both participant and risk facilitator, and explanation by the risk facilitator of the objective of the interview.
- "Gaps" in the project: "What do you see as gaps in knowledge, skills, and experience for this project?" and "Can these gaps be bridged within the time and resource constraints of the project?"
- The reference list with potential risks: "What other gaps might be difficult to bridge?"
- Closing the interview: "Did we forget something?"
- Next steps: The risk facilitator briefly explains again what the interviewee can expect next, especially the risk questionnaire and the risk management session.

Processing the Interviews: Design of a Risk Questionnaire

After having interviewed all the participants, the risk facilitator analyzes the interview notes and clusters the critical issues according to the risk categories distinguished in Appendix 2.9.A (e.g., product technology risk, manufacturing technology risk, and project team risk). Then the risk facilitator designs a risk questionnaire, in which the critical issues from the interviews are translated into positive statements of "objectives to be realized" (see Figure 2.9.3). For example, if in one of the interviews a risk team member says, "We will be using a new ingredient in our product solution, and I have read in a

Risk Statements:	What is the *level of certainty* that the statement will be true? Very Very Low High					Ability of team to *influence* course of actions within time & resource limits Very Very Low High					Relative *importance* of statement for obtaining project success Very Very High Low				
	1	2	3	4	5	1	2	3	4	5	1	2	3	4	5
1. The new product will be safe to use for people with a sensitive skin.	O	O	O	O	O	O	O	O	O	O	O	O	O	O	O
2. With the trade customer clear after sales arrangements have been agreed.	O	O	O	O	O	O	O	O	O	O	O	O	O	O	O
3. For localized dye damage we have an appropriate solution.	O	O	O	O	O	O	O	O	O	O	O	O	O	O	O

Figure 2.9.3   Example of a risk questionnaire

journal that this material sometimes causes skin irritations," the statement would be for-mulated thus: "The new product formulation will be safe to use also for people with sensitive skin."

Answering the Risk Questionnaire

Respondents are asked individually to score the risk statements that they have developed on three five-point scales (see Figure 2.9.3):

- The level of certainty that the objective formulated in the risk statement will be realized.
- The ability of the team to reach an appropriate solution using the project's allotted time and resources.
- The relative importance of the objective to project performance.

Respondents are asked to answer the questionnaire as completely as possible, but not to respond to those issues about which they have no idea or opinion. The typical number of risk statements in these questionnaires is 50–60, and it takes 45–60 minutes to complete.

Constructing the Risk Profile

After the respondents have completed the risk questionnaire, the risk facilitator constructs a risk profile from their scores. Every risk statement is reported with its scoring for the three evaluation parameters (see Figure 2.9.4 for an example).

The risk profile presents both the degrees of risk perceived by the majority of the respon-dents and the distribution of their perceptions. Although the criterion can be chosen differ-ently, we have chosen to mark with a dot the column in which a support of a minimum of 50% (the average of the scores) is reached. This will give an initial view of the thinking of the majority of the respondents. Next, the risk facilitator classifies the risk statements in two ways. First, every risk statement is classified along the three parameters into four groups by the following decision rules:

- ("*"): At least 50% of the scores are 1 or 2 on the 5-point scale (1 being *very risky*), and there are no scores of 5 on the 5-point scale.
- ("0"): At least 50% of the scores are 4 or 5 on the 5-point scale, and there are no scores of 1 on the 5-point scale.
- ("m"): At least 50% of the scores are 3 on the 5-point scale, and there are no scores of 1 or 5 on the 5-point scale.
- ("?"): For all remaining cases. There exists a lack of consensus, visible in a wide distribu-tion of opinions. After discussion with the interviewees, the "?" scores may be changed to one of the other three.

Next, the risk facilitator classifies each risk statement into a "risk class" by examining the questionnaire responses. RDM uses five risk classes: S = safe; L = low; M = medium; H = high; and F = fatal. For example, a combination of scores "*,*,*" on a given risk state-ment would result in its classification as so risky that not lessening this risk would be fatal for the project (which would then be assigned a risk class of F), while the combination "0,0,0" would result in a classification as safe (risk class S). The total number of possible

Risk Statements:		What is the *level of certainty* that the statement will be true? 'C' Very Very Low High					Ability of team to *influence* course of actions within time & resource limits 'A' Very Very Low High					*Relative importance* of statement for obtaining project success 'I' Very Very High Low					Score for each dimension of risk			Risk Class
		1 5	2	3	4		1 5	2	3	4		1 5	2	3	4		C	A	I	
1. The new product will be safe to use for people with a sensitive skin.	N resp. ≥ 50%	0	0	1	3	5	0	0	1	3	5	0	2	5	2	0	0	0	m	L
2. With the trade customer clear after sales arrangements have been agreed.	N resp. ≥ 50%	3	2	4	0	0	0	2	5	2	0	3	5	1	0	0	*	m	*	H
3. For localized dye damage we have an appropriate solution.	N resp. ≥ 50%	1	3	2	0	2	2	2	0	4	0	4	0	1	1	2	?	*	?	L-F

N resp: Number of team members who scored in certain column
≥ 50%: Column in which at least 50% of team response is first met
'*': At least 50% of the scores are 1 or 2 on the five point scales and there are no scores of 5
'0': At least 50% of the scores are 4 or 5 on the five point scales and there are no scores of 1
'm': At least 50% of the scores are 3 on the five point scales and there are no scores of 1 or 5
'?': Lack of consensus: There is a wide distribution of opinions.
Risk Classes used: **S** = Safe; **L** = Low Risk; **M** = Medium Risk; **H** = High Risk and **F** = Fatal risk.

Figure 2.9.4   Example of a project risk profile

combinations of risk scores is 64 (see Keizer, Halman, and Song, 2002). If there is a distribution of opinions, the risk score can be represented by a range between the lowest and highest risk classes that can be reached if the respondents achieve consensus (e.g., L-M or H-F). For example, in Figure 2.9.4, the scores indicate a lack of consensus within the team for risk statement number 3. If, after discussion and clarification, the team as a whole is convinced that the statement "finding an appropriate solution for localized dye damage" is very uncertain and very important to project success, the risk range will change from "L-F" to "F." It should be stressed that this lack of consensus in the risk profile is very valuable information and should not be "swept under the carpet." It has happened more than once that a member of a team had a clearly divergent opinion that appeared, after discussion and clarification, to be right!

Preparing a Risk Management Session

In this RDM step, the project manager reaches agreement with the risk facilitator on the agenda for the risk management session. Certain risks will be better tackled in a plenary session, and others in subgroups. The choice will depend on the number and difficulty of

- Every one's viewpoint is valid!
- No holding back – Say what's worrying you!
- No management hierarchy
- The things we don't want to hear are probably the key issues
- Explain from your area of expertise

Figure 2.9.5   Rules of engagement for a risk management session

risks requiring a solution. Experience suggests that a risk management session will require a one-day meeting where the team can work in plenary as well as syndicate (subgroup) meetings.

Risk Management Session

The objective of the risk management session is to achieve consensus on action plans for dealing with the high risks and on procedures for dealing with the medium and lower risks. In addition to the project manager and the risk facilitator, all persons who participated in the RDM process are invited to this session, which the project manager usually leads. A typical session includes an introduction to the objectives of the meeting, the program, and some "rules of conduct."

Our experience has shown that observing these rules of conduct (see Figure 2.9.5) helps participants to increase the effectiveness of the process and foster breakthroughs in problem solving. Not only do these rules enforce the requirements generally agreed on for brain-storming (Ivancevich and Matteson, 1996) but also they limit as far as possible the potential negative effects of group dynamics discussed in this section. After the introduction and an agreement to follow the rules of conduct, the risk facilitator presents the risk profile: What are the high risks everyone agrees on? What are the risks about which opinions differ but that could potentially turn out to be high? The risk facilitator also shows how certain issues are related to each other.

The first part of the risk management session is designed to create a common understanding of the risks and to generate ideas for managing them. In the second part of the risk management session, the group is split up into subgroups, which are asked to work further on the suggested ideas and to formulate action plans specifying what needs to be done, by whom, and when. Appendix C presents some trigger questions that might help the subgroups design these plans. In the third part of the session, the subgroups present the outcomes of their respective discussions. After further clarification and discussion, the project team decides on which follow-up actions should be taken to manage the diagnosed risks, and on how to present the results to senior management.

Drawing Up and Executing a Risk Management Plan

The risks and corresponding action plans are brought together in a risk management plan. In addition to documenting the risk assessment results and the outcome of the risk management session, the risk management plan states who is responsible for each of the diagnosed risks,

Project: Golden Eagle	Risk issue #: T07
Project number: 01A2552	**Project Leader**: Tom Jefferson

Risk Issue:
Deformation of the product due to overexposure

Date of assessment: 13 June 2001	**Action Responsible**: Marc Erlich
Risk Type: Manufacturing Technology	**Start Date**: 18 June 2001
Risk Class: F **H** M L S	**Due Date**: 3 Sept 2001

Clarification of risk issue:
During production an uneven cooling down of the product surface causes instability in product surface structure

Actions Agreed on:
1. Investigate alternative mould options
2. Investigate how GE has solved this problem

Follow-Up agreements:
1. Report results and present proposal in project review session of 14 September 2001
2. In case of satisfying new mould option make supplementary work package proposal
3. In case GE-solution is applicable and alternative mould options don't work out,
develop cross-over proposal to negotiate with GE

Mitigation plan status:
PT-meeting 6/25: Drafts for mould options a/b/c/d are ready
PT meeting 7/23: Prototypes for mould options a/b/c/d casted
PT meeting 8/7:   Prototypes a/b/c/d tested, prototype c seems satisfactory
PT meeting 8/21:
PT meeting 9/3:

Figure 2.9.6   Example of a risk-tracking form

how much time and resources are needed to deal with these risks, and how progress will be monitored and reported. This plan enables management to decide upon the feasibility of the project and make a "go" or "no go" decision. The action plans drawn up by the subgroups are documented in risk-tracking forms (see Figure 2.9.6). These provide a framework for recording information about the status and progress of each diagnosed risk. To guarantee follow-up, besides regular monitoring and control of the project risks in project team meetings, senior management should also require formal approval of the risk management plan and verify the progress of the risk actions plans in all subsequent gate reviews.

It may be necessary to repeat the RDM for some product innovation projects. In particular, in cases where the project newness and complexity are great, modifications and unforeseen issues are almost certain to arise; this might demand reassessment of the overall risks at later stages of the project. Senior management in consultation with the project team therefore should reconsider at each stage whether to repeat the RDM process or simply to have the project team update the existing risk management plan.

2.9.4   Added Value of RDM

Evaluation studies show that professionals in the field of new product development are very satisfied with the way RDM allows them to identify, confront, and manage risks in their projects (Vuuren, 2001). Factors causing this satisfaction are, for example, that RDM helps to pull out the key risk issues and actions and to clear misunderstandings and varying interpretations about the existing project risks. Similarly the limited time that is needed to conduct the whole process, the application of the rules of engagement, and the appeal to think cross-functionally contribute a lot to this satisfaction. Also the use of the risk reference list is evaluated positively: It stimulates the team to think more in detail about the project and to identify risks the team hadn't thought of before. RDM provides the project team with the following:

- A list of classified risks identified while there is still time.
- A risk evaluation on scenarios that can be followed.
- A plan on how the major risks will be tackled.
- A solid basis for communicating the risks, project planning, and required resources with senior management and other stakeholders.

Appendix 2.9: Reference List with Potential Risk Issues in the Innovation Process

I.	**Product family and brand-positioning risks**
1.	New product helps to achieve business strategy,
2.	Project is important for project portfolio,
3.	New product contributes to brand name position,
4.	Project includes global rollout potential and schedule,
5.	New product fits within existing brand.
6.	New product fits with brand image.
7.	New product enhances the potential of product family development.
8.	New product provides opportunities for platform deployment.
9.	New product supports company reputation.
10.	New product has brand recovery potential.
11.	New product has brand development potential.
12.	New product's platform will be accepted by consumers.

II.	**Product technology risks**
1.	New product's intended functions are known and specified.
2.	New product fulfills intended functions.
3.	In-use conditions are known and specified.
4.	Interactions of product in use with sustaining materials, tools, and so on are understood.
5.	Components' properties, function, and behavior are known.
6.	Correct balance between product components is established.
7.	Assembled product meets safety and technical requirements.
8.	Alternatives to realize intended product functions are available.
9.	New product shows parity in performance compared to other products.
10.	New product shows stability while in storage (in factory, shop or warehouse, and transportation, and at home).
11.	New product format meets functional requirements.

III.	**Manufacturing technology risks**
1.	Raw materials are available that meet technical requirements.
2.	Process steps to realize the new product are known and specified.
3.	Conditions (temperature, energy, safety, etc.) to guarantee processing of good product quality are known and specified.
4.	Production means (equipment and tools) necessary to guarantee good product quality are available.

5. Scale = up potential is possible according to production yield standards.
6. Production system requirements (quality and safety standards, training of human resources, facilities, etc.) will be met.
7. Product-packaging implications are known and specified.
8. Manufacturing efficiency standards will be met.
9. Alternative approaches to process the intended product will be available.
10. Adequate production capacity is available.
11. Adequate production start-up is assured.
12. The reusability of rejects in production is foreseen.

IV. Intellectual property risks
1. Original know-how will be protected.
2. Required external licenses or know-how is known and available.
3. Relation to legal and patent rights of competitors is known and arranged.
4. Relevant patent issues are understood.
5. Patent-crossing potential is known and arranged.
6. Trademark registration potential is known and arranged.

V. Competitor risks
1. Product will provide clear competitive advantages.
2. Introduction of new product will change existing market share positions.
3. Introduction of the new product will have an impact on market prices.
4. New product will be launched before competitors launch a comparable product.
5. Response actions toward public and media expected from competitors will be anticipated.
6. New product enables the creation of potential barriers for competitors.
7. Implications of being technology leader or follower for this project have been identified.
8. Competitor's actions will be monitored and followed with adequate response.
9. Competitor's challenges will be monitored adequately.

VI. Supply chain and sourcing risks
1. Suppliers will meet required quality.
2. Capacity available to meet peak demands.
3. Appropriate after-sales services are available.
4. Contingency options are available for each of the selected suppliers.
5. Financial position of each supplier is sound.
6. Past experiences with each of the suppliers are positive.
7. Suppliers are ready to accept modifications if required.
8. Supply contracts can be canceled.
9. Each supplier will be reliable in delivering according to requirements.
10. Required quantities will be produced for acceptable prices.
11. Appropriate contract arrangements with suppliers will be settled.

VII. Consumer acceptance risks
1. Product specifications meet consumer standards and demands.
2. New product fits consumer habits and/or user conditions.
3. New product offers unique features or attributes to the consumer.
4. Consumers will be convinced that they get value for money, compared to competitive products.
5. New product appeals to generally accepted values (e.g., health, safety, nature, and environment).
6. New product offers additional enjoyment, compared to competitive products.
7. New product will reduce consumer costs, compared to competitive products.
8. Non-intended product use by consumers is adequately anticipated.
9. Target consumers' attitudes will remain stable during the development period.
10. New product will be communicated successfully with target consumers.
11. New product will provide easy-in-use advantages, compared to competitive products.

12.   Primary consumer requirements are known.
13.   Target consumers will accept the new product's key product ingredients.
14.   Niche marketing capabilities are available if required.
15.   Communication about new product is based on realistic product claim.
16.   Advertising will be effective.
17.   Product claims will stimulate target consumers to buy.
18.   New product has repeat sales potential.

VIII.   Trade customer risks
1.   Product specifications will meet trade customer standards and demands.
2.   Trade customers will welcome the new product from the perspective of potential sales.
3.   Trade customers will welcome the new product from the perspective of profit margin.
4.   Trade customers will welcome the new product given required surface and volume on shelf and storage facilities.
5.   Trade customers' attitudes will remain stable during the development period.
6.   New product will be communicated successfully to trade customers.
7.   Right distribution channels will be used.
8.   Trade will give new product proper care.
9.   Trade-supporting persons will endorse the new product.

IX.   Commercial viability risks
1.   The market target is clearly defined and agreed.
2.   Market targets are selected based on convincing research data.
3.   Capital cost projection for new product is feasible.
4.   Delays in product launch will leave the commercial viability of the new product untouched.
5.   Sales projections for new product are realistic.
6.   Estimated profit margins are based on convincing research data.
7.   Profit margin will meet the company's standards.
8.   The estimated return on investment will meet the company's standards.
9.   Volume estimates are based on clear and reliable estimates.
10.   Product viability will be supported by repeat sales.
11.   Suppliers will get attractive purchasing agreements.
12.   Knowledge of pricing sensitivity is available.
13.   Adequate investments to secure safety in production will be made.
14.   Long-term market potential is to be expected.
15.   Financing of capital investment is secured.
16.   Fallback to prior product concept is feasible.

IX.   Commercial viability risks
1.   The market target is clearly defined and agreed.
2.   Market targets are selected based on convincing research data.
3.   Capital cost projection for the new product is feasible.
4.   Delays in product launch will leave the commercial viability of the new product untouched.
5.   Sales projections for new product are realistic.
6.   Estimated profit margins are based on convincing research data.
7.   Profit margins will meet the company's standards.
8.   The estimated return on investment will meet the company's standards.
9.   Volume estimates are based on clear and reliable estimates.
10.   Product viability will be supported by repeat sales.
11.   Suppliesr will get attractive purchasing agreements.
12.   Knowledge of pricing sensitivity is available.
13.   Adequate investments to secure safety in production will be made.
14.   Long-term market potential is to be expected.

15.   Financing of capital investment is secured.
16.   Fallback to prior product concept is feasible.
17.   New product is commercially viable in case of market restrictions.

X.   Organizational and project management risks

1.   Internal political climate is in favor of this project.
2.   Top management actively supports the project.
3.   Project goals and objectives are feasible.
4.   Project team is sufficiently authorized and qualified for the project.
5.   Project team will effectively utilize the knowledge and experience of (internal) experts.
6.   Roles, tasks, and responsibilities of all team members are defined and appropriate.
7.   Decision-making process in the project is effective.
8.   Communication between members in the project team is effective.
9.   Required money, time, and (human) resources estimations are reliable and feasible.
10.   Required money, time, and (human) resources will be available when required.
11.   External development partners will deliver in time, and conform to budget and technical specifications.
12.   Sound alternatives are available to external development partners.
13.   Collaboration within the project team is effective.
14.   Project will effectively be organized and managed.
15.   Collaboration with external parties is effective.
16.   Collaboration between project team and the parent organization is effective.
17.   Project team is highly motivated and committed.
18.   Project team is paying attention to the right issues.
19.   Project has effective planning and contingency planning.
20.   Project team is learning from past experiences.

XI.   Public acceptance risks

1.   It is clearly understood who is responsible for the public relations of the project.
2.   The key opinion formers for the new product are known.
3.   Support of key opinion formers will be assured.
4.   Legal and political restrictions will be adequately anticipated.
5.   Environmental issues will be adequately anticipated.
6.   Safety issues will be adequately anticipated.
7.   Possible negative external reactions will be effectively anticipated.
8.   In case of new technology, prior (external) experience will be consulted.

XII.   Screening and appraisal

1.   New product performance targets will be tested and measured adequately.
2.   Trade customer appreciation will be tested and measured adequately.
3.   Consumer appreciation will be tested and measured adequately.
4.   Adverse properties as a consequence of the technological change will be tested and measured adequately.
5.   Credibility of the (internal) measures to external agencies is warranted.
6.   Tests will provide reliable evidence.

References

Cooper, R.G. (1993) *Winning at New Products*, Addison Wesley, Reading, MA.

Halman, J.I.M. and Keizer, J.A. (1994) Diagnosing risks in product innovation projects. *Int. J. Project Management*, **12**(2), 75–81.

Ivancevich, J.M. and Matteson, M.T. (1996) *Organizational Behavior and Management*, Irwin, Chicago.

Keil, M., Cule, P.E., Lyytinen, K., and Schmidt, R.C. (1998) A framework for identifying software project risks. *Commun. Assoc. Comput. Mach.*, **41**(11), 76–83.

Keizer, J.A., Halman, J.I.M., and Song, M. (2002) From experience: applying the risk diagnosing methodology RDM. *J. Prod. Innovat. Manag.*, **19**(3), 213–233.

Montoya-Weiss, M.M. and Calantone, R.J. (1994) Determinants of new product performance: A review and meta analysis. *J. Prod. Innovat. Manag.*, **11**(1), 31–45.

Sitkin, S.B. and Pablo, A.L. (1992) Reconceptualizing the determinants of risk behavior. *Acad. Manage. Rev.*, **17**(1), 9–38.

Smith, P.G. (1999) Managing risk as product development schedules shrink. *Res. Technol. Manage.*, **42**(5), 25–32.

Vuuren, W. (2001) An evaluation study on the use of RDM in Unilever, Report for the Brite Euram III program, Project Risk Planning Process (RPP), Contract no. BPRPR CT 98 0745, Project no BE97-5052, Report TUE_6_04_01.

Wheelwright, S.C. and Clark, C.B. (1992) *Revolutionizing Product Development, Quantum Leaps in Speed, Efficiency and Quality*, Free Press, New York.

2.10 A Multilevel Design Model Clarifying the Mutual Relationship between New Products and Societal Change Processes

Peter Joore

NHL University of Applied Sciences, P.O. Box 1080, The Netherlands

2.10.1 Introduction

Designers working on the development of sustainable and energy-efficient products inevitably run into issues such as user acceptance, infrastructural integration, or governmental regulations. Although such topics may not be part of their specific expertise, these issues cannot be ignored during the design process. First, this is because societal issues largely determine the development of many energy-efficient products, for instance through public awareness regarding the depleting resources of fossil fuels and the availability of sustainable energy technologies. Secondly, these issues cannot be neglected because the successful implementation of many energy-efficient products is largely determined by existing frameworks such as infrastructures and legal regulations, as well as other aspects that cannot be directly influenced by an individual designer. This means that the broader context in which new products will be functioning has to be considered during the design process (Joore, 2008, 2010). To support designers and to demarcate their efforts, it is necessary to structure the design process and the role of designers in such a way that the mutual relationship between new products and the sociotechnical or societal context in which these products function is taken into account in a systematic manner.

2.10.2 A Multilevel Design Model

In 2010 a new multilevel design model (MDM) was published (Joore, 2010) that may help to clarify this relationship between physical artifacts, on the one hand, and more intangible societal topics, on the other. The MDM combines two types of models that will be described in this section. The first group of models originates from the field of industrial design engineering and systems engineering, for instance the V-Model that is often used during the development of complex technological and software systems (KBST, 2004). Although these models may offer sufficient insight into the technological aspects of a design process, they are often formulated around the development of one single product or system, such that the broader sociotechnical and societal aspects are not sufficiently addressed.

The second group of models originates from the area of sustainable systems innovations and transition management. Here we refer to a dynamic multilevel model developed by Geels (2005) and methods like constructive technology assessment (CTA) and the innovation journey, which are described in detail in elsewhere in this section. These models offer a detailed insight into the interrelationships between innovations and their sociotechnical and societal contexts, However, these models often are based on qualitative and descriptive research and a high abstraction level, whereas the embedding of the model's outcomes toward the practice of industrial product development has not yet been formalized by a standardized design method.

Combining both groups of models leads to the MDM, which can be considered a descriptive multilevel systemic approach combined with a prescriptive design process. The design process consists of four phases: (1) experience, (2) reflection, (3) analysis, and (4) synthesis. This process is applied to four aggregation or system levels, being described as the (1) product–technology system (indexed by P), (2) the product–service system (indexed by Q), (3) the sociotechnical system (indexed by R), and (4) the societal system (indexed by S). Figure 2.10.1 shows a scheme with these four design steps in relation to the system levels and the multilevel design model. This figure also shows for each system level the envisioned transformation processes (T_S, T_R, T_Q, and T_P) between objectives at each system level ($S2'$, $R2'$, $Q2'$, and $P2'$) toward new situations, systems,

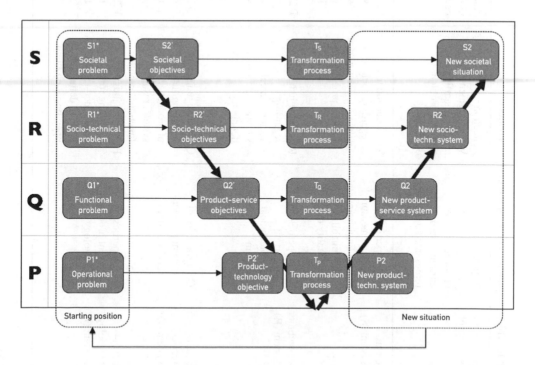

Figure 2.10.1   Multilevel design model (Joore, 2010, 88)

Table 2.10.1 Explanation of the multilevel design model

Design phase / System level	Experience Starting position, characteristics of current system (not represented in Figure 2.0.1)	Reflection Value judgment regarding starting position, problem definition	Analysis Objectives and criteria for new (sub)system	Synthesis Design, creation of new (sub)system	Experience Characteristics of new (sub)system
Societal system (S)	S1: properties of society, measured with Environmental Sustainability Index or Quality of Life Index	S1*: value judgment regarding societal situation, definition of societal problem	S2': preferences regarding social order, based on worldview and values, resulting in objectives for ideal new societal situation	Ts: vision development process, resulting in future vision for new societal situation	S2: living in society, executing societal experiment
Sociotechnical system (R)	R1: properties of current sociotechnical system	R1*: value judgment regarding sociotechnical situation and system deficiency	R2': dominant interpretative framework, leading to objectives for new sociotechnical system	Tr: system design process, leading to proposal for new sociotechnical system	R2: experiencing new sociotechnical system, for example by means of niche experiment
Product–service system (Q)	Q1: properties of current product–service system	Q1*: value judgment regarding functioning of product–service system, resulting in functional problem	Q2': determining functional demands and requirements to be met	Tq: design of a new product–service system	Q2: using and experiencing new product–service system
Product–technology System (P)	P1: properties of current product–technology system	P1*: value judgment regarding functioning of product–technology system, definition of operational problem	P2: target definition regarding new product and technology, leading to program of demands	Tp: product design process, leading to (prototype of) new product–technology system	P2: simulation, testing, using and experiencing new product

and products and services (S2, R2, Q2, and P2). The objectives are based on the experience of and reflection on existing problems at each level (S1*, R1*, Q1*, and P1*). The meaning of the symbols used in the MDM is further explained in Table 2.10.1.

To emphasize the difference in size or dimension of the system at the various levels, the model resembles a V-shape. Arrows in this V-shape indicate the relations that occur at and between various levels. At each level an identical process is presented. Only the width of the various layers differs, creating the characteristic V-shape. The main difference compared to the V-model used in systems engineering is the fact that sociotechnical and societal issues are explicitly part of the MDM.

2.10.3   Example Based on the Development of an Electrical Transport System

The MDM can be clarified by comparing the design of an electric transport vehicle, a transport service, and a regional transport system. The choice to use examples from this domain is made because the field of transport is complex enough to visualize the various aspects of the MDM, while the area of transportation also covers many energy issues. The example is visually represented in Figure 2.10.2.

Level P: The Product–Technology System

Products form the basic level of the MDM. These can be defined as "physical objects that originate from a human action or a machine process." As these objects are made up of technical components, the term *product–technology system* is being used. However, to improve readability we will generally refer to these as *products*. Products refer to tangible, inextricably linked technical systems, physically present in place and time. With most of these artifacts, you could "drop them on your toes." Product–technology systems generally fulfill one clearly distinguishable function. A system dysfunction occurs as soon as one or more technical components are missing.

An electric vehicle or a battery-charging station is an example of a product–technology system. The vehicle is discernible in place and time and fulfills a clearly defined primary function aimed at transporting people or things. As soon as certain technical components are missing, the product ceases to function as such, for example with a flat tire or an engine that is out of order. The direct relationship with the vehicle as a product–technology system is limited to individual persons, such as the driver, passengers, and maintenance mechanic.

Level Q: The Product–Service System

The second level of the MDM is formed by product–service systems. These can be defined as "a mix of tangible products and intangible service designed and combined so that they jointly are capable of fulfilling final customer needs" (Tukker and Tischner, 2006, p. 24). A product–service system is built up of physical as well as organizational components, which form a united and cohesive whole that together fulfills a specific function, usually definable in time and place. The system fulfills one or more clearly defined functions that can no longer be performed if one of the technical or organizational components is missing. The product–service system can indeed be compatible with certain policy, legal, social,

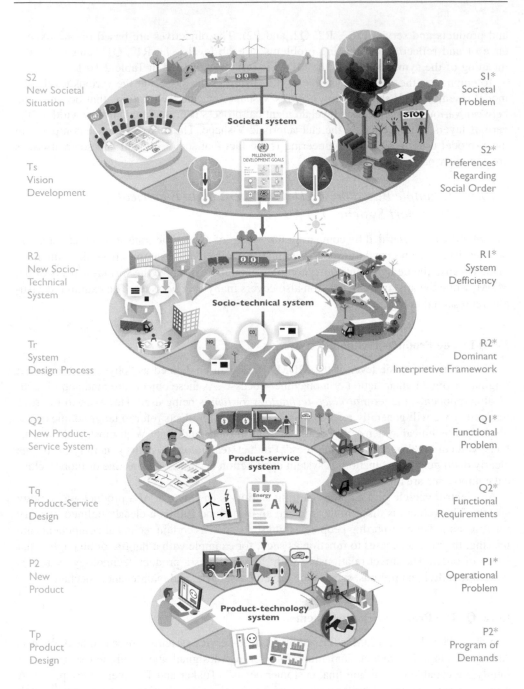

S2
New Societal
Situation

Societal system

S1*
Societal
Problem

S2*
Preferences
Regarding
Social Order

Ts
Vision
Development

R2
New Socio-
Technical
System

Socio-technical system

R1*
System
Defficiency

R2*
Dominant
Interpretive Framework

Tr
System
Design Process

Q2
New Product-
Service System

Product-service
system

Q1*
Functional
Problem

Q2*
Functional
Requirements

Tq
Product-Service
Design

P2
New
Product

Product-technology
system

P1*
Operational
Problem

P2*
Program of
Demands

Tp
Product
Design

Figure 2.10.2 Multilevel representation of the electric transport case (See Plate 19 for the colour figure)

cultural, or infrastructural elements, but these do not form an inextricable part of the product–service system.

An example of a product–service system is an electric transport service, which is made up of technical as well as organizational components. If, for example, the truck driver is missing or business problems occur, the transport service may no longer work. The product–technology system "electric vehicle" may still be able to function perfectly well, but the product–service system "transport service" no longer works. To function properly, good roads and corresponding traffic regulations are necessary. When using electric vehicles, battery-charging units may be essential. However, these do not form an inseparable component of the product or the service itself, but they are part of the even larger "sociotechnical system."

Level R: The Sociotechnical System

The third level of the multilevel design model is the sociotechnical system. This can be defined as "a cluster of aligned elements, including artifacts, technology, knowledge, user practices and markets, regulation, cultural meaning, infrastructure, maintenance networks and supply networks, that together fulfill a specific societal function" (Geels, 2005). Changes that take place at this level are often referred to as a *system innovation*, which can be defined as "a large scale transformation in the way societal functions are fulfilled. A change from one socio-technical system to another" (Elzen, Geels, and Green, 2004, p. 19). At this level a large number of components are combined that are not necessarily formally related to each other. Several elements together form a joint system that fulfills a combination of functions that have a narrow, joint relationship with each other. Product–service systems, accompanying infrastructure, government legislation, and cultural as well as social aspects may form a mutually interdependent whole. In contrast to the first two levels, the sociotechnical system continues to function if one or more elements are missing, and elements may even assume each other's function.

In this way, road transport can be considered a sociotechnical system, where transport vehicles, rental trucks, freight trains, and other means of transport meet each other on public roads. They are joined there by buses, pedestrians, and cyclists. Other elements that are part of this system are the traffic rules, the insurance and licenses that a company must have, the fuel stations, the price that is paid for that fuel, the availability of parking places, and the attitude of citizens toward the various forms of transportation. In case of the introduction of electric transport, electric battery-charging points may need to be introduced as new parts of the sociotechnical system.

In case one of these subsystems fails, its function can be taken over by another subsystem. If the buses stop running, people will take the bicycle. If diesel becomes too expensive, people will buy a car that runs on gasoline. However, switching to electric transport may not be so easy, as battery-charging points are currently hardly available. Even so, the current position of fuel stations is not suitable to place these battery-charging points, as they are often located in rather remote areas. While this is no problem when filling up a tank of gasoline in a few minutes, these remote areas are not very attractive when waiting several hours for a battery to be charged. Here, government may play an intermediary role, for instance when deciding to support the placement of battery-charging points in inner-city areas. However, even when supported by policy regulations it will still take a substantial amount of time until these battery-charging stations are as readily available as regular fuel stations. This example

shows that changes at the sociotechnical level often take more time and have a greater societal impact than changes at the level of individual product–service systems.

Level S: The Societal System

The highest level of the MDM is being defined as the *societal system*, being "the community of people living in a particular country or region and having shared customs,
laws, and organizations" (*Oxford Dictionary*). This is, just like the previous level, built
up from a combination of material, organizational, policy, legal, social, cultural, or infrastructural elements. Changes that take place at this level are often referred to as a *transition*, which can be considered "a gradual, continuous process of societal change, where
the character of society (or of one of its complex subsystems) undergoes structural
change" (Rotmans *et al.*, 2000, p. 11).

Whereas the sociotechnical system can more or less be defined and demarcated, at the
societal system level a complete summary can no longer be made of which elements do or
do not make up the components of the system. It extends over several influence spheres and
domains, and the boundary between these areas cannot easily be determined. Also the societal system does not fulfill one distinct function, but is made up of functions that are not
necessarily related.

An example of development on the society level is the influence of the sociotechnical system "road transport" on other sectors. Noise pollution and toxic emissions as a consequence
of road transport affect the health of people, including when these people are not part of the
transport system. The transport system can function perfectly even when everybody who
lives along highways becomes ill. This indicates that this problem is apparently located at
the societal system level and can no longer be resolved within the boundaries of one delim-
ited sociotechnical system.

2.10.4   Benefits for the Design Process

The examples given in this section show that the development of new energy-efficient products, for example a new electric transport vehicle, is very much interwoven with development
in the broader system context in which the new product will be functioning. If no battery-
charging points are available, the acceptance of electric vehicles will be hindered. At the same
time, companies will invest in the development and placement of battery-charging points only
when a substantial amount of electric vehicle owners are willing to pay for their use. In other
words, especially at the higher system levels there may often arise a "chicken–egg" situation
that may hinder the introduction of new sustainable and energy-efficient products.

The use of the MDM may not solve these chicken–egg dilemmas. However, they may help
the designer to demarcate the scope of a specific design process in which he or she is
involved. Consciously distinguishing between the various system levels may help designers
to determine what projects may suit their specific expertise. For instance, the designer of a
new electric vehicle must know everything about engineering, materials, production processes, draft angles, and other technological details. The designer of a transport service
doesn't necessarily need to bother about the design of the physical artifact. Here it is more
about business model generation and developing innovations in which the vehicle has become

part of a broad "transport solution." At a still higher level it can then be about the development of a future vision of the way mobility will develop in a wider sense during the next 10
years, and the way that policy measures may influence this development. It seems obvious
that the designer who is skilled in designing the technical details of a new vehicle does not
necessarily require the same qualities as the designer who is skilled in the development of a
"transport solution" or a future vision aimed at what mobility will look like in the year 2020.

Secondly, distinguishing between the various system levels may help designers to determine the design methods that are most suitable for a specific project. The manner in which
the design process progresses at the level of the product–technology system appears to be
mostly in keeping with the various models in the areas of industrial design engineering and
system engineering. At the product–service level, the way in which the design process progresses appears to be mostly in keeping with the various models in the areas of sustainable
product development, as these models have a rather strong focus on the organizational aspect.
Change processes that take place on the sociotechnical level appear to be mostly in keeping
with the various models in the field of sustainable system innovations and transitions. Here it
is usually a matter of slowly progressing, difficult-to-direct developments. At the societal
level, it is also usually a matter of progressing, difficult-to-direct developments, so the question is of course if it is at all possible to speak about a "design process" at all.

Thirdly, distinguishing between the various system levels may help designers to determine
the way that a certain design should be tested. Testing a new technological product can often
be done in a laboratory setting, with users commenting on the design in a protected environment. Trying a new transport service probably needs a broader setting, perhaps introducing the
service for several weeks or months in a small-scale experiment in a dedicated environment.
Finding out the effect of certain sociotechnical or societal system changes would need even
more time, so that the impact of specific interventions (like the introduction of a certain policy
regulation) can be measured over a longer period of time. Here the concept of a strategic niche
experiment of bounded sociotechnical experiment (Brown *et al.*, 2003) may be useful.

Fourthly, distinguishing between the various system levels may help to determine which
actors to involve during the design process. As for the involvement of actors and designer, at
the product–technology level it is generally a matter of a limited group of actors who are in
direct contact with the product. In most cases, one organization can be identified that delivers
the product. At the product–service level, the relationship with actors is mostly restricted to a
limited number of parties that are usually in a formal or legal relationship, for example as
consumer-suppliers or as formally cooperating partners. At the sociotechnical level, agreements between actors are less tightly defined, although they can be formalized collectively,
for example in the form of legislation, regulation, or collective standardization. At the societal level, the influence of the system extends to all sorts of parties that do not maintain any
deliberate relationship with each other, but become implicitly related as developments touch
several sectors of society.

2.10.5 *Conclusions*

In this section the MDM has been described, and I have clarified the design specified process
by distinguishing four separate system or aggregation levels. Distinguishing between these
levels may help designers of sustainable and energy-efficient products to execute the design

process more effectively. Firstly, it may help them to determine the specific skills that may be required for a certain design project, asking themselves if a certain project indeed matches their specific expertise. Secondly, it may help them to determine the design methods that are most suitable for a specific project. Thirdly, it may help them to determine the way that a new design should be tested, choosing for instance between short-term in-house laboratory experiments or long-term sociotechnical experiments. Fourthly, it may help them to determine which actors to involve during the design process.

References

Brown, H., Vergragt, P., Green, K., and Berchicci, L. (2003) Learning for sustainability transition through bounded social-technical experiments in personal mobility. *Technol. Anal. Strateg.*, **15**, 291–315.

Elzen, B., Geels, F., and Green, K. (eds) (2004) *System Innovation and the Transition to Sustainability: Theory, Evidence and Policy*, Edward Elgar Publishing Ltd, Cheltenham, UK.

Geels, F. (2005) The dynamics of transitions in socio-technical systems: A multi-level analysis of the transition pathway from horse-drawn carriages to automobiles (1860–1930). *Technol. Anal. Strateg.*, **17**, 445–476.

Joore, P. (2008) Social aspects of location-monitoring systems: The case of Guide Me and of My-SOS. *Soc. Sci. Inform.*, **47**, 253–274.

Joore, P. (2010) New to Improve: The Mutual Influence between New Products and Societal Change Processes. PhD thesis, Delft University of Technology, Delft.

KBST (2004) V-Modell XT, Version 1.2, Koordinierungs- und Beratungsstelle der Bundesregierung für Informationstechnik in der Bundesverwaltung.

Rotmans, J., Kemp, R., Van Asselt, M. *et al.* (2000) *Transities & Transitiemanagement. De casus van een emissiearme energievoorziening*, International Centre for Integrative Studies (ICIS), Maastricht Economic Research Institute on Innovation and Technology (Merit), Maastricht.

Tukker, A. and Tischner, U. (eds) (2006) *New Business for Old Europe: Product-Service Development, Competitiveness and Sustainability*, Greenleaf Publishing Ltd, Sheffield, UK.

3

Energy Technologies

3.1 Introduction

There are many reasons to consider an increased utilization of renewable energy sources to supplement and replace part of the conventional fossil fuel-based energy production types that are most prevalent today. Besides the security of supply, an important issue is the future availability of coal, oil, and natural gas; finite reserves of these fossil fuels will be available for a period ranging from 50 to a maximum of 300 years. However, part of this supply of fossil fuels will provide society not only with electrical energy but also with an environmental impact as a result of the emission of carbon dioxide, a greenhouse gas. The emission of carbon dioxide decreases in the sequence of C (coal) $\rightarrow$ CH_2 – (oil) $\rightarrow$ CH_4 (natural gas), and these fossil fuels provide about 80% of the world's annual primary energy use. However, the current renewable energy sources (i.e., solar photovoltaics (PV), solar heat, hydropower, geothermal, wind, and biomass) provide only 2.7% of the alternative energy.

Looking from the perspective of industrial design engineering at the possibilities and opportunities to integrate sustainable energy technologies in products, several requirements toward product and integration can be identified. An energy technology is suitable for product applications if it meets the following criteria:

- It is sizable, preferably modular, allowing customization to its application and hence product integration.
- It can deliver power in a low power range, that is, less than 3 kWp.
- It doesn't require significant modifications of the environment of use once integrated in a product.
- It doesn't emit pollutants during use.
- If heat flows during use, it is at a level that is acceptable for users.

For the reasons mentioned here, we won't describe the following sustainable energy technologies in this chapter: biomass conversions, geothermal systems, or hydropower systems.

Hence in this chapter of *The Power of Design* the functional principles of various sustainable energy technologies that are relevant for product integration will be explained, namely, batteries (Section 3.2), PV solar energy (Section 3.3), fuel cells (Section 3.4), small wind

The Power of Design: Product Innovation in Sustainable Energy Technologies, First Edition.
Edited by Angèle Reinders, Jan Carel Diehl and Han Brezet.
© 2013 John Wiley & Sons, Ltd. Published 2013 by John Wiley & Sons, Ltd.

turbines (Section 3.5), human power (Section 3.6), energy-saving lighting (Section 3.7), energy-saving technologies in buildings (Section 3.8), and piezoelectric energy conversions (Section 3.9).

3.2   Rechargeable Batteries for Energy Storage

Joop Schoonman

Delft University of Technology, Faculty of Applied Sciences, Department of Chemical Engineering (ChemE), Section Materials for Energy Conversion and Storage, Juliana-laan 136, The Netherlands

3.2.1   Introduction

A shift toward the use of renewable energy sources is increasingly attracting attention. The renewable energy sources (i.e., solar PV, solar heat, geothermal, wind, and biomass) provide to date only a few percent of the alternative energy carriers electricity and hydrogen (Goldemberg, 2007). In particular, the decentralized conversion of solar and wind energy is being studied and engineered worldwide. Solar PV and wind energy produce electrons and can be coupled directly to the electricity grid, thereby reducing the emission of greenhouse gases (van Geenhuizen and Schoonman, 2010). However, these renewable energy sources are discontinuous and, therefore, require adequate storage systems.

While present and future energy storage technologies fall in the broad categories of fly wheel, compressed air, superconducting coil, hydrogen, supercapacitor, and rechargeable battery, advanced materials for supercapacitors and rechargeable batteries are attracting widespread research and development attention in academia and industry. This chapter will focus on rechargeable batteries and, in particular, on the rechargeable lithium-ion (Li-ion) battery. The types of Li-ion batteries will be presented, along with the Li-ion battery principle and the state-of-the-art battery anode, cathode, and electrolyte materials. A new class of soft-matter electrolytes is obtained via heterogeneous doping of non-aqueous (polymer) salt solutions. These composite electrolytes behave as a solid and are referred to as *soggy sand* electrolytes. Nanostructured battery components are also the focus of widespread research, and the nanoscale will give Li-ion batteries benefits in terms of capacity, power, cost, and materials sustainability. The utilization of advanced rechargeable batteries to design emission-free mobility is a challenge, next to the option of hydrogen and, in this case, the polymer electrolyte membrane fuel cell (PEMFC) for emission-free vehicles, the major problem then being the safe large-scale storage of hydrogen. The safe storage of hydrogen is possible in the form of a metal hydride, an example being the nickel–metal hydride battery.

Energy carrier hydrogen can be obtained by (photo-)electrolysis of water using renewable energy sources. In fact, hydrogen is a means of storing renewable energy and can be converted back into electrical energy (as mentioned) in a fuel cell, thereby setting the scene for a hydrogen economy.

Here, the main focus will be on rechargeable batteries, in particular rechargeable Li-ion batteries and the design criteria for electric vehicle (EV) requirements and technology comparison.

3.2.2   Rechargeable Batteries

To date, more and more devices and tools are being invented and created that require energy, and usually this energy is provided by primary or rechargeable batteries. Regarding the types of batteries, there are at least five electrochemistries available today, that is, lead–acid (Pb–acid), nickel–cadmium (Ni–Cd), nickel–metal hydride (Ni–MH), lithium-ion (Li-ion), and zinc–air (Zn–air). The Ni–Cd battery is replaced by the Ni–MH battery, because cadmium is no longer accepted in the environment.

The nominal cell voltage, specific energy, energy density, cycle life, and efficiency of selected batteries are presented in Table 3.2.1.

Several metals and alloys can absorb hydrogen to form a hydride as has been referred to in Section 3.2.1. In these metal hydrides the hydrogen ions (protons) are mobile and, therefore, can also participate in an electrode reaction. Here, the nickel–metal hydride battery, which contains a nickel positive electrode and a metal hydride as the negative electrode, is presented as an example. The electrode reactions are as follows:

Positive electrode reaction:

$$NiOOH + H_2O + e^- \rightarrow Ni(OH)_2 + OH^- \qquad (3.2.1)$$

Negative electrode reaction:

$$MH + OH^- \rightarrow M + H_2O + e^- \qquad (3.2.2)$$

Overall battery reaction:

$$NiOOH + MH \rightarrow Ni(OH)_2 + M \qquad (3.2.3)$$

The open circuit voltage (OCV) of this battery is 1.32 V.

A cross-section of a nickel–metal hydride battery is presented in Figure 3.2.1.

Figure 3.2.2 presents a comparison of the gravimetric and volumetric energy densities of the various rechargeable battery systems.

Table 3.2.1   Practical specifications of selected batteries; data from Flipsen (2006) and Kan (2006)

Battery type	Nominal cell voltage (V)	Specific energy (Wh/kg)	Energy density (Wh/L)	Cycle life, 20% fading (cycles)	Efficiency (%)
Sulfuric lead–acid	2.0	30–50	80–90	200–500	85
Nickel–metal hydride	1.2	75–120	240	300–500	
Lithium-ion	4.1	110–160	400–500	500–1000	95–98
Lithium–manganese dioxide	3.0	100–135	265–500	300–500	
				2000	

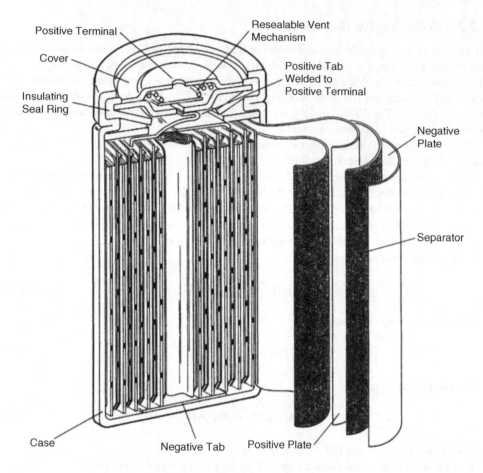

Figure 3.2.1 Internal structure of the cylindrical Ni–MH battery

The development of the rechargeable batteries differs a lot from a few years old for metal–air systems to more than 150 years old for lead–acid technology. The properties of these batteries also differ significantly. The application, therefore, will primarily determine which battery to use. Figure 3.2.3 presents Ragone diagrams of these different applications, where the requirements for two specific applications, electric and hybrid vehicles, are indicated in detail.

It seems that these requirements have not yet been reached for any battery technology, but the lithium battery technology is the closest in terms of both power and energy density. Indeed, to achieve high energy and power density, a combination of a large specific capacity and current and a high output potential is required. Since lithium is one of the lightest elements in nature, lithium batteries have a very high specific capacity. Moreover, because it has the lowest electrochemical potential, $EoLi+/Li = -0.3045$ V/NHE (normal hydrogen electrode), the output potential will also be very high (Simonin, 2009). Therefore, the lithium battery is one of the best candidates for high-energy applications.

The sulfuric lead–acid, nickel–metal hydride, lithium-ion, and lithium–manganese dioxide batteries have different designs, and the design aspects of these batteries are presented in Table 3.2.2.

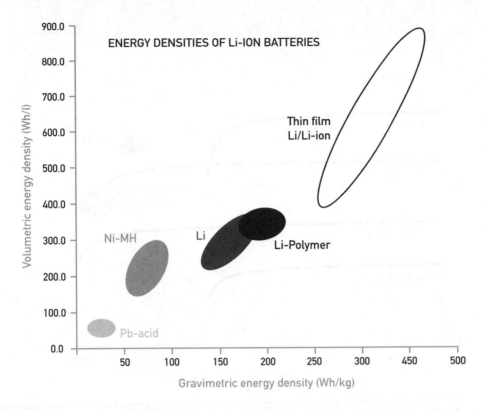

Figure 3.2.2 The gravimetric and volumetric energy densities of selected rechargeable battery systems

3.2.3 Lithium Batteries

The first lithium-ion battery was commercialized by Sony in 1991 and was based on the exchange of lithium ions between lithiated carbon (LiC6) and a layered oxide (Li1-xMO2), with M being a transition metal ion, usually cobalt, nickel, or manganese. The energy it stores is about 180 Wh/kg and only five times higher than that stored by the lead–acid battery, but it took a revolution in materials science to achieve it (Armand and Tarascon, 2008).

The primary lithium battery comprises a lithium anode (Li) and a manganese dioxide (MnO2) cathode, and has an open-circuit voltage of about 3 V. More interesting are the rechargeable lithium batteries, and these are classified as follows:

- *Lithium battery*: metallic lithium versus an insertion compound.
- *Lithium-ion battery*: insertion compound versus an insertion compound and a liquid electrolyte.
- *Lithium–polymer battery*: insertion compound versus an insertion compound and a polymer electrolyte.
- *Lithium–ceramic battery*: insertion compound versus an insertion compound and a ceramic electrolyte.

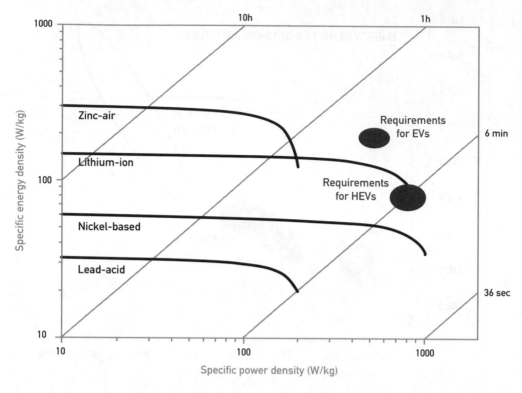

Figure 3.2.3 Ragone diagrams for different types of rechargeable batteries. Requirements for emission-free electric vehicle and hybrid vehicle applications are indicated

Table 3.2.2 Design aspects of different battery technologies; data from Flipsen (2006) and Kan (2006)

Battery type	Shape	Flexible	Safety and environmental considerations	Operating temperature (°C)	Maintenance	Specific costs (€/Wh)
Sulfuric lead–acid	Cubic	Fluid	Hazard of shock Release of gas	−20 to 60°C	3–6 mo	0.50
Nickel–metal hydride	Cylindrical	No		−20°C to 60°C	60–90 d	3.80
Lithium-ion	Cylindrical, prismatic, and pouch cells	Yes	Flammable	−20°C to 45°C	Not necessary	9.50
Lithium–manganese dioxide	Cylindrical, prismatic, and pouch cells	Yes		0–45°C	Not necessary	19.00

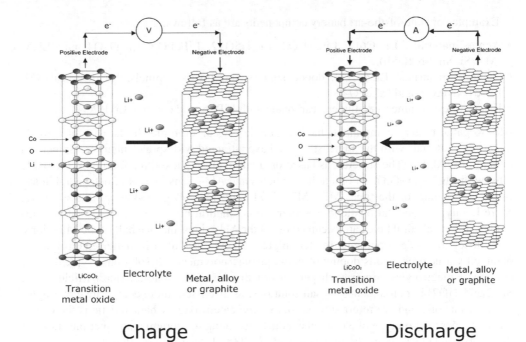

Figure 3.2.4 The lithium-ion battery principle

Several metal oxide electrode materials can incorporate lithium atoms during discharge of the battery and release lithium atoms during charge (positive electrode), or can release lithium atoms during discharge and incorporate lithium atoms during charge (negative electrode). Such metal oxides are referred to as *insertion compounds*.

In Figure 3.2.4 the lithium-ion battery principle is shown; the anode (negative electrode) is, for example, lithiated carbon (LiC6) and the cathode (positive electrode) is a metal oxide that can incorporate lithium atoms during discharge and release lithium atoms upon charging the battery. Usually these metal oxides comprise oxides of the transition metals. Characteristic of the insertion compounds is that they are mixed ionic–electronic conductors (MIECs), that is, they exhibit lithium-ion conductivity and electronic conductivity. The liquid, polymer, or ceramic electrolytes exhibit only lithium-ion conductivity. Any electronic conductivity would internally short-circuit the battery.

In the Periodic Table of the Elements, there are 10 elements between Ca ($Z = 20$) and Ga ($Z = 31$), whose occurrence at this position is due to the filling of the 3D orbitals. These elements are called the *transition elements* or sometimes the *d-block elements*. Their common characteristic is that the neutral atom, some important ion it forms, or both have an incomplete set of 3D electrons and exhibit variable valence, which leads to many characteristic physical and chemical properties that are important for application in, for example, rechargeable lithium-ion batteries.

Examples of state-of-the-art battery components are as follows:

- Anode materials: Li, C6Li, Li0.5TiO2, Li7Ti5O12, LiFePO4, LiCrTiO4, and LixM (M = Si, Sn, Sb, SnSb).
- Cathode materials: LiMn2O4, doped spinels and inverse spinels, LiCoO2, LiNiO2, LiCo1-xNixO2, and LiFePO4.
- Electrolytes: commercial ethylene carbonate with LiPF6, and ceramic LixBPO4.

All inorganic battery components have a specific crystal structure. In the case of cathode materials, materials with the spinel structure have attracted widespread attention. Spinel is a mineral, MgAl2O4. The structure is based on a cubic close-packed (ccp) array of the oxide ions, with Mg2+ ions in a set of tetrahedral holes and Al3+ ions in a set of octahedral holes. Many compounds of the types M2 + M3 + 2O4 and M4 + M2 + 2O4 adopt this structure. More highly charged cations tend to prefer the octahedral holes so that in the latter compounds, the octahedral holes are occupied by all the M4+ ions and one half of the M2+ ions.

With regard to developments in electrolytes, a new class of soft-matter electrolytes is obtained via heterogeneous doping of non-aqueous (polymer) salt solutions. Typically, the composite electrolytes comprise dispersions of fine inorganic acidic oxide particles (e.g., SiO2 and TiO2) in non-aqueous Li–salt solutions; as discussed, these composite electrolytes behave as a solid and are referred to as *soggy sand* electrolytes. Characteristic is the occurrence of a percolation type of electrical conductivity along the ceramic–polymer interfaces in the soggy sand electrolytes (Bhattacharyya *et al.*, 2006; Edwards *et al.*, 2006).

To date, research and development are focused on improved anode, cathode, and electrolyte materials, as well as on interfacial problems between the electrodes and the electrolyte. While many attempts to improve the properties of lithium-ion batteries have tackled the problem at the macroscopic scale, current battery materials research is focusing on the nanoscale (Chan *et al.*, (2007, 2008, 2009); Kim *et al.*, 2008). Nanostructured components will give lithium-ion batteries benefits in terms of capacity, power, cost, and materials sustainability that are still far from being fully exploited.

Recent examples are reported on fast and completely reversible Li insertion in vanadium pentoxide nanoribbons (Chan *et al.*, 2007) and on spinel LiMn2O4 nanorods as lithium-ion battery cathodes. In addition, electrode kinetic issues can be circumvented by combining nanomaterials with nanopainting, in which the nanoparticles are coated with a very thin layer of carbon to provide the required electrical conductivity to the individual particles, whose small sizes shorten the diffusion pathways of lithium ions and electrons (Armand and Tarascon, 2008). Moreover, the insertion and removal reactions of lithium in nanostructured LixM anode materials cause the volume to expand or contract several-fold, and accommodating the strains associated with the lithium insertion and removal reactions has been studied in detail with practical solutions having been reported (Chan *et al.*, 2008, 2009; Cui *et al.*, 2009; Simonin, 2009).

Examples of commercially available Li-ion batteries are presented in Figure 3.2.5 and were taken from the Panasonic *Lithium Ion Batteries Technical Handbook* (Panasonic, 2007).

In general, lithium-ion batteries are used in battery packs that contain both lithium-ion batteries and battery safety circuits. Both items are sealed in a container made of a material such as resin, so that the battery pack cannot easily be disassembled.

The "constant voltage–constant current" method is used to charge Li-ion batteries, as is illustrated in Figure 3.2.6 for a single cell (Panasonic, 2007).

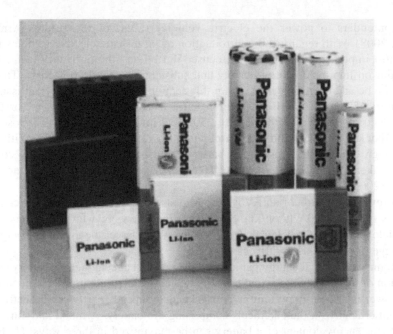

Figure 3.2.5 Lithium-ion batteries, dedicated to support various types of mobile equipment that require small size, light weight, and high performance (Panasonic, 2007)

3.2.4 Electric Vehicles

Ni–MH batteries are used to power cheaper electronics and are used in hybrid vehicles. Lithium-ion batteries have conquered high-end electronics and are now being used in power tools. Lithium-ion batteries are also entering the hybrid vehicle market and are

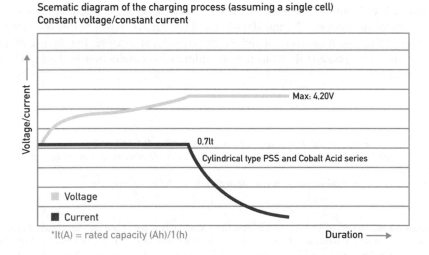

Figure 3.2.6 The diagram to illustrate the "constant–voltage current" of the charging process

serious contenders to power the electric vehicles (EVs) of the future (Armand and Tarascon, 2008). With regard to the introduction of large-scale electric vehicles into society, the most important development is focused indeed on lithium battery technology, and in particular on improved energy density and substantial reduction of costs. The issue of cost will not be addressed in detail within this section, but it is an important issue, as in a recent report of the German Bank stating that a 25 kWh battery package for a battery-powered electric vehicle will cost US$11,250. SB Limotive (Samsung and Bosch) anticipate €350 per kWh in 2015 and €250 around 2020 (van Ginneken, 2011). With regard to battery design criteria, the present EV battery requirements and technology comparison have been taken in part from Broussely (2009) of Saft, Besenhard (1999), and Balbuena and Wang (2004).

The energy consumption of an EV is proportional to its weight. The energy consumption of an EV is typically 120 Wh/ton/km. Since the allowable weight for the battery onboard a given vehicle is limited, the better the energy density, the larger the available energy and the vehicle range (Besenhard, 1999; Broussely, 2009).

The amount of electrical energy per mass or volume that a rechargeable battery can deliver is a function of the cell's voltage and capacity, which are dependent on the chemistry of the system. Power is also an important parameter, which depends partly on the battery's engineering, but crucially on the chemicals comprising the battery. Armand and Tarascon state that the stored energy content of a battery can be maximized in three ways: (1) by a large OCV, (2) by making the mass (or volume) of the active electrode materials per exchanged electrons as small as possible, and (3) by ensuring that the electrolyte is not consumed in the chemistry of the battery. This latter condition holds only for the Ni–MH and the lithium-ion batteries.

In a comparison of different technologies (energy density) of EV batteries, that is, lead–acid (33 Wh/kg), Ni–Cd (45 Wh/kg), Ni–MH (70 Wh/kg), and Li–ion (120 Wh/kg) for a typical size of 250 kg, as well as power characteristics for a 250 kg battery, that is, 75 W/kg, 120 W/kg, 170 W/kg, and 370 W/kg, respectively. Here, lithium-ion batteries exhibit superiority over traditional electrochemistries.

For pure EV applications, the battery manufacturers have developed ranges of cells with individual capacities between 25 and 100 Ah. Each EV battery maker has its own proprietary choice of negative electrode material and electrolyte solvent mixture, but the use of the lithium hexafluoro–phosphate (LiPF6) to provide lithium-ion conduction in the liquid electrolyte is standard.[*]

References

Armand, M. and Tarascon, J.M. (2008) Building better batteries. *Nature*, **451**, 652–657.
Balbuena, P.B. and Wang, Y. (eds.) (2004) *Lithium-Ion Batteries: Solid-Electrolyte Interphase*, Imperial College Press, London.
Besenhard, J.O. (ed.) (1999) *Handbook of Battery Materials*, Wiley-VCH, Weinheim.

[*] The current manufacturers for EV batteries are Shin Kobe Machinery in collaboration with Hitachi Maxwell (with Nissan equipping these EV batteries), Japan Storage Battery, Matsushita Batteries Industrial Co. Ltd, JSB high-energy cells for EV, Panasonic, and NGK Insulators. All these rechargeable batteries are based on the combination of an insertion compound versus an insertion compound.

Bhattacharyya, A.J., Maier, J., Bock, R., and Lange, F.F. (2006) New class of soft matter electrolytes obtained via heterogeneous doping: Percolation effects in "soggy sand" electrolytes. *Solid State Ionics*, **177**, 2565–2568.

Broussely, M. (2009) Lithium-ion batteries for EV, HEV and other industrial applications, in *Lithium Batteries – Science and Technology* (ed. G.-A. Nazri and G. Pistoa), Springer, Berlin.

Chan, C.K., Peng, H., Twesten, R.D. *et al.* (2007) Fast, completely reversible Li insertion in vanadium pentoxide nanoribbons. *Nano Lett.*, **7**, 490–495.

Chan, C.K., Peng, H., Liu, G. *et al.* (2008a) High performance lithium battery anodes, using silicon nanowires. *Nature Nanotechnology*, **3**, 31–35.

Chan, C.K., Zhang, X.F., and Cui, Y. (2008b) High capacity Li-ion battery anodes using Ge nanowires. *Nano Lett.*, **8**, 307–309.

Chan, C.K., Ruffo, R., Hong, S.S. *et al.* (2009) Structural and electrochemical study of the reaction of lithium with silicon nanowires. *J. Power Sources*, **189**, 34–39.

Cui, L.-F., Ruffo, R., Chan, C.K. *et al.* (2009) *Nano Lett.*, **9**, 491–495.

Kim, D.K., Muralidharan, P., Lee, H.-W. *et al.* (2008) Spinel LiMn$_2$O$_4$ nanorods as lithium ion battery cathodes. *Nano Lett.*, **8**, 3948–3952.

Edwards, W.V., Bhattacharyya, A.J., Chadwick, A.V., and Maier, J. (2006) An XAS study of the local environment of ions in soggy sand electrolytes. *Electrochem. Solid-State Letters*, **9**, A564–A567.

Flipsen, S.F.J. (2006) Power sources compared: The ultimate truth? *J. Power Sources*, **162**, 927–934.

Goldemberg, J. (2007) Ethanol for a sustainable energy future. *Science*, **315**, 808–810.

Kan, S.Y. (2006) Energy matching – key towards the design of sustainable powered products. Thesis, Technical University of Delft, Design for Sustainability Program, Delft.

Panasonic (2007) Lithium ion batteries technical handbook. http://industrial.panasonic.com/www-data/pdf/ACI4000/ACI4000PE5.pdf (accessed April 20, 2012).

Simonin, L. (2009) Synthesis and characterisation of tin and antimony nano-compounds for lithium battery applications. PhD thesis, Delft University of Technology, Delft.

van Geenhuizen, M. and Schoonman, J. (2010) Renewable energy technologies and potentials for adoption, in *Energy and Innovation: Structural Change and Policy Implications* (ed. M. van Geenhuizen, W.J. Nuttall, D.V. Gibson, and E.M. Oftedal), Purdue University Press, West Lafayette, IN.

van Ginneken, R. (2011) Accu elektrische auto nog niet goed genoeg. *Technisch Weekblad*, July 10 (in Dutch), http://www.technischweekblad.nl/accu-elektrische-auto-nog-niet-goed-genoeg.80165.lynkx (accessed April 20, 2012).

3.3 Photovoltaics and Product Integration

Angèle Reinders[1] and Wilfried van Sark[2]

[1]*Delft University of Technology, Faculty of Industrial Design Engineering, Design for Sustainability, Landbergstraat 15, The Netherlands*
[2]*Utrecht University, Faculty of Science, Science, Technology and Society (STS), Budapestlaan 6, The Netherlands*

3.3.1 Introduction

It was more than a century since the discovery of the photovoltaic (PV) effect – the conversion of photons to electricity – by Becquerel in 1839 (Becquerel, 1839) before solar cells were developed. In the 1950s, PV solar cells were developed at Bell Telephone Laboratories in the United States with the purpose to apply them in products that lacked permanent electricity supply from the mains (Chapin, Fuller, and Pearson, 1954). The solar cells were called *silicon solar energy converters*, commonly known as the *Bell Solar Battery* (Prince, 1955). Furnas (1957) reports,

"The Bell Telephone Laboratories have recently applied their findings in the transistor art to making a photovoltaic cell for power purposes . . . and exposed to the sun, a potential of a few volts is obtained and the electrical energy so produced can be used directly or stored up in a conventional storage battery. . . . The Bell System is now experimenting with these devices for supplying current for telephone repeaters in a test circuit in Georgia. As to cost, one radio company has produced a power pack using this type of photoelectric cell for one of its small transistorized radios."

The *Journal of the Franklin Institute* mentioned as early as 1955 that "these (solar) batteries can be used as power supplies for low-power portable radio and similar equipment" (Anonymous, 1955). Expectations regarding the applicability of photovoltaic cells in products were high; Sillcox (1955) reports on predictions by researchers of New York University "that small household appliances like toasters, heaters or mixers using the sun's energy might be in fairly widespread use within the next five years (i.e., 1960)." These predictions have not become reality, because at that time the costs of silicon PV cells were about $200 per Watt for high-efficiency cells (Prince, 1959) – where 12% was considered high efficiency – and the costs of a dry cell to operate a radio for about 100 hours would be less than a dollar (Furnas, 1957). Therefore, it was believed that the solar battery could be an economical source for all except the most special purposes. As such by the end of the 1950s silicon PV solar cells were applied as a power supply for satellites (Zahl and Ziegler, 1960). The Vanguard TV-4 test satellite that launched on March, 17 1960 was the first satellite ever equipped with a solar-powered system and it announced a new area of space technology with solar-powered satellites.

At present, various applications of photovoltaic solar cells exist ranging from stand-alone PV systems, satellites, grid-connected PV, building integrated PV systems, and a very large system with a power of tens of megawatts. The focus of this chapter, however, is on product-integrated PV (PIPV) from the perspective of industrial design engineering. PIPV can be applied in consumer products, lighting products, boats, vehicles, business-to-business applications, and arts (Reinders and van Sark, 2012).

3.3.2   PV Cells

The efficiency of a PV solar cell is an important variable in the design of product-integrated PV, because it determines the power that can be produced for the application designed. The efficiency depends on the PV material and technology of the cell and the intensity of irradiance that impinges on a PV cell surface. In addition, the temperature of the PV cell and the spectral distribution of the light affect the efficiency.

If a solar cell is illuminated, a photocurrent I_{ph} is generated. This photocurrent is in most cases linearly related to the intensity of irradiance. Because of the semiconductor materials in the PV cell, the electric behavior of a PV cell can be represented by a current source in parallel with two diodes, D_1 and D_2. A series resistance, R_s, and a parallel resistance, R_{sh}, add to this electric circuit, as shown schematically in Figure 3.3.1.

The electric behavior or current–voltage characteristic (I–V curve) of a PV cell is described by the following (Green, 1982):

$$I = I_{ph} - I_{s1}\left(e^{\frac{q(V+IR_s)}{n_1 kT}} - 1\right) - I_{s2}\left(e^{\frac{q(V+IR_s)}{n_2 kT}} - 1\right) - \frac{V + IR_s}{R_{sh}} \qquad (3.3.1)$$

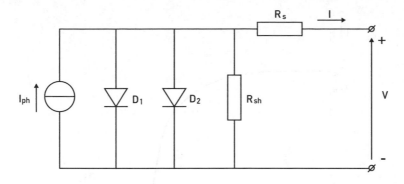

Figure 3.3.1   Equivalent circuit of a solar cell represented in the two-diode model

Here I_{s1} and I_{s2} are the saturation currents of the two diodes, respectively, and n_1 and n_2 are the quality factors of the two diodes. In general, n_1 will not deviate much from 1, and usually $n_2 = 2$ if no imperfections occur. V is the voltage over the circuit, T is the temperature, and k is the Boltzmann constant. Figure 3.3.2 shows the I–V curve of a solar cell at a certain irradiation. It crosses the y-axis at the open circuit voltage, V_{oc}, and the x-axis at the short circuit current, I_{sc}.

In Figure 3.3.2 it is shown that the I–V curve of a PV cell has one point that delivers maximum power. This point is called the *maximum power point*, P_{mp}, and is characterized by V_{mp} and I_{mp}. This maximum power point is used to determine the efficiency, η:

$$\eta = \frac{I_{mp} V_{mp}}{AG} \tag{3.3.2}$$

Here, AG is the optical power falling onto the solar cell, with A the solar cell area and G the irradiance. Please note that from the two-diode model, it follows that if the operational voltage, V, of the cell deviates from the maximum power point settings, the efficiency will drop accordingly.

Single values of efficiencies of solar cells are usually efficiencies at so−called standard test conditions (STC), η_{STC}, which means that it is measured at $1000\,\text{W/m}^2$ irradiance, AM 1.5 spectrum, and 25°C cell temperature.

PV cells are made from several semiconductor materials using different technologies. These materials yield various efficiencies. The most well-known solar cell technology is the single-crystal silicon wafer–based solar cell, indicated by c-Si. It has improved significantly for nearly 60 years, and it has always been and is still today the dominant solar cell technology. Crystalline silicon solar cell technology also represents multicrystalline silicon solar cells, indicated by m-Si. Both c-Si and m-Si are so-called first-generation solar cells (Green, 2003).

Besides this, second-generation solar cells have been developed with the intention to find a cheaper alternative for crystalline solar cell technology by using less material. They can usually be characterized as thin-film solar cells. Several semiconductor materials allow for the production of thin films, namely, copper indium gallium diselenide ($CuInGaSe_2$, abbreviated to CIGS), cadmium telluride (CdTe), hydrogenated amorphous silicon (a-Si:H), and thin-film polycrystalline silicon (f-Si). Also, thin film PV cells can be made from organic materials. Firstly, dye-sensitized cells (DSCs) consist of titanium oxide nanocrystals covered

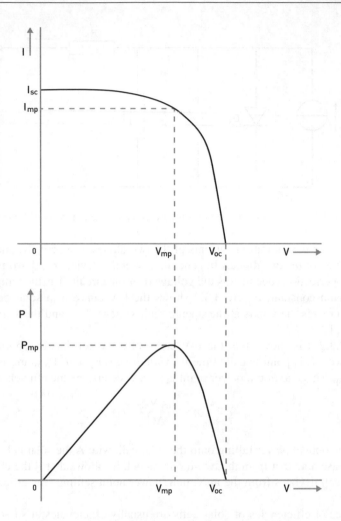

Figure 3.3.2 Above: The current–voltage characteristic of a solar cell. Below: The power–voltage characteristic of a solar cell

with light-absorbing organic molecules. Secondly, polymer organic solar cells are made from conducting polymers with, for example, light-absorbing C_{60} molecules. A third group of solar cells is made from compounds from the elements Ga, As, In, P, and Al, also denoted as III–V technology. The specific cells are named after their compounds, for instance GaAs, GaInP, or InP. These PV cells have been developed for space applications because of their high efficiency. Nevertheless, they are increasingly being applied in terrestrial concentrator systems. Table 3.3.1 shows typical efficiencies for the PV technologies mentioned in this section.

The sensitivity of each solar cell strongly depends on the wavelength of the light falling onto the solar cell. The sensitivity as a function of wavelength is called the *spectral response* (expressed in $[(A/m^2)/W/m^2)]$ or in $[A/W]$), and it is quantified by measuring the short-circuit current occurring at illumination with a monochromatic light beam. Table 3.3.1 shows

Table 3.3.1 Characteristic efficiencies and spectral response range of several PV technologies adapted from Kan (2006), updated with Kazmerski (2010) and Green *et al.* (2011)

Type of photovoltaic (PV) cell	Record lab cells η_{STC} (%)	Commercially available η_{STC} (%)	Spectral range (nm)
c-Si	25	14–17 24.2	350–1200
m-Si	20.4	14–16	350–1200
a-Si	12.5	8–10	300–800
Nano-, micro-, or poly-Si	16.5	11–13	300–800
CIGS	20.3	12–17	300–1200
CdTe	16.7	10,7	350–850
III–V three junction	43.5 (under 300 suns)	27–30	300–1250
III–V single junction, thin film	26	21–23	300–1000
DSC	11.1	8–10	300–800
Polymer	8	3–5	300–800

the range of spectral response of different cell technologies, which is determined by the band gap of the semiconductor material and the charge generation and recombination processes that internally take place in a solar cell. Figure 3.3.3 presents spectral response curves of samples of a-Si and c-Si under an AM1.5 spectrum by Reich *et al.* (2005).

3.3.3   Irradiance and PV Cell Performance

Outdoor Irradiance in Relation to Solar Cell Performance

Irradiance (G) is the power density of light expressed in W/m^2. In daytime, outdoor irradiance is predominantly determined by sunlight. In Figure 3.3.4, the spectral distribution

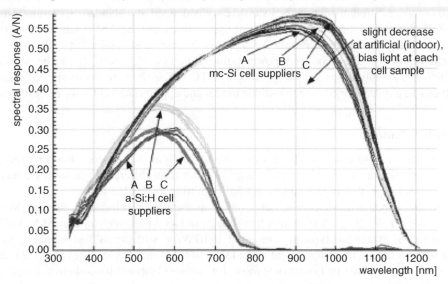

Figure 3.3.3   Spectral response of a-Si samples and m-Si samples from Reich *et al.* (2005)

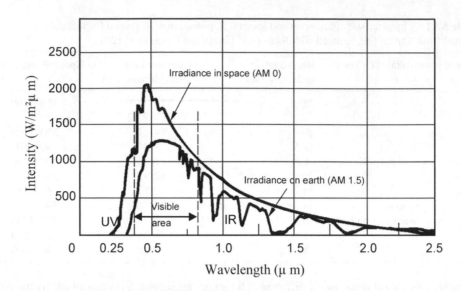

Figure 3.3.4 The wavelength-dependent AM 0 and AM 1.5 spectra of sunlight

of the sunlight falling onto the earth's surface is shown. This spectrum resembles the spectrum of a black body with a temperature of 5700 K and is determined by the path length of the sunlight through the atmosphere (the air mass, or AM): AM 0 is the extraterrestrial spectrum. The various "dips" in the spectra arise because of absorption in the atmosphere, by water vapor among other things. In addition, the spectrum is influenced by scattering taking place in the atmosphere.

The response of a solar cell depends on the spectral composition of incident light. To be able to compare different solar cells, its efficiency is defined at a standard spectrum. This is the AM 1.5 standard spectrum for terrestrial applications and the AM 0 standard spectrum for space travel applications. Figure 3.3.5 shows for c-Si and m-Si solar cells measured efficiency curves in the maximum power point (Reich *et al.*, 2009). It can be seen that the efficiency steeply drops with respect to the STC efficiency with decreasing irradiance, which can be explained using the two-diode model and appropriate values for its parameters.

Indoor Irradiance

Indoor irradiance usually consists of a mixture of sunlight that enters a building through windows and skylights, and artificial light originating from different light sources, such as incandescent lamps, fluorescent lamps, and light-emitting diodes (LEDs).

Müller has conducted extensive research on the measurement of indoor irradiance (Müller, 2010; Müller *et al.*, 2009). She reports that though indoor irradiance can exceed 500 W/m^2, basic orders of magnitude typically are about 1–10 W/m^2 with worst-case scenarios in the winter without the use of artificial light in the range of 0.1 W/m^2. For surfaces oriented toward windows, solar radiation will contribute most; for surfaces oriented toward electric light, the latter will contribute most of the radiation. These features can be important concerning the dominating spectral distribution of irradiance that can be converted to electricity by PV solar

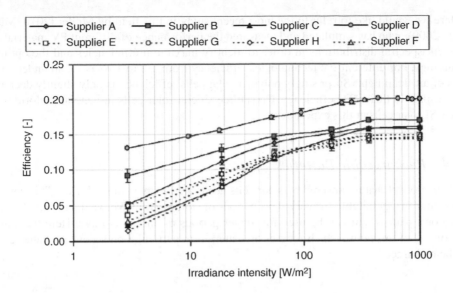

Figure 3.3.5 Measured irradiance intensity-dependent efficiencies of various Si solar cells by Reich *et al.* (2009)

cells. In other words, the spectral range of irradiance from sunlight is from 300 up to 4000 nm, whereas the spectral distributions of artificial light are narrower. For instance, artificial light emitted by incandescent lamps has a spectral range from 350 up to 2500 nm, by LEDs from 400 to 800 nm, and by fluorescent lamps from 300 to 750 nm (Kan, 2006). In Figure 3.3.6, measured spectra of incandescent light sources and fluorescent lamps are shown.

Solar Cell Performance at Indoor Irradiance

Several studies have been devoted to solar cell performance under weak light or indoor irradiance conditions. These studies were conducted by Randall (2006), Randall and Jacot (2003), Reich *et al.* (2005, 2009), Girish (2006), and Gong *et al.* (2008).

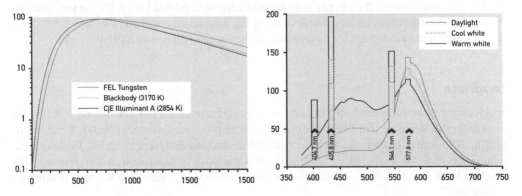

Figure 3.3.6 Spectra of incandescent lamps and typical fluorescent lamps from Ryer (1997)

Here we would like to refer to the most recent study in this field by Müller (2010; Müller *et al.*, 2009), who has simulated and measured the performance of different PV materials for different spectral distributions in order to identify maximum efficiencies and the optimum technology. For different PV technologies, efficiency from 4% to 16% is found under conditions of artificial light. Surprisingly polymer solar cells' efficiency is only slightly decreased with respect to their STC efficiency. Crystalline silicon solar cells in general perform better than a-Si solar cells in this experiment.

3.3.4   Rechargeable Batteries

A PIPV product that is powered by solar cells needs an energy storage device. This could be a capacitor that is suitable for very short periods of storage (Kan, Verwaal, and Broekhuizen, 2006) or a battery that can be used for longer periods of energy storage. Here we refer to Section 3.2 of this book, which focuses on batteries, in particular on secondary batteries that can be recharged.

3.3.5   System Design and Energy Balance

PV system design in a product context is a rather complex task because of the interdisciplinary character of product development. It not only concerns an appropriate energy balance of the system components, but also addresses issues related to manufacturability, costs, safety, operating temperatures, and environmental aspects. Finally, the integration of PV systems in products should result in customer benefits like (1) a better functionality, increased comfort, or autonomy of a product; (2) less dependency from the electricity grid; (3) smaller batteries; and (4) fewer user interaction for recharging batteries (Reinders, 2002). Figure 3.3.7 aims to represents system design during product development of PIPV by showing the relationships between the user, the product, and the integrated PV system and indicators that might be relevant for the decision making in the conceptual design stage. These indicators will be related to energy matching, final weight of the product, area required, volume, customer benefits, costs of the product, and safety and environmental indicators. Apart from this it should be possible to produce a product with existing manufacturing processes. Within this vast context the energy balance of a certain combination of PV system components, product, and user can be estimated, using the flow scheme shown in Figure 3.3.8. Next we will give some guidelines for estimations of the energy balance of a PIPV.

Irradiance

Irradiance can be measured on locations of expected use of a PIPV. However, it might be convenient to simulate irradiance during the design process: Reinders (2007), Reich *et al.* (2008, 2009), and Tiwari and Reinders (2009) developed different methods to determine irradiance in CAD tools. Moreover, at present the widely used design environment 3D Studio Max comprises a module with the established Perez model for irradiance simulation.

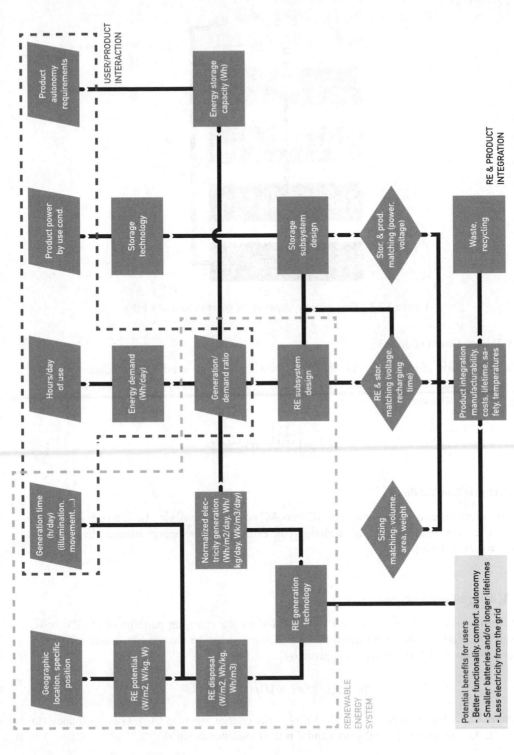

Figure 3.3.7   Issues involved in the integration of renewable energy sources, in particular PV technologies in a product context Adapted from P. Diaz (2010)

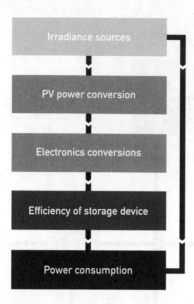

Figure 3.3.8   Flow scheme showing the energy chain of a PIPV

PV Power Conversion

The conversion of irradiance, G, should at least include the irradiance dependency of the efficiency, η_{PV}, of solar cells. In a certain period of time, the energy E_{PV} that can be produced by the solar cells is given by

$$E_{PV} = \int G(t) \cdot \eta_{PV}(G)dt \qquad (3.3.3)$$

Electronic Conversions

Power electronics needed to convert DC into AC power in the device have a certain conversion efficiency that should be taken along in the estimates of the energy balance, in particular in low-power devices.

Efficiency of the Storage Device

The efficiency of a battery in a PIPV is affected by the charging patterns of the PV cells. Therefore, some recommend determining the solar energy to charge conversion efficiency and to optimize it. This efficiency is given by

$$\eta_{PV\text{-}CC} = (Vav \times Q)/(G \times A \times T) \qquad (3.3.4)$$

Where V_{av} is the average voltage (in V), Q is the charge increase (Ah), G is irradiance (in W/m^2), A is the area of PV cells (m^2) and T is the time interval (in s).

Power Consumption

A reliable estimate of the power consumption of future users of PIPVs can be based on questionnaires and a quantitative insight of the power required for the different functions of a PIPV. The variability between user behaviors can be used to set minimum and maximum values for the area of PV cells needed to fulfill the energy demand of the future users.

3.3.6 Design and Manufacturing of Product-Integrated PV

Opportunities of the integration of PV technology in the final design of a product that can be manufactured partly depend, on one hand, on the visual appearance of PV cells in the context of the product design and, on the other hand, on the possibility to shape and form PV cells and to attach them to the product. Roth and Steinhueser (1997), Reinders and Akkerman (2005), researchers in the EU PV-Accept project (2005), Kan (2006), and Gorter *et al.* (2010) have addressed these issues.

For PV cells, characteristic visual features of different PV technologies and possibilities to shape and form them have been collected in Table 3.3.2. It can be found that visual appeal, flexibility, and the number of possible operations vary considerably among the different technologies. Coloring is possible for c-Si and m-Si technologies, although it will adversely affect the efficiency. For other solar cells the use of colored glass sheets or colored plastics as a cover could bring in more variations with respect to their dark brown and grayish colors. a-Si and CIGS can be produced with customized patterns; c-Si and m-Si can be decorated by grid designs that are different from the common H-pattern. Newly developed back-contact cells may be more appealing from a visual perspective.

PV cells or series of PV cells can be attached on product surfaces in the following ways:

1. Attachment on a surface and covering with a glass sheet or a plastic sheet.
2. Attachment on a surface and covering with resin, for instance epoxy resin.
3. Lamination in between plastic sheets and next attachment on a surface.
4. Encapsulation in fiber-reinforced plastic and next attachment on a surface.

The fourth option has the advantage of preshaping of a series of PV cells in a mold that is injected with the fluid plastic, so-called *injection molding*.

The selection of glass or plastics – such as epoxies, fluorides, polyolefins, and silicones – that could be applied for the protection of solar cells in products is based on variables like transparency, transmittance, service temperature, thermal expansion coefficient, glass temperature costs, weight, and expected lifetime in relation to the product's expected lifetime. For the last reason, we assume that for PIPV in consumer products, less stringent criteria are set to UV stability and impact resistance compared to PIPV in business-to-business products that must be able to endure harsh outside environments for many years.

3.3.7 Conclusions

PIPV can be applied well in different product categories and for various markets. With the steep decrease of prices of conventional PV technologies such as c-Si, the emergence of low-cost PV technologies such as polymer solar cells, and a slight decrease of the power demand

Table 3.3.2 Design aspects of different PV technologies, adapted from Kan (2006) and extended with manufacturing data

Type of PV cell	Maturity	Color, surface, and other	Typical area (mm)	Typical thickness (μm)	Flexibility of cell	Operations on cells during design and manufacturing
c-Si	Highly commercially available	Blue, dark-gray, or black-smooth surface with silver grid patterns on top; cells can be colored (gold, orange, pink, red, green, or silver) by variable Si_3N_4 layer; decorative grid patterns possible	156 × 156 cell area	>180–220	Low	Bending only to a limited extent
m-Si	Highly commercially available	Shiny blue, dark blue-shiny grains, smooth surface with silver grid patterns on top; cells can be colored (gold, orange, pink, red, green, or silver) by variable Si_3N_4 layer; decorative grid patterns possible	156 × 156 cell area	>180–220	Low	Laser cutting Heating Injection transfer molding in plastics Lamination in plastics Bending only to a limited extent Laser cutting Heating Injection transfer molding in plastics Lamination in plastics

a-Si	Commercially available	Dark brown or black-smooth surface with thin lines (cell interconnects); patterned deposition is possible	Customizable from 10 × 10 to 1000 × 2000 module area	<1	High	Bending Lamination in plastics Deposition on curved surfaces Cutting not possible
CIGS	Commercially available	Gray or black-smooth surface with light lines (cell interconnects); patterned screen printing is possible	Customizable from 10 × 10 to 1000 × 2000 module area	1–3	High	Bending Lamination in plastics Deposition on curved surfaces Cutting not possible
CdTe	Highly commercially available	Brownish-smooth	Customizable from 10 × 10 to 1000 × 2000 module area	1–3	Low	Heating
III–V three junction	Mainly available for space applications, concentrators	Black-smooth	40 × 80 80 × 80 Cell area	140–200	Low	Connection to ceramics
III–V single junction, thin film	Commercially available for terrestrial applications	Black-smooth	40 × 80 80 × 80	5	High	Bending Heating

(continued)

Table 3.3.2 (*Continued*)

Type of PV cell	Maturity	Color, surface, and other	Typical area (mm)	Typical thickness (μm)	Flexibility of cell	Operations on cells during design and manufacturing
			Cell area			Injection transfer molding in plastics
						Lamination in plastics
DSC	Available	Red or brown-transparent and smooth; cells can be colored by dye molecules	Customizable	1–10	High	Bending
			Cell and module area			Lamination in plastics
Polymer	Limitedly available	Orange, red, or brown-smooth	Long strips, customizable	<1	High	Bending
			Cell and module area			Lamination in plastics

of products induced by technological advances, PIPV might become a respectable widely applied energy source. For instance, PIPV might become common in public-lighting products and other urban furniture in public spaces. Also we expect that the market of PV-powered LED lamps in developing countries might grow in the forthcoming years. In this chapter, it was shown that PIPV can offer energy to products with a wide range of power demand. Therefore we believe that PIPV will be further developed in the field of micro-energy harvesting, that is, for sensors and security, and that it will be applied more often in boats and cars.

References

Anonymous (1955) Improvements in solar battery. *J. Franklin I.*, **260** (1), 87–88.

Becquerel, E. (1839) Mémoire sur les effets électriques produits sous l'influence des rayons solaires. *C. R. Seances Acad. Sci. D*, **9**, 561–567.

Chapin, D.M., Fuller, C.S., and Pearson, G.L. (1954) A new silicon p-n junction photocell for converting solar radiation into electrical power. *J. Appl. Phys.*, **25**, 676.

Diaz, P. and Reinders, A.H.M.E. (2010) Sustainable energy design support tool. Department Design Production and Management, University of Twente.

Furnas, C.C. (1957) The uses of solar energy. *Sol. Energy*, **1**, 68–74.

Girish, T.E. (2006) Some suggestions for photovoltaic power generation using artificial light illumination. *Sol. Energ. Mat. Sol. C*, **90**, 2569–2571.

Gong, C., Posthuma, N., Dross, F. *et al.* (2008) Comparison of n- and p-type high efficiency silicon solar cell performance under low illumination conditions. In Proceedings of 33rd IEEE PVSC, San Diego, pp. 1–4.

Gorter, T., Voerman, E.J., Joore, P. *et al.* (2010) PV-boats: Design issues in the realization of PV powered boats. In Proceedings of 25th European Photovoltaic Solar Energy Conference, Valencia, pp. 4920–4927.

Green, M.A. (1982) *Solar Cells: Operating Principles, Technology and Systems Application*, Prentice Hall, Englewood Cliffs, NJ.

Green, M.A. (2003) *Third Generation Photovoltaics: Advanced Solar Energy Conversion*, Springer, Berlin.

Green, M.A., Emery, K., Hishikawa, Y. *et al.* (2011) Solar cell efficiency tables (version 38). *Prog. Photovolt: Res. Appl.*, **19**, 565–572.

Kan, S.Y. (2006) Energy matching – key towards the design of sustainable powered products, thesis, Technical University of Delft, Design for Sustainability Program, Delft.

Kazmerski, L. (2010) *Best Research Cell Efficiencies*, NREL, Golden, CO.

Müller, M. (2010) Energieautarke Mikrosysteme am Beispiel von Photovoltaik in Gebäuden. PhD thesis, University of Freiburg.

Müller, M., Hildebrand, H., Walker, W.D., and Reindl, L.M. (2009) Simulations and measurements for indoor photovoltaic devices. In Proceedings of 24th European Photovoltaic Solar Energy Conference, Hamburg, pp. 4363–4367.

Prince, M.B. (1955) Silicon solar energy converters. *J. Appl. Phys.*, **26**, 68–74.

Prince, M.B. (1959) The photovoltaic solar energy converter. *Sol. Energy*, **3**, 35.

PVAccept (2005) Final report EU PVACCEPT/IPS-2000-0090 (ed. I. Hermannsdörfer and C. Rüb). http://www.pvaccept.de (accessed April 20, 2012).

Randall, J. (2006) *Designing Indoor Solar Products: Photovoltaic Technologies for AES*, John Wiley and Sons, Chichester, UK.

Randall, J.F. and Jacot, J. (2003) Is AM1.5 applicable in practice? Modelling eight photovoltaic materials with respect to light intensity and two spectra. *Renew. Energ.*, **28**, 1851–1864.

Reich, N.H., van Sark, W.G.J.H.M., Alsema, E.A. *et al.* (2005) Weak light performance and spectral response of different solar cell types. In Proceedings of the 20th European Photovoltaic Solar Energy Conference (ed. W. Palz, H. Ossenbrink, and P. Helm), WIP-Renewable Energies, Munich, Germany, pp. 2120–2123.

Reich, N.H., van Sark, W.G.H.M., Reinders, A.H.M.E., and de Wit, H. (2009) Using CAD software to simulate PV energy yield: Predicting the charge yield of solar cells incorporated into a PV powered consumer product under 3D-irradiation conditions. In Proceedings of 34th IEEE PVSC, Philadelphia, pp. 1291–1296.

Reich, N.H., van Sark, W.G.H.M., Alsema, E.A. *et al.* (2009) Crystalline silicon cell performance at low light intensities. *Sol. Energ. Mat. Sol. C*, **93**, 1471–1481.

Reinders, A. (2002) Options for photovoltaic solar energy systems in portable products. In Proceedings of the TMCE, Wuhan.

Reinders, A.H.M.E. (2007) A design method to assess the accessibility of light of PV cells in an arbitrary geometry by means of ambient occlusion. In Proceedings of 22nd European Photovoltaic Solar Energy Conference, Milan.

Reinders, A.H.M.E. and Akkerman, R. (2005) Design, production and materials of PV powered consumer products: The case of mass production. In Proceedings of 20th European Photovoltaic Solar Energy Conference, Barcelona.

Reinders, A.H.M.E. and van Sark, W.G.J.H.M. (2012) Product integrated photovoltaics, in *Photovoltaic Technology*, Vol. 1 (ed. W.G.J.H.M. van Sark), in Comprehensive Renewable Energy (ed. A. Sayigh), Elsevier, London, 709–732.

Roth, W. and Steinhueser, A. (1997) *Photovoltaische Energieversorgung von Geraeten und Kleinsystemen*, OTTI – Technologie-Kolleg, Fraunhofer Institut Solare, Energiesysteme ISE, Freiburg.

Ryer, A. (1997) *The Light Measurement Handbook*, International Light, Newburyport, MA.

Sillcox, L.K. (1955) Fuels of the future. *J. Franklin I.*, **259** (3), 183–195.

Tiwari, A. and Reinders, A. (2009) Modeling of irradiance in a CAD environment for the design of PV powered products. In Proceedings of PVSEC-19, Kolkata.

Zahl, H.A. and Ziegler, H.K. (1960) Power sources for satellites and space vehicles. *Sol. Energy*, **4**, 32–38.

Further Reading

Geelen, D., Kan, S.Y., and Brezet, H. (2008) Photovoltaics for product designers: How to design a photovoltaic solar energy powered consumer product. In Proceedings of 23rd European Photovoltaic Solar Energy Conference, Valencia, pp. 3329–3333.

Gorter, T., Reinders, A.H.M.E., Pascarella, F. *et al.* (2009) LED/PV lighting systems for commercial buildings: Design of a sustainable LED/PV symbiotic system. In Proceedings of 24th European Photovoltaic Solar Energy Conference, Hamburg, pp. 4250–4255.

Kan, S.Y. and Silvester, S. (2004) Synergy in a smart photo voltaic (PV) battery: SYN-ENERGY. *J. Sustainable Product Design*, **3**, 29–43.

Kan, S.Y., Verwaal, M., and Broekhuizen, H. (2006) The use of battery-capacitor combinations in photovoltaic powered products. *Short Commun. J. Power Sources*, **162**, 971–974.

Kan, S.Y. and Strijk, R. (2006) Towards a more efficient energy use in photovoltaic powered products. *Short Commun. J. Power Sources*, **162**, 954–958.

Reich, N.H., van Sark, W.G.H.M., Alsema, E.A. *et al.* (2008) A CAD based simulation tool to estimate energy balances of device integrated PV systems under indoor irradiation conditions. In Proceedings of 23rd European Photovoltaic Solar Energy Conference, Valencia, pp. 3338–3343.

Reich, N.H., van Sark, W.G.H.M., Turkenburg, W.C., and Sinke, W.C. (2010) Using CAD software to simulate PV energy yield: The case of product integrated photovoltaic operated under indoor irradiation. *Sol. Energy*, **84**, 1526–1537.

3.4   Fuel Cells

Frank de Bruijn

Nedstack fuel cell technology B.V., Westervoortsedijk 73, The Netherlands
Energy and Sustainability Research Institute Groningen University of Groningen,
Nijenborgh 4, Groningen, the Netherlands

3.4.1   Fuel Cell Principles and Characteristics

Fuel cells are electrochemical devices that convert chemical energy into electricity and heat. Like batteries, fuel cells are so-called galvanic cells: Because of a negative change in Gibbs free energy, the electrochemical reaction is a spontaneous process. The counterpart of the fuel cell is the electrolyzer, which is a forced process.

The core of the fuel cell consists of an electrolyte and two electrodes, the anode and the cathode. Oxidation takes place at the anode, and reduction at the cathode. The electrolyte separates the two electrodes, while at the same time it facilitates the transport of ions to prevent the accumulation of charge at either side of the fuel cell.

In comparison to batteries, fuel cells are open systems: A continuous feed of reactants is needed to sustain the electrochemical process, while the products have to be expelled. The increase of power density is often a combination of optimizing the activity of the electrodes, minimizing ohmic resistances, and maximizing the rate at which reactants can reach the reaction interface. A schematic drawing of a series of three cells in a fuel cell stack is shown in Figure 3.4.1, including the flows of gas, ions, and electrons.

The basic requirements to the separate components become obvious when observing Figure 3.4.1. The electrolyte not only needs to be ion conductive but also needs to separate the reactant gases hydrogen and air, and it needs to be an electronic insulator. At the electrodes, the electrochemical reactions take place. Access for gas, ion, and electrons to the reaction interface is required at both electrodes, which leads to the need for a careful design of the electrode structure at the submicrometer scale. The flow plates connect the individual cells electronically, contain flow patterns to distribute the reactant gas over the cell area, and contain a flow field for cooling to dissipate the heat; the heat generation is typically in the same order as the electric power generation, the electric efficiency being 50–60% in most fuel cells.

3.4.2   Comparison of Fuel Cell Types

Six basic types of fuel cells have emerged, with a number of subtypes that consist of minor variations. The main distinction between these basic types is set by the electrolyte. Both cation- and anion-conducting electrolytes are used, and the temperature at which these electrolytes provide sufficient conductivity is the determining factor for the temperature window of the fuel cell, the materials that can be used for the electrodes, and the fuels that can be fed to the anode. The six fuel cell types and the most important subtypes are given in Table 3.4.1.

The PEMFC is split into three different varieties: air-cooled, water-cooled, and high-temperature polymer electrolyte membranes (PEMs), as they have distinct characteristics and all are being developed toward separate products. The other fuel cell types have not

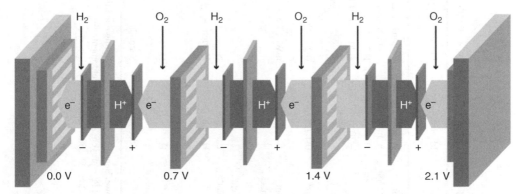

Figure 3.4.1   Schematic drawing of a fuel cell stack containing three repetitive units, using a polymer electrolyte membrane fuel cell (PEMFC) as an example (See Plate 17 for the colour figure)

Table 3.4.1 Overview of fuel cell types and their most important characteristics (De Bruijn, 2005; Stolten, 2010)

	PEMFC water cooled	PEMFC air cooled	HT PEMFC	DMFC	AFC	PAFC	MCFC	SOFC
Electrolyte or conducting ion	Proton exchange membrane H^+	Proton exchange membrane H^+	H_3PO_4 doped PBI H^+	Proton exchange membrane H^+	KOH OH^-	H_3PO_4 H^+	$Li_2/K_2\,CO_3$ CO_3^{2-}	Yttrium-ZrO_2 O^{2-}
Operating temperature (°C)	−20°C to 90°C	−20°C to 70°C	100–160°C	30–80°C	40–200°C	200°C	650–700°C	600–1000°C
Power density[a] $W.cm^{-2}$	0.4–0.7	0.2–0.4	0.2–0.5	0.1	0.1–0.3	0.14	0.12	0.15–0.7
Typical output range of stack	1–100 kW	mW–1 kW	100 W–10 kW	mW–1 kW	1–5 kW	25–125 kW	50–125 kW	mW–125 kW
State of development[b]	Pr	Pr	D	Pr	Pr	Pr	Pr	D
Quality of heat[c]	L	N	M	L	L	M	H	H
Fuels	H_2 rich	H_2 rich	H_2 rich + maximum 2% CO	Methanol	H_2 rich	H_2 rich + maximum 2% CO	CH_4 and biogas	CH_4, biogas, H_2, and coal gas
Sensitivity to contaminants	H	H	M	H	H	M	L	L
Startup time	Seconds	Seconds	Minutes	Minutes	Seconds	Hours	Hours	Hours

[a]Practical power densities under normal operating conditions.
[b]Status 2011: Pr = proven; and D = development.
[c]N = nil; L = low; M = medium; and H = high.

been diversified to this extent. As one can conclude from Table 3.4.1, there is no single fuel cell type that scores high on all operational properties. A key advantage of high-temperature fuel cells is their tolerance toward impurities. It makes them suitable for operation on a wide variety of fuels, either directly or indirectly using a fuel processor that converts the primary fuel into, for example, synthesis gas (a mixture of hydrogen and carbon monoxide). The same high-temperature operation leads to long start-up times, which makes them unsuitable for a variety of applications such as transport and backup power. While the PEMFC has many advantages over the other fuel cell types, a clear disadvantage is its intolerance toward impurities; it needs hydrogen of high quality. High quality is not necessarily high purity, as inert components such as water, nitrogen, and methane can be tolerated.

3.4.3   Key Characteristics of Fuel Cells

Current–Voltage–Power–Efficiency Relations

Figure 3.4.2 shows the typical relation between current density, voltage, power, and efficiency of a single fuel cell, in this case for a PEM fuel cell. Similar characteristics can be drawn for the other fuel cell types as well, albeit that the current and power densities vary considerably for the different fuel cell types; see Table 3.4.1. As one can see, there is a clear trade-off between efficiency and power output, which can be translated into a trade-off between fuel efficiency and fuel cell investment cost.

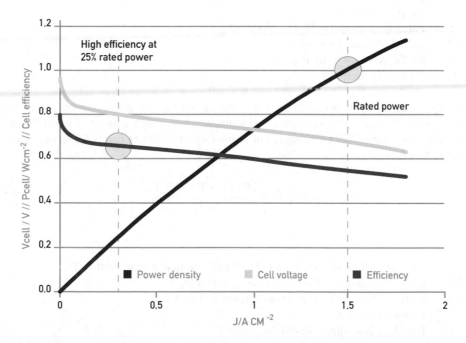

Figure 3.4.2   Cell voltage (green), power density (black), and efficiency (red) versus cell current density. The points of rated power and 25% rated power are indicated. Follow the vertical lines to find the corresponding cell characteristics at these points

The electrical efficiency η_{el} of the fuel cell stack, referred to the *lower heating value of the fuel used*, is determined by the voltage at which the individual cells are operated and the utilization (U_{fuel}) of the fuel:

$$\eta_{elFC, stack} = \frac{\text{Cell voltage}}{1.254} \times U_{fuel} \qquad (3.4.1)$$

The maximum theoretical voltage of a fuel cell is reached at zero current, and is referred to as the *Nernst voltage*. This Nernst voltage depends on temperature, pressure, and the ratio of reactant and product partial pressures. At 25°C and at partial pressures of hydrogen and oxygen of 101 325 Pa, the Nernst voltage amounts to 1.23 V. For a hydrogen–oxygen (air) fuel cell, the relation between the Nernst voltage and partial pressure of hydrogen, oxygen, and water at a given temperature is given by Equation (3.4.2):

$$Nernst\ Voltage = E°(T) + \frac{RT}{2F} \times \ln\frac{\sqrt{PO2 \times PH2)}}{PH2O} \qquad (3.4.2)$$

in which:

E° (T) the standard potential (V) of the fuel cell reaction at temperature T, 1.23 V for a
         hydrogen–oxygen cell at 298 K
   R   gas constant, $8.314\,J.mol^{-1}\,K^{-1}$
   T   temperature in K
   F   Faradays constant $= 96\,484.56\,C.mol^{-1}$
  PO2   partial pressure of oxygen
  PH2   partial pressure of hydrogen
 PH2O   partial pressure of water

From equation (3.4.2), it becomes clear that increasing the (partial) pressures of hydrogen and oxygen leads to a higher reversible cell voltage, while a high content of water in the gas phase leads to a lower Nernst voltage. For temperatures different from 298 K, one needs to calculate E° (T), using:

$$E°(T) = -\frac{\Delta H(T) - T \times \Delta S(T)}{2F} \qquad (3.4.3)$$

in which

E° (T) the standard potential (V) of the fuel cell reaction at temperature T
  ΔH   enthalpy change of the reaction at temperature T
  ΔS   entropy change of the reaction at temperature T
   T   temperature in K
   F   Faradays constant $= 96\,484.56\,C.mol^{-1}$

As with many fuel cell reactions, the entropy change is negative, and the Nernst potential generally diminishes with an increase in temperature.

In principle, the fuel cell stack power P_{stack} is the simple multiplier of cell current I_{cell}, cell voltage V_{cell}, and the number of cells n_{cells}:

$$P_{stack} = \eta_{cells} \times I_{cell} \times V_{cell} \tag{3.4.4}$$

From Figure 3.4.2, it can be concluded that at cell and stack levels, efficiency is highest at zero power, and in general decreases with increasing power.

A fuel cell stack needs to be fed with hydrogen and air to generate power, and generally the power needs to be conditioned to obtain the desired voltage level and quality. For maintaining the stack temperature in the optimal range, a cooling circuit is needed. Figure 3.4.3 gives a simple systems layout for a hydrogen-fed fuel cell system. Hydrogen can be either fed from a hydrogen storage device or generated by a so-called fuel processor. The PEM fuel cell especially needs to remain hydrated; a complex water management system is needed that is not included in Figure 3.4.3.

On a systems level, efficiency is lowered by parasitic losses of so-called balance-of-plant components (BOP in equation (3.4.5)), such as compressors, humidifiers, and power conditioning:

$$\eta_{elFC,system} = \eta_{elFC,stack} \times \left(1 - \frac{Power\ consumption\ BOP}{Net\ power\ output\ system + power\ consumption\ BOP} \right)$$

$$\tag{3.4.5}$$

Net system power refers to the useful power produced by the fuel cell system available for propulsion. This net system power is obtained by subtracting the power use of the balance-of-plant components from the power output of the fuel cell stack.

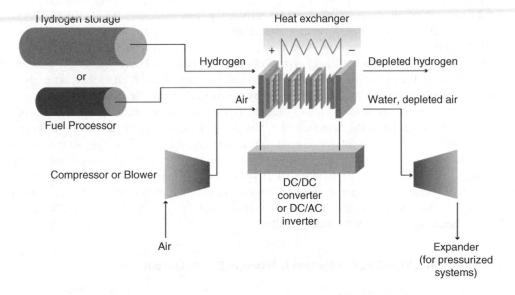

Figure 3.4.3 Simple systems layout for a PEMFC system

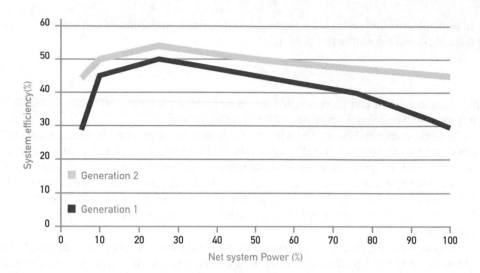

Figure 3.4.4 PEMFC vehicle systems efficiency based on composite data generated from fuel cell vehicles taking part in the demonstration fleet in the United States, for two generations (NREL, 2009)

As the power consumption of these components does not necessarily vary proportionally with the net system power output, system efficiency often follows a trend as illustrated in Figure 3.4.4 for two succeeding generations of automotive fuel cell systems.

Combustion engines generally show an increase of efficiency with increasing power output, with especially low efficiency at partial load. Whereas fuel cell systems of generation 1, displayed in Figure 3.4.4, had difficulties to beat combustion engines at high loads, today's systems, represented by generation 2 in Figure 3.4.4, have improved such that over the whole output range, fuel cell systems offer superior efficiency (Froeschle and Wind, 2010). This is due to both improved cell performance and balance-of-plant components with low power consumption.

It can be concluded that stack and system efficiency depend on many factors. At constant fuel utilization, the operating voltage determines the electrical stack efficiency. At a given current–voltage relation, which is determined by the fuel cell properties and the stack design, higher power generally leads to lower electrical efficiency at the stack level. The proper design of the system and the selection of balance-of-plant components are decisive for net system efficiency. For the total well-to-wheel efficiency, the complete fuel supply chain should be taken into account. Given all these complications, fuel cell efficiency numbers are not included in Table 3.4.1. Equations (3.4.1) and (3.4.4) give the basic relation between the fuel cell efficiency when operating on hydrogen and oxygen and the cell voltage. A good insight in complete chain efficiencies for vehicles can be obtained from the "Well-to-Wheels" study published by the JRC (2007).

Modular Setup, Enabling Continuous Increase of Power Output

A single fuel cell has a typical power density of 0.1–0.7 W.cm^{-2}, depending on fuel cell type, operating conditions, and cell voltage. The voltage, current, and power requirements set the design of the fuel cell and fuel cell stack. At given cell characteristics, such as those displayed in Figure 3.4.2, the point of operation can be chosen and renders the cell current density and power density. Cell and stack current are then determined by the cell area, while the stack voltage and power are set by the number of individual cells put in series. With a given design, one can vary the power output of a fuel cell stack by a factor of around 5, just by the variation of the number of individual cells put in series. A fuel cell system can contain a number of stacks, put in series, in parallel, or a in combination of these, leading to the desired system voltage and current. With a single fuel cell stack design, a power range between 2 kW and 1 MW can thus be covered. This offers a flexibility that is unsurpassed by competing technologies.

Low or No Emissions

The key benefit of fuel cells is the low level of harmful emissions when compared to many other conversion devices. All fuel cells operate below the temperature where nitrogen and oxygen combine to nitrogen oxides, which lies above 1500°C. When hydrogen is used as a fuel, water is the only byproduct. Due to the intolerance of all fuel cells to sulfur, no emissions of sulfur oxides are formed.

High-temperature fuel cells can be fed with simple hydrocarbons, mostly methane or methane-rich gases, or with synthesis gas, a mixture of hydrogen and carbon monoxide. More complex hydrocarbons often lead to carbon deposition, limiting the fuel cell life to far shorter than the required 40,000–90,000 hours.

Fuel cells are thus especially considered where air quality is compromised by the combustion of fuels, such as in cities and buildings. In these cases, fuel cells compete with the only other zero-emission technology, which is the battery. The absence of noise production is a prime benefit of fuel cells and batteries as well.

Split between Energy Storage and Power

The key difference between fuel cells and batteries is the split between power and energy. The fuel cell, being an open system, contains no energy itself. The energy is contained in the fuel, which is stored separately. As an oxidant, both air and oxygen can be used. Oxygen is used primarily in space and submarine applications. In most terrestrial applications, the benefit of higher power densities does not compensate the higher costs and space requirements of oxygen storage.

Whether expressed in driving distance or operating hours, fuel cell systems beat batteries in volume, weight, and cost to a point, because the storage of fuel per unit of energy is lighter, more compact, and cheaper than in batteries (Wagner et al., 2010). Below that point, batteries are more compact and lighter because of system simplicity.

3.4.4 Cost

The cost of fuel cell systems and stacks heavily depends on the power rating of the system and the production volume. As of today, only two fuel cell types are sold on a commercial base: the PEMFC and the DMFC.

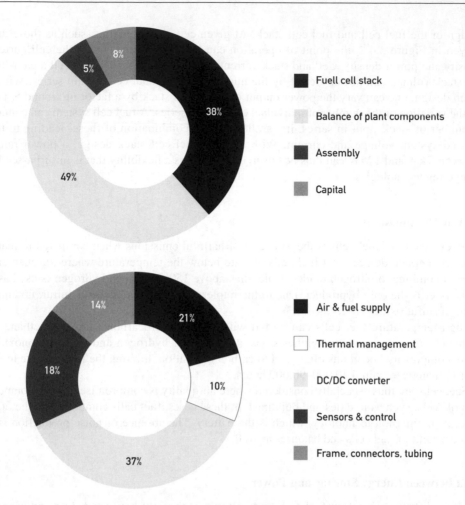

Figure 3.4.5 Estimated cost breakdown of a 5 kW PEMFC system for backup power. Left: Cost breakdown on system level. Right: Cost breakdown of balance of plant components Analysis made for the US Department of Energy (2010)

The cost of PEM fuel cell systems has come down impressively since the turn of the twenty-first century. An economic analysis that is relevant for today's applications in telecommunications has recently been published by the US Department of Energy in their *2010 Annual Progress Report* (Mahadevan *et al.*, 2010). Figure 3.4.5 shows the estimated cost breakdown of a 5 kW PEMFC system, both for the system as a whole and as an estimated cost breakdown of the balance-of-plant components. The cost of a 5 kW system at an annual sales volume of 2,000 pieces was estimated to amount to $7,000 in 2010, and could be reduced with increasing annual production volume of 100,000 units to $4,200 in 2015.

Although the actual cost is at present higher than estimated in this report, the analysis provides a valuable insight into the major cost components of a fuel cell system of 5 kW output with limited production volume. For example, the high cost of the DC/DC conversion is an issue that is overlooked by many integrators.

3.4.5   Fuel Cell Applications and Basic Requirements

Potential Applications and Their Requirements

Although most fuel cell types have been considered for a wide range of applications, one can clearly come to a selection of preferred solutions for single applications, taking into account the fuel availability, power density, window of operation, durability, and costs. In the rest of this section, an overview of the main applications is given in conjunction with the current status of various technologies being developed or already applied. Table 3.4.2 gives an overview of the basic requirements for various applications where fuel cells are considered.

3.4.6   Passenger Vehicles

Requirements:

Power range: 40–100 kW; lifetime: 5,000 hrs; system cost: 30–50 €/kW; fuel: hydrogen; shock and vibration resistant; cold start; freeze proof; high power density; and high cycle ability

The application of fuel cells in passenger cars has undoubtedly drawn the most public attention. Already since the early 1970s, prototype fuel cell passenger cars have been showcased. Only since the threat of zero-emission regulation for road vehicles to be introduced in California by 2004, car OEMs have been actively developing passenger cars running on fuel cells that are able to meet consumer expectations, that is, offering functional space, power, and driving range comparable to those of internal combustion engine–powered cars. PEM fuel cells are the only fuel cell type being considered nowadays. State-of-the-art fuel cell vehicles, such as those demonstrated by Honda, Toyota, General Motors, and Daimler, contain fuel cell stacks with a power density of around $2\,kW.l^{-1}$ that can be used over the expected range of ambient conditions, have a proven lifetime of over 100,000 km, and have a driving range of 500–800 km without refueling (De Bruijn, 2009). The fuel of choice has emerged to be hydrogen; cars using onboard reforming of gasoline or methanol could not meet efficiency, start-up time, start-up energy use, and power density requirements.

Although the latest generation of passenger vehicles is regarded to be fit for the first market introduction, further developments are devoted to the reduction of vehicle cost over its useful life, being a combination of initial procurement costs and durability. Both the costs of platinum per kW output as well as the costs of balance of system components receive the most attention. The latter is thought to diminish considerably when heat and water management can be simplified, for which new fuel cell components are needed that can operate at low humidity level and higher temperature (De Bruijn, 2009).

3.4.7   City Buses

Requirements:

Power range: 75–150 kW; lifetime: 20,000 h; system cost: 1,000 €/kW; fuel: hydrogen; shock and vibration resistant; cold and rapid start; freeze proof; and high cycle ability

In comparison to passenger cars, fuel cells for city buses have longer lifetime requirements, but at the same time are not pushed to the same high power output as fuel cells for automotive are. As in automotive use, only PEM fuel cells are being considered. In practice,

Table 3.4.2 Basic requirements for various applications where fuel cells are considered

	Automotive and urban bus	Portable	Stationary backup power	Stationary base load	Stationary power and CHP
Power range	40–160 kW	Micro: < 5 W Small: 5–500 W Large: > 500 W	4–20 kW	4–20 kW	1–1000 kW
Systems volume	123 l for 80 kW	0.5 l for 50 W	Not relevant	Not relevant	Not relevant
Systems weight	123 kg for 80 kW	0.5 kg for 50 W	Not relevant	Not relevant	Not relevant
Lifetime	5,000–20,000	5,000	1,500–4,000	10,000–40,000	40,000–90,000
Startup time	30 s at −20°C 5 s at +20°C	5 ms	5 ms	30 min	30 min
Startup energy	5 MJ at −20°C (for 80 kW) 1 MJ at +20°C (for 80 kW)	Not specified	Not relevant	Not relevant	Not relevant
Preferred fuel cell types	PEMFC	PEMFC DMFC	PEMFC	PEMFC AFC SOFC	SOFC PAFC MCFC PEMFC
Competing technologies	ICE Battery–electric Hybrids	Batteries	Batteries Batteries + gensets	Gensets	Stirling Gas engine Diesel engine Gas turbine

similar technology is used, for example the 150 kW Ballard FC velocity–HD6 stack module contains two 75 kW PEMFC stacks that are also used to power passenger cars. The development status is similar to that of automotive use: Fuel cell buses are demonstrated in large numbers across the world and offer functionality comparable to that of internal combustion engine buses, and fuel cell lifetime is over 12,000 hours (Froeschle and Wind, 2010). With respect to development goals, these are similar to those for automotive use, although total cost of ownership is more important than for automotive use, where first ownership costs dominate.

3.4.8   Materials Handling

Requirements:

Power range: 4–20 kW; lifetime: > 10,000 h; system cost: 600 €/kW; fuel: hydrogen; shock and vibration resistant; and high cycle ability

Fuel cell systems for materials-handling vehicles are primarily introduced as replacement for battery-powered vehicles used indoors, where emissions are generally not allowed. PEM fuel cells are primarily used, although direct methanol is being considered as well. Fast refueling, compared to battery charging over hours or labor-intensive swapping of batteries, makes the application of fuel cells economically viable in the short term (US Department of Energy, 2011). As with batteries, weight is not an issue; in fact, forklifts often need a counterweight for safe operation. Size matters, as fuel cell systems are often required to be a drop-in replacement of existing battery packages. Hundreds of fuel cell–powered forklifts have been put into service and meet most important end user requirements, such as lifetime, durability, and power density. Tax credits or subsidies are often still needed to convince industries, partly because fuel costs are higher in the case of fuel cells. Because of many similarities with fuel cell systems for automotive applications, materials-handling vehicles are regarded as their forerunner.

3.4.9   Portable Applications

Requirements:

Power range: micro: < 5 W; small: 5–500 W; large > 500 W; lifetime: 2,000–5,000 h; system cost: 500–2,000 €/kW; fuel: hydrogen, methanol; cold and rapid start; and high cycle ability

Portable fuel cells are in many applications in direct competition with batteries. Energy density is the most important property on which fuel cell systems can offer an advantage, so it is no surprise that the direct methanol fuel cells are most often developed for low-power applications (Garche, 2010). The most successful commercialization of portable fuel cells, by SFC energy, is based on direct methanol fuel cells, having sold over 20,000 fuel cells by 2011 for leisure and defense applications in the 25–90 W range. A lifetime of over 5,000 hours is claimed for a 55 W battery charger system for defense applications.

PEM fuel cells running on hydrogen have a market share as well, and for very small applications even solid-oxide fuel cells have been showcased. The application in the Watt range, for example, for consumer electronics seems to have become commercially out of reach with the successful development of new generations of lithium-based batteries.

3.4.10   Stationary Fuel Cells: Backup Power

Requirements:

Power range: 4–20 kW; lifetime: 1500–4000 h; system cost: 1,000–2,000 €/kW; fuel: hydrogen; cold and rapid start; and high cycle ability

Together with materials handling and portable power, backup power is a market already in full development. The need for high reliability in the telecom sector leads to the installation of backup power systems in both landline and wireless telecom networks. Operating hours can vary highly per location, depending on the reliability of the local grid (Colbow, 2010). Operating hours are spread over 5–10 years with frequent startup and shutdown cycles, and can resemble automotive use, where fuel cell stacks are operated at high current density (1 A.cm^{-2} and higher) during relatively short times, and are in cold stand still for the largest share of their life. Although voltage cycling leads to higher voltage decay rates when compared to continuous operation, PEMFC systems can meet the application requirements successfully.

3.4.11   Stationary Fuel Cells: Base Load Power

Requirements:

Power range: 5–1000 kW; lifetime: 20,000–90,000 h; system cost: 1,000–2,000 €/kW; fuels: hydrogen, methanol, ammonia, and natural gas; an high efficiency

Base load power generation using fuel cells in the power range of 5–20 kW refers to locations where a power grid is absent, for example, for powering telecom stations in remote locations. On the high end of the range, at 200–1000 kW, power is generated at industrial sites where hydrogen is generated as a byproduct. In both cases, fuel cell stacks are mostly operated at moderate current densities (0.2–0.6 A.cm^{-2}), and consequently high cell voltages (0.7 V and higher), to obtain high fuel efficiency. Using practical hydrogen gas qualities, voltage decay rates as low as 2 μV.hr^{-1} are realistic for continuous operation using PEMFC. In these applications, PEM, alkaline, and phosphoric acid fuel cells are applied, the first dominating the market.

Alkaline fuel cells have been used especially onboard space missions, operating on hydrogen and oxygen, and supplying both electricity and drinking water for astronauts. The power output of these systems is typically around 6 kW. Being intolerant to carbon dioxide, terrestrial application requires the application of CO_2 absorbents, such as limestone (De Bruijn, 2005). Most efforts by companies to launch alkaline fuel cells for stationary applications, especially those advocating the ability of using nonnoble metals for the conversion of hydrogen and oxygen, have failed so far.

3.4.12   Stationary Fuel Cells for Combined Heat and Power Generation

Requirements:

Power range: 1–2000 kW; lifetime: 40,000–90,000 h; system cost: 1,000–2,000 €/kW; fuel: natural gas, biogas, coal gas, kerosene, and propane; and high efficiency

Combined heat and power generation from a very small scale for residential use to a large scale for commercial and industrial uses is a challenging application that has been pursued for the last three decades. While in the 1970–2000 period molten carbonate fuel cells (MCFCs) and phosphoric acid fuel cells were primarily used in the 100–200 kW range (Moreno, McPhail, and Bove, 2008), in the 2000–2010 period the focus was shifted to small systems for residential use. Energy saving is the main driver for combined heat and power, preventing the waste of heat that is typical for central power production at the GW scale. Because operating on the fuel readily available at the location of use is preferred, systems need to operate on natural gas, propane, and even kerosene. The ability to internally reform carbon-containing fuels and the generation of high-temperature heat gives high-temperature fuel cells a clear advantage. Especially solid-oxide fuel cells (SOFCs) have emerged in the 1–50 kW range as the fuel cell of choice (Föger, 2010). For very large systems, where SOFCs seem to lose on cost and manufacturability of large modules, MCFC systems are being applied. Use of biomass-derived gas is in development to reduce overall emissions of greenhouse gases even further.

Because temperature and redox cycles are severe stress factors, SOFCs are not considered for the aforementioned backup applications. For the same reason, previously considered applications as auxiliary power units for transportation, running on pre-reformed liquid fuels such as diesel and gasoline, seem to have become out of reach.

3.4.13 Conclusions

Fuel cells are in development for a wide range of applications, from small sub-Watt portable applications converting methanol directly to power to large systems converting hydrogen- and methane-rich fuels up to the MW scale. Their key advantage lies in their capability to efficiently deliver amounts of power and energy comparable to those of combustion engines, but without the emissions of carbon dioxide, non–greenhouse gas emissions and noise during use. The aforementioned emission over the full energy chain depends on the energy source and the efficiency of the process to convert the primary fuel to the fuel fed to the fuel cell. Depending on fuel and applications, different fuel cell types might qualify as the best choice. Even without legislation on air quality and greenhouse gas emissions, fuel cell systems are entering markets on a commercial basis, by offering a cost and durability that meet customer requirements.

References

Colbow, K. (2010) Fuel cells in extended duration emergency backup power. Paper presented at 2010 Fuel Cell Seminar, San Antonio, TX.
De Bruijn, F.A. (2005) The current status of fuel cell technology for mobile and stationary applications. *Green Chem.*, **7** (3), 132–150.
De Bruijn, F.A. (2009) PEM fuel cells for transport applications: State of the art and challenges. *AIP Proc.*, **1169**, 3–12.
Föger, K. (2010) BlueGen: Ceramic fuel cells first product for commercial roll-out. Paper presented at 2010 Fuel Cell Seminar, San Antonio, TX.
Froeschle, P. and Wind, J. (2010) Fuel cell power trains, in *Hydrogen and Fuel Cells: Fundamentals, Technologies and Applications* (ed. D. Stolten), Wiley-VCH, Weinheim, Chapter 38.
Garche, J. (2010) Portable applications and light traction, in *Hydrogen and Fuel Cells: Fundamentals, Technologies and Applications* (ed. D. Stolten), Wiley-VCH, Weinheim, Chapter 35.
Joint Research Centre of the European Commission (2007) Well-to-Wheels analysis of future automotive fuels and powertrains in the European context, Well-to-Wheels Report Version 2c. http://ies.jrc.ec.europa.eu/WTW (accessed April 20, 2012).

Mahadevan, K., Contini, V., Goshe, M. *et al.* (2010) Economic analysis of stationary PEM fuel cell systems. FY 2010 annual progress report, DOE Hydrogen Program, pp. 693–698. http://www.hydrogen.energy.gov/annual_progress10_fuelcells.html#h (accessed April 20, 2012).

Moreno, A., McPhail, S., and Bove, R. (2008) *International Status of Molten Carbonate Fuel Cell Technology*, ENEA, Rome.

NREL (2009) Hydrogen and fuel cells research: Composite data products by topic. www.nrel.gov/hydrogen/cdp_topic.html (accessed April 20, 2012).

Stolten, D. (2010) *Hydrogen and Fuel Cells: Fundamentals, Technologies and Applications*, Wiley-VCH, Weinheim.

US Department of Energy (2011) Information brochure on the application of fuel cells in materials handling vehicles. www1.eere.energy.gov/hydrogenandfuelcells/education/pdfs/early_markets_forklifts.pdf (accessed April 20, 2012).

Wagner, F., Lakshmanan, B., and Mathias, M.F. (2010) Electrochemistry and the future of the automobile. *J. Phys. Chem. Lett.*, **1**, 2204–2219.

3.5 Small Wind Turbines

Paul Kühn

Fraunhofer Institute for Wind Energy and Energy System Technology IWES, Königstor 59, Germany

3.5.1 Introduction

In the last two decades, wind power utilization has emerged from a niche industry to an industrial sector with global significance. In the mid-1980s, wind turbines had an average rated power of 30 kW and rotor diameters of less than 15 m. Since then, wind turbines with a rated power of 5 MW and more and rotor diameters of more than 125 m have been developed. Today, multi-megawatt wind turbines of 2 MW and more dominate the market. But in the last few years the interest in small wind turbines has also grown again. Small wind turbines are used in a broad field of application ranging from very small mobile battery chargers rated at only a few hundred Watts to grid-connected systems with up to about 100 kW.

3.5.2 Turbine Size and Applications

Some useful parameters for small wind turbine classification are their physical size, that is, the rotor diameter or the rotor swept area as well as their electrical properties, usually the rated electrical power. According to the international standard for the design requirements, small wind turbines have a rotor swept area equal to or smaller than 200 m^2, generating at a voltage below 1000 V AC or 1500 V DC (IEC, 2006). This corresponds to a rotor diameter of up to 16 m and a rated power of up to about 75 kW. Furthermore, it is useful to distinguish between larger commercial systems and smaller residential systems. Figure 3.5.1 depicts the size of three small wind turbine classes in comparison to a large wind turbine.

Another way to differentiate small wind turbines is by their application, for example battery charger, grid-tied systems, or water pumper; type of installation, for example, free-standing or building mounted, or mobile design characteristics, for example, horizontal or vertical rotor axis, or number of rotor blades.

Figure 3.5.2 shows four examples of available small wind turbine models of different size and design used for different applications.

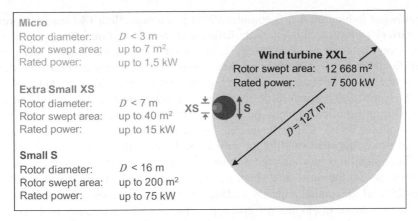

Micro
Rotor diameter:	$D < 3$ m
Rotor swept area:	up to 7 m²
Rated power:	up to 1,5 kW

Extra Small XS
Rotor diameter:	$D < 7$ m
Rotor swept area:	up to 40 m²
Rated power:	up to 15 kW

Small S
Rotor diameter:	$D < 16$ m
Rotor swept area:	up to 200 m²
Rated power:	up to 75 kW

Wind turbine XXL
Rotor swept area:	12 668 m²
Rated power:	7 500 kW

$D = 127$ m

Figure 3.5.1 Small wind turbines divided into three size classes (See Plate 18 for the colour figure)

Model	Superwind 350	Quietrevolution qr5	Fortis Montana	Gaia-Wind 133-11 kW
Rotor diameter	1,2 m	3,1 m x 5 m	5 m	13 m
Rotor swept area	1,1 m²	13,6 m²	19,6 m²	133 m²
Axis of rotation	horizontal (upwind)	vertical	horizontal (upwind)	horizontal (downwind)
Rated power output	0,35 kW	2,3 kW	5,8 kW	11 kW
Rated wind speed	12 m/s	11 m/s	17 m/s	9,5 m/s
Rotor speed	500... 1300 min⁻¹	100... 260 min⁻¹	max. 450 min⁻¹	56 min⁻¹
Type of generator	permanent magnet	permanent magnet	permanent magnet	induction with gearbox
Tower head mass	11,5 kg	450 kg	230 kg	900 kg
Tower height	2,6... 15 m	6 m	12... 24 m	15... 27 m
Application	off-grid, remote, mobile, battery charger	on-grid, building mounted	on- and off-grid	on-grid farms, small businesses

Figure 3.5.2 Example of small wind turbine models of different sizes and designs

The small wind turbine market is complex. Worldwide more than 180 manufacturers offer a huge variety of turbine models with different technical features. The quality of these turbines is very heterogeneous. Whereas some available turbines do not comply with international standards and are unsafe to operate, others have a successful track record. Studies show that small wind turbines can achieve high reliability and a life expectancy of 20 years and more (Kühn, 2007). According to current market reports, about 20,000 small wind systems with an installed capacity of about 40.000 kW are installed every year (AWEA, 2010; Fraunhofer IWES, 2011; Renewable UK, 2011). It is estimated that approximately 75% of all newly installed small wind turbines fall within the *micro category* (cf. Figure 3.5.1). These small turbines are predominantly employed in off-grid systems and used as battery chargers. At the same time small wind turbines of the micro category contribute to less than 25% of the annual capacity added. Small wind turbines that fall within the XS category are deployed in on- and off-grid applications, whereas turbines within the S category are mainly grid tied.

On-grid systems dominate the market in terms of installed capacity, with the United States, United Kingdom, and China being the biggest markets. Typical grid-connected small wind turbines are tower mounted and provide electricity for private homes, small businesses, and farms. Yet the use of small wind turbines in cities and the built environment is controversial and challenging for reasons of safety, structural engineering, vibration, noise, and the complex flow conditions on rooftops and around buildings.

Regardless of their size, small wind turbines have huge potential in remote applications, especially to be used to supply off-grid telecommunication towers and to be employed for rural electrification in hybrid power supply systems and mini-grids. But so far, these future markets are largely untapped. Other small wind turbines target niche applications such as research and measurement stations, water pumps, heating, and advertisement.

3.5.3   Turbine Design and Technology

Despite the variety of design concepts, many small wind turbines are surprisingly similar. Common features are the three-bladed rotor made from fiberglass-reinforced plastics with a horizontal axis directly driving a permanent magnet synchronous generator.

One of the most fundamental design characteristics is the *number of rotor blades*. For large wind turbines, three blades are state of the art. The same is true for most small wind turbine models, although two-bladed and multibladed machines are also available. The choice for a three-bladed rotor can be attributed to being the best compromise between cost, aerodynamic characteristics, and dynamic behavior (Manwell, McGowan, and Rogers, 2006). Some design criteria related to the number of rotor blades are briefly considered in this section.

The ratio of blade area to the rotor swept area is called *solidity*. Small wind turbines with fewer rotor blades have a lower solidity and use less blade material accordingly. This is a major advantage as it results in less tower head mass and lower production costs. Contrariwise rotors with fewer blades are more challenging from the engineering point of view because they require a more advanced airfoil design and slender blade shape. In addition, fewer blades allow more compact generators and gearboxes as rotor speed and torque are linked to the number of rotor blades. For example, a rotor with three blades instead of four compensates for the missing blade by rotating faster at lower torque. At the same time, the loss in efficiency is comparatively small. The difference between three and four blades is only about 1%. A two-bladed rotor is about 3–4% less efficient than a three-bladed rotor (Hau, 2005).

Generally, rotors with fewer blades also emit more noise due to higher rotor speeds. Two-bladed small wind turbines are generally easier to assemble and install than machines with three or more blades (Gipe, 2004). A particular drawback associated with one- and two-bladed rotors is structural dynamic problems induced by nonsymmetric forces acting over the rotor plane. Three-bladed machines run more smoothly. In some applications a multi-bladed rotor might be favored. Examples include mechanical wind-driven water pumpers using 24 or more blades to provide a high torque. Also some battery chargers in the micro category have a six-bladed rotor design to allow a good starting behavior, that is, a high starting torque at very low wind speeds.

Small wind turbines with *permanent magnet generators* usually come in different voltage options. Typical voltages of battery chargers are between 12 and 48 V, whereas turbines for grid connection generate at voltages higher than 100 V. The variable voltage output from the permanent magnet generator is rectified and fed into a battery via a charge controller. In grid-tied systems, an inverter performs the conversion of the rectified generator voltage to the frequency of the grid, that is, 50 or 60 Hz. Instead of a directly driven permanent magnet generator, some small wind turbine systems in the S category use an induction generator and a gearbox. In these cases, the generator is directly connected to the grid.

Similar to all current turbine models in the multi-megawatt class, the majority of small wind turbines are *upwind machines*, that is, the rotor position is in front of the tower. A characteristic feature of these small *horizontal-axis* machines is the passive yaw system with a tail vane to orient the rotor into the wind (see Figure 3.5.2). But there are also small wind turbines that do have *downwind rotors*. An advantage of this rotor concept is the possibility to free yaw without the need for a tail vane. Special attention should be paid to the so-called *tower shadow effect*. That is when the downwind rotor blades pass through the disturbed flow in the wake of the tower. If not properly designed, it can result in having a significant impact on the dynamic behavior of the complete small wind turbine as well as on noise emissions.

The *overspeed control* ensures that a wind turbine remains within design limits at anytime, especially in extreme wind conditions. There are two distinctive and popular overspeed control options with small wind turbines, both being passive mechanisms. The first is the *furling* mechanism, which is common with turbines that have a wind vane. These machines are controlled by means of minimizing the projected rotor swept area. The second control mechanism is *passive pitch*, which functions by turning the leading edge of the rotor blades into the wind (Gasch and Twele, 2002). The pitch angle of the blades is changed by means of spring assemblies and induced by the wind thrust. There are also small wind turbines in the XS and S categories that use *active control* and protection systems similar to those in larger turbines, for example active pitch and active yaw. The control systems of these turbines are more complex and require additional electronics, actuators, and sensors.

The current market share of small wind turbines with a *vertical rotor axis* is rather small. An advantage of the three-dimensional rotor is that it does not need to be aligned to the wind, that is, a yaw system is not required. Another more subjective criterion is the more appealing design of vertical axis machines. Drawbacks of vertical-axis machines include a more complex rotor construction and a relatively high specific tower head mass. Both facts lead to higher efforts for manufacturing, transport, and transportation and result in higher specific costs (€/kW or €/m^2). They also require more massive support structures when compared to horizontal-axis turbines.

The *tower* supports the nacelle and rotor and puts the small wind turbine into the wind out of the effects of the terrain. The choice of the right tower type and of the optimal height strongly depends on the wind turbine specification and on the site conditions. The most common tower types are tube or lattice towers that are either freestanding or guyed. Up to about 30 m, height tilt-up towers are very popular with small wind turbines in the micro and XS category. Tilt-up towers allow easy installation and maintenance. The most common tower type for turbines in the S category is a freestanding tower with a concrete gravity foundation. Usually freestanding towers can be climbed for maintenance and smaller repairs. However, they require a crane for installation and major repairs.

3.5.4 Performance

A wind turbine independent of its design, number of rotor blades, and rotor geometry can capture a maximum of 16/27 or about 59% of the kinetic energy in the wind. This is known as the *Betz limit*. The theoretical maximum power output of a wind turbine is as follows:

$$p_{\text{Betz}} = C_{\text{P,Betz}} \frac{1}{2} \rho A V^3 \tag{3.5.1}$$

where:

$C_{\text{P,Betz}}$ is the maximum theoretical efficiency of the turbine, that is, 59% of the available wind power can be converted into electrical power;

ρ is the air density in kg/m^3, and at standard atmosphere the air density ρ is 1225 kg/m^3 (temperature is 15°C, air pressure is 1013,35 hPa); and

A is the rotor swept area of the wind turbine in m^2.

The wind power is proportional to the air density and to the rotor swept area or squared to the diameter of circular shaped rotor, respectively. As air is about 800 times less dense than water, wind turbine rotors need to be comparatively large. The formula also shows that the power available in wind is proportional to the cube of the wind speed, that is, when the wind speed doubles, the power in the wind increases eightfold. Typical operating wind speeds of small wind turbines are between cut-in at about 3.5 m/s to cut-out at about 15 m/s. At higher wind speeds, the wind power becomes too high and the wind turbine needs to be protected from destructive wind forces. The engineering challenge is to convert the maximum of the highly fluctuating wind power over the whole operating wind speed range. The so-called *method of bins* is helpful to describe the power performance of wind turbines. Using wind speed bin-widths of 0.5 m/s or 1 m/s, the power curve of a wind turbine gives the mean power output for each wind speed bin i. The power output of a wind turbine $P_{\text{WT,i}}$ in wind speed bin i is represented by the following equation:

$$P_{\text{WT,i}} = C_{\text{P,i}} \frac{1}{2} \rho A V_i^3 \tag{3.5.2}$$

where:

$C_{\text{P,i}}$ is the average power coefficient in wind speed bin i; and
V_i is the average wind speed in wind speed bin i.

Typical maximum power coefficients of small wind turbines usually range between 0.25 and 0.35, whereas modern large wind turbines achieve maximum power coefficients of more than

0.5. The generally lower power coefficients mainly result from simpler design and control systems. A problem with the power curves of small wind turbines is that they cannot be trusted at any rate. In many cases, nonstandard procedures and test conditions as well as inadequate measurement equipment are used to determine the power curves. It is therefore advisable to check for the plausibility of a small wind turbine's power curve before performing yield estimations.

The average specific yield for small wind turbines is considerably lower than that of large wind turbines. Besides the generally lower power coefficients, low towers (i.e., hub heights of typically 10–40 m) and the influence of obstacles on flow and turbulence are reasons for this. Most small wind turbines are installed near the place of electricity consumption, that is, near buildings that act as obstacles slowing and distorting the flow of air. Experience shows that besides choosing the wrong turbine model, there are three main reasons for disappointed operators: inadequate sites, overestimated wind resources, and too-short towers (Kühn, 2010). This highlights that proper planning is essential to ensure good performance and longevity, as bad siting can considerably reduce overall power output and high turbulent flow can significantly shorten the life expectancy of a small wind turbine.

Wind resource assessment, yield estimation, and siting are fundamental tasks when planning a small wind system. The average wind speed is the most important parameter for the characterization of the wind resource and is ideally measured at the site and at the hub height of the planned small wind turbine. In many cases a wind measurement campaign, common with large wind projects, is not carried out. The accuracy of the yield estimation, that is, the annual electricity production highly depends on the available data about the local wind regime. Because these data are often unknown (e.g., the site characteristics), the wind speed must itself be estimated, allowing only very rough estimations of the annual electricity production. Wind maps can give a useful hint about the general wind resources of a region. From Figure 3.5.3 it becomes clear that Scotland, for example, has a much better wind regime than the Tuscany region in Italy. However, local conditions and obstacles cannot be accounted for. Furthermore, the resolution of these maps is not high enough to be used for yield estimations. Also, most wind maps show the wind regime for greater heights that are not relevant for small wind turbines with relatively small hub heights. It is therefore advisable to conduct a measurement campaign whenever possible.

The wind resource can be assessed by measurement campaigns lasting at least over a period of one year. Wind data can be obtained by means of a mast equipped with an anemometer and wind vane measuring wind speed and wind direction. Wind speed increases with height above ground. This is called *vertical wind shear*. If data from the measurement height z_1 are available, the wind speed V at height z_2 can be roughly approximated by the Hellmann power law:

$$V(z_1) = V(z_2) \left(\frac{z_1}{z_2}\right)^{\alpha} \tag{3.5.3}$$

Where:

$V(z_1)$ is the measured or reference wind speed at height z_1 in m/s;
$V(z_2)$ is the wind speed at height z_2 in m/s; and
α is the wind shear exponent.

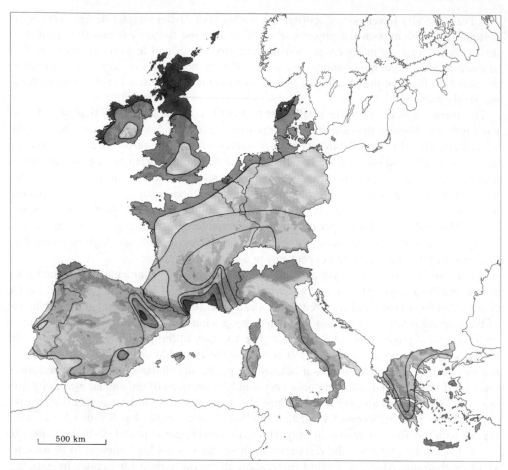

Wind resources[1] at 50 metres above ground level for five different topographic conditions										
Sheltered terrain[2]		Open plain[3]		At a sea coast[4]		Open sea[5]		Hills and ridges[6]	
$m\,s^{-1}$	Wm^{-2}	$m\,s^{-1}$	Wm^{-2}	$m\,s^{-1}$	Wm^{-2}	$m\,s^{-1}$	Wm^{-2}	$m\,s^{-1}$	Wm^{-2}
> 6.0	> 250	> 7.5	> 500	> 8.5	> 700	> 9.0	> 800	> 11.5	> 1800
5.0-6.0	150-250	6.5-7.5	300-500	7.0-8.5	400-700	8.0-9.0	600-800	10.0-11.5	1200-1800
4.5-5.0	100-150	5.5-6.5	200-300	6.0-7.0	250-400	7.0-8.0	400-600	8.5-10.0	700-1200
3.5-4.5	50-100	4.5-5.5	100-200	5.0-6.0	150-250	5.5-7.0	200-400	7.0- 8.5	400- 700
< 3.5	< 50	< 4.5	< 100	< 5.0	< 150	< 5.5	< 200	< 7.0	< 400

Figure 3.5.3 Wind resources at 50 m above ground level for five different topographic conditions from the *European Wind Atlas* (Troen and Petersen, 1989) (See Plate 20 for the colour figure)

The wind shear exponent varies with the roughness of the terrain and with the stability of the atmosphere, among other factors. Another parameter to describe the surface roughness is the roughness length z_0, which is equivalent to the height in m at which the wind speed is 0 m/s. Table 3.5.1 shows some values for α and z_0.

For yield estimation, it is relevant to know the distribution of wind speeds, for example, over a year. This is important because very high wind speeds are equivalent to very high powers available in the wind (compare Equation 3.5.2). However, their contribution to the overall available wind energy is usually comparatively low. This is firstly because very high

Table 3.5.1 Wind shear exponent α and surface roughness lengths z_0 for different terrain types (Gipe, 2004, p. 41)

Terrain	Wind shear exponent α	Roughness lengths z_0 in m
Cut grass	0,14	0007
Hedges	0,21	0085
Trees, hedges, and a few buildings	0,29	0,3
Suburbs	0,31	0,4

wind speeds occur generally less frequently. Secondly, small wind turbines will cut out at very high wind speeds or will operate at limited power.

The Weibull distribution is applied in wind data analysis to model the probability distribution of the wind speed, especially if no time series of the wind speeds are available. The Weibull distribution can approximate a variety of different wind regimes, using only two parameters. The scale factor A is a measure of the wind speed. The shape factor k describes the form of the wind speed distribution. A special case of the Weibull distribution is when the shape factor $k = 3.4$, resulting in almost normal distribution, that is, the wind speeds would be equally distributed around the average wind speed. In reality, however, this distribution is not very common. The Rayleigh distribution is another special case of the Weibull function. It has a shape factor of $k = 2$. This distribution offers a fair approximation for many wind regimes, for example, for locations in Continental Europe. It is also used by many manufacturers of wind turbines as well as in international standards to calculate an annual reference electricity production. The Rayleigh probability distribution has the advantage that it requires only the average wind speed in order to be calculated:

$$F(V) = dV \frac{\pi}{2} \left(\frac{V_i}{V_{ave}^2} \right) \exp\left(-\frac{\pi}{2} \left(\frac{V_i}{V_{ave}} \right)^2 \right) \tag{3.5.4}$$

Where:

dV is the width; and
V_i is the average wind speed in wind speed bin i.

The annual yield or annual electricity production can simply be estimated by applying the power curve of a small wind turbine to the wind speed distribution calculated with the Rayleigh distribution. Based on the Rayleigh distribution, the annual electricity production can also be estimated without power characteristics of a wind turbine. First the total specific power density of the wind in W/m^2 can be calculated for each wind speed bin i by equation (3.5.3), without applying the Betz factor. Now, by using the Rayleigh distribution the annual specific power density and the annual wind energy density can easily be derived. Some example values are given in Table 3.5.2 for Rayleigh distributed wind speeds and standard atmospheric conditions. In the fourth column is the specific annual electricity output of a theoretical conversion rate of 20% (e.g., a constant average power coefficient of 0.2). The estimation of the annual electricity of a small wind turbine with a rotor diameter of 5 m (i.e., about 20 m^2 rotor swept area) is 2620 kWh at 4 m/s average wind speed and almost doubles to 5140 kWh at 5 m/s average wind speed.

Table 3.5.2 Estimates of the annual electricity output of small wind turbines with an overall wind energy conversion rate of 20% based on a Rayleigh wind speed distribution and standard atmospheric conditions. Adapted from Gipe (2004, p. 58)

Average annual wind speed in m/s	Specific annual wind power density in W/m^2	Specific annual wind energy density in kWh/m^2	Specific annual electricity output in kWh/m^2
4	75	657	131
5	146	1283	257
6	252	2212	442

References

AWEA (2010) *Small Wind Turbine Global Market Study*, American Wind Energy Association, Washington, DC.

Fraunhofer IWES (2011) *Small Wind Turbine Market Survey 2011*, Fraunhofer Institute for Wind Energy and Energy System Technology, Kassel, Germany.

Gasch, R. and Twele, J. (2002) *Wind Power Plants: Fundamentals, Design, Construction and Operation*, Solarpraxis, Berlin.

Gipe, P. (2004) *Wind Power: Renewbale Energy for Home, Farm, and Business*, Rev. edn, Chelsea Green, White River Junction, VT.

Hau, E. (2005) *Wind Turbines: Fundamentals, Technologies, Application, Economics*, 2nd edn, Springer, Berlin.

International Electrotechnical Commission (IEC) (2006) *International Standard IEC 61400 2:2006: Wind Turbines – Part 2: Design Requirements for Small Wind Turbines*, International Electrotechnical Commission, Geneva.

Kühn, P. (2007) Big experience with small wind turbines – 235 small wind turbines and 15 years of operational results. In European Wind Energy Conference EWEC, Milan, Italy. http://www.iset.uni-kassel.de/abt/FB-I/publication/2007-073_Big_Experience_with_Small_Wind_Turbines.pdf (accessed April 20, 2012).

Kühn, P. (2010) Small wind turbine market and performance in Germany. In European Wind Energy Conference EWEC, Warsaw, Poland. http://www.iset.uni-kassel.de/abt/FB-I/publication/2010-029_Small_wind_turbine-Paper.pdf (accessed April 20, 2012).

Manwell, J.F., McGowan, J.G., and Rogers, A.L. (2006) *Wind Energy Explained*, John Wiley & Sons Ltd, West Sussex, UK.

RenewableUK (2011) Small Wind Systems UK Market Report RenewableUK, London.

Troen, I. and Petersen, E.L. (1989) *European Wind Atlas*, Risø National Laboratory, Roskilde.

3.6 Human-Powered Energy Systems

Arjen Jansen

Delft University of Technology, Faculty of Industrial Design Engineering, Landbergstraat 15 The Netherlands

3.6.1 Introduction

Human-powered energy systems have been around since the beginning of humankind and can be seen in many forms. Section 3.6 focuses on products characterized by the presence of a technical system converting muscular work into electricity. Some examples of this type of human-powered products are the dyno-torch (a flashlight powered by squeezing it), hand-cranked flashlights, and human-powered radios. Literature provides multiple synonyms for the conversion process involved in human-powered energy systems, that is, *energy scavenging, energy harvesting, power from the people*, and *self-powering*. Human power is generally

Figure 3.6.1   From left to right: the Philips dyno torch, Seiko kinetic watch, and BayGen Freeplay radio

defined as a *nonconventional power source* and is seen as a battery alternative for electrical products.

The first human-powered product (as defined within the boundaries of this section), the human-powered flashlight or dyno-torch, was produced by Philips as early as the pre–World War II era. Although the design and engineering have evolved, similar products are produced even today. Traditionally, these products were aimed at use scenarios where the user wanted to be independent from the grid. In 1988 the first human-powered quartz wristwatch was marketed by Seiko (Auto-Quartz, later named Seiko Kinetic). A next generation of human-powered products was introduced by the BayGen company in 1995, including the BayGen radio (Figure 3.6.1).

Product characteristics or unique selling points, as used in the sales descriptions of human-powered products, are as follows:

- Environmental benefit or eco-friendliness; human-powered products are widely perceived as more sustainable than their battery-powered alternatives.
- No batteries; consumers are attracted to the idea of not having to buy batteries, although many human-powered products do contain nonreplaceable secondary batteries or capacitors.
- Independence from the grid; since human-powered products can be used everywhere and anytime, the user experiences independence from the grid.
- Emergency preparedness; due to the self-discharge behavior of batteries, products for emergency preparedness cannot rely on them.
- Freedom in choice of power source; especially with products featuring a combination of photovoltaic solar cells, human power, batteries, and grid power, the user is able to choose the power source that best fits his or her specific requirements.

Next to products in the market (see examples in Figure 3.6.2), graduation projects at Delft University (see Jansen, 2011, chap. 8) have resulted in a plethora of ideas and concepts for the use of human power. These projects started from an environmental perspective and gradually merged toward exploiting the general user benefits of human-powered products. On the Internet, numerous examples of human-powered product concepts can be found as well (Figure 3.6.3). Many of them made it no further than a computer drawing or the prototype phase; this might be due to an absence of consumer adoption, lack of a sense of engineering reality, or even lack of knowledge of basic physics principles.

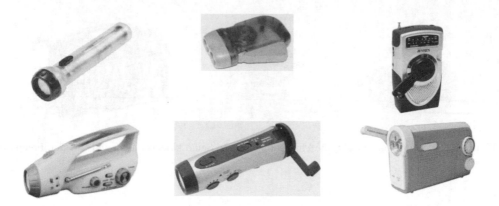

Figure 3.6.2 Examples of human-powered products from the 2010 study: Eternity Flashlights I pow-ered by shaking (top left), an Eternal Torch small squeeze torch (middle right), a Jensen MR-550 radio (top right), an XLN288s Dynamo Flashlight (bottom left), a Kaito KA404 flashlight radio (middle bot-tom), and a Kaito KA218 flashlight radio (bottom right)

3.6.2   The Human Body as a Power Source

The human body can be used in many ways to generate electric power. A first overview of this was provided in Starner (1996). It presents first estimates of the power available from exhalation, breathing bands, blood pressure, footfalls, finger motion, body heat, and arm motion. The largest power potential, however, is generated by using the muscular groups of both arms and legs to convert muscular power into electricity. Examples of various types of muscular output are provided in Table 3.6.1. Note that these values are muscular power out-put values; the efficiency of converting them into electricity is not included.

The total energy output is determined by multiplying the power output and duration of the task. The duration of the task, however, strongly influences the maximum power available from the user, as can be seen in Figure 3.6.4. As a rule of thumb, it can be stated that

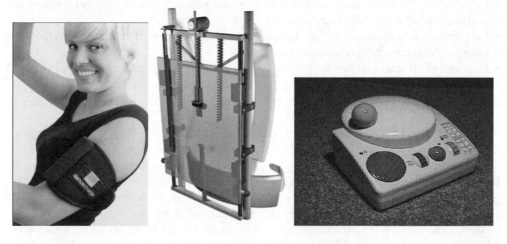

Figure 3.6.3 (Left to right): Orange dance charger, backpack charger (Rome *et al.*, 2005), and My First Wind-Up Sony (DUT student project by Yannic Dekking) (See Plate 21 for the colour figure)

Table 3.6.1 Examples of muscular output

Type of input and input variables	Muscular power output (W)
Push button with thumb (as in ball point pen): 16 N @ 8 mm travel and 2 Hz repetition rate[1]	0.56
One-hand squeeze: 400 N @ 30 mm travel and 1 Hz repetition rate[1]	12
One-hand cranking of five different radios; average power input required[2]	5.15–21.2
One-hand ergometer cranking with 175 mm crank length; sustainable cranking output for 95% of the population[3]	31
One-hand ergometer cranking with 175 mm crank length; maximum power exertion from five subjects[3]	109–168

Notes: (1) Adapted from Jansen and Stevels (1999); (2) adapted from Jansen (2004); and (3) adapted from Jansen and Slob (2003).

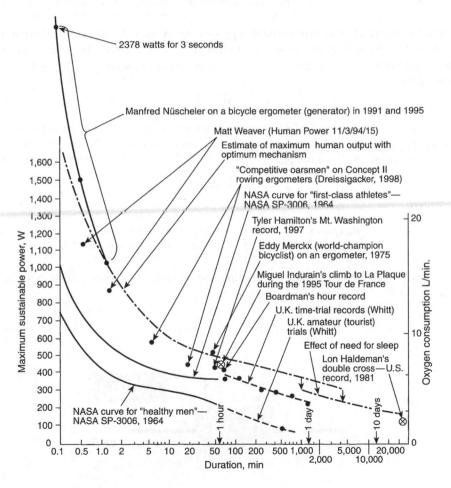

Figure 3.6.4 Human power output, principally by pedaling. Curves connect the terminations through exhaustion of constant-power tests. (Wilson, 2004)

comfortable (sustainable) power levels are – in general – 15% of the maximum power output reached during short (all-out) output bursts.

3.6.3   Kinetic Energy Formulas: From General Models to Specific Models

General descriptive models from literature only partly suffice in describing human-powered energy systems in consumer products. This section presents the results of adapting existing general models in order to apply them to human-powered energy systems.

Figure 3.6.5 presents a descriptive model (closed model) of a human-powered energy system in a consumer product, focusing at the flow of energy with specified system borders, energy input (E_{in}) and energy output (E_{out}), in which

E_{in} (Q, W)	Energy input; supplied by the human body
ΔU (Q, W)	Change in internal energy (temperature, pressure, etc.)
E_{out} (Q, W)	Energy output; electric, pneumatic, hydraulic, or mechanical work

Given the nature of human-powered energy systems, work will be performed not only *by* the system but also mainly *on* the system. In this case, W is negative. In order to make this formula applicable for a description of models of human-powered energy systems, it is rephrased into

$$E_{in} = \Delta U + E_{out} \tag{3.6.1}$$

This equation shows how the total amount of energy into the human-powered energy systems equals the change in internal energy plus the energy output of the system. The energy input can be represented in this equation:

$$E_{in} = P_{human}(t)dt \tag{3.6.2}$$

In which

E_{in} (joule)	Energy from human work
$P_{human}(t)$ (Watt or J·s^{-1})	Power from muscular work

The power from muscular work for (continuous) translational movements is described by this equation:

$$P_{human}(t) = F(t) \cdot S(t) \cdot t^{-1} \tag{3.6.3}$$

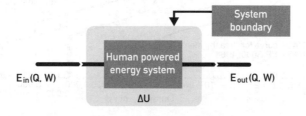

Figure 3.6.5   First law of thermodynamics for a closed system

In which

$P_{human}(t)$ (Watt or J·s^{-1})	Human power as a function of time
$F(t)$ (N)	Force exerted by the human
$S(t)$ (m)	Distance traveled

The power from muscular work for (continuous) rotational movements is described by this equation:

$$P_{human}(t) = F(t) \cdot r(t) \cdot \omega(t) \tag{3.6.4}$$

In which

$P_{human}(t)$ Watt or J·s^{-1})	Human power as a function of time
$F(t)$ (N)	Human force exerted at the crank
$r(t)$ (m)	Crank length
$\omega(t)$ (rad·s^{-1})	Angular velocity of the crank

The energy balance as presented earlier can therefore also be captured by this equation:

$$E_{product} = \eta_a \cdot W_{human} \tag{3.6.5}$$

In which

$E_{product}$ (joule)	Energy consumed by the product at time considered
η_a (%)	Efficiency of the total energy systems = the product of the efficiency of the various transformation steps
W_{human} (joule)	Energy input from the user

This equation shows that the total amount of energy provided by the human body should be equal to the amount of energy dissipated by the conversion system and the human powered product. This energy balance provides a critical condition in assessing the feasibility of a human-powered energy system.

Practically, the energy balance will consist of the difference of the integral of both the muscular power exerted by the user and an integral of the power consumption by the product, as presented in this equation:

$$\eta_a \cdot \int_{t_1=0}^{t_1=t} P_{human}(t)dt - \int_{t_2=0}^{t_2=t} P_{product}(t)dt = 0 \tag{3.6.6}$$

In which

$\eta_{a \, (a=1 \text{ to } n)}$ (%)	Efficiency of the total energy system ($E_{out}/E_{human\,in}$), consisting of n steps
P_{human} (Watt)	Power input from human muscular work
$P_{product}$ (Watt)	Power consumption of the product

The power input considered in this mathematical model is based upon active muscular power input by the user and can also be applied to "passive" power input by the user.

3.6.4   The Design of Human-Powered Energy Systems

A simple method to design a human-powered energy system is provided in this subsection. It consists of the following three steps, which will be explained in detail here:

1. Identifying opportunities.
2. System definition.
3. Feasibility check of the matching of human energy and power requirements.

Step 1: Identifying opportunities

Opportunities for the design of human-powered products are set by their unique characteristics as described here. Whenever a fit between these characteristics and the expected product's behavior exists, an opportunity is identified.

Human-powered products

- are always and everywhere available, and have a long shelf life;
- reduce the cost of ownership of products (no battery costs on the long run and lower maintenance cost because battery replacement is not required);
- reduce a product's environmental impact over its life cycle;
- can be used in drawing attention to other environmental initiatives (e.g., the Dance Floor project, (Randag, 2007)); and
- provide unique and innovative solutions.

The advantages of human-powered products are evident, although in some cases they become visible only when system boundaries are enlarged beyond the product itself. They then provide environmental benefits or cost benefits not visible on the product level alone. Specific opportunities for introducing human-powered energy systems can be found in replacing current systems that damp the motion of (parts of) consumer products.

Step 2: System definition

A draft of the human-powered product is set up by combining the product functionality and human movement and the conversion steps required. Putting this in a schematic overview helps to communicate and clarify the design proposal (see Figure 3.6.6).

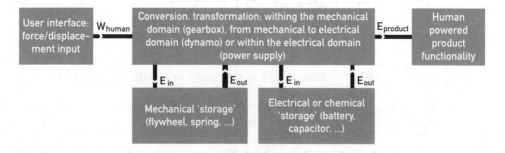

Figure 3.6.6 Product functions, human movement, and conversion steps

Defining the energy system takes a qualitative approach: The feasibility of the human-powered energy system now has to be assessed by taking a more quantitative approach. This is done in step 3.

Step 3: Feasibility check of the matching of human energy and power requirements

A first-order approximation of the total energy harvested from the human body and the energy consumption of the product will show the idea's feasibility. By definition, the input of human energy should exceed the product's power consumption times the conversion system's efficiency. This is done by quantifying the efficiency of the individual conversion steps and determining the overall energy system efficiency. Visualizing the human muscular work and the product's energy consumption by drawing a power graph helps in communicating the energy balance.

Figure 3.6.7 is an example of a power graph showing an energy balance calculation. Human power input is 10 W during 30 s (t_1). This means the total amount of human work is 300 J. When this energy is "stored" in a battery and the energy conversion and storage system has an efficiency of approximately 40%, 120 J is available for the product. This means that a product with a 1.8 W power consumption can now be powered during 66.7 s (t_2) (Figure 3.6.7).

Sankey diagrams can also be very illustrative in presenting a quantitative view on the product's conversion efficiency, as shown in Figure 3.6.8, presenting the energy system of the first BayGen Freeplay radio.

Design recommendation do's and don'ts: During the research project into human power and also as a result of 20 graduation projects at Delft University, valuable practical insights were developed. These design recommendations are fully documented in Jansen (2011), and some of them are as follows:

1. Whenever the variation of input force is very large (i.e., due to the use of body mass as input), the conversion device can be decoupled from (and therefore protected from

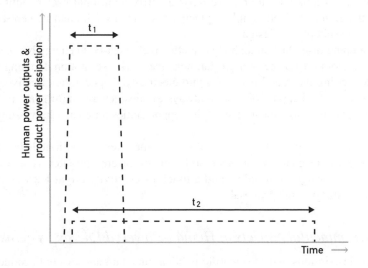

Figure 3.6.7   Power graph of human power and product consumption

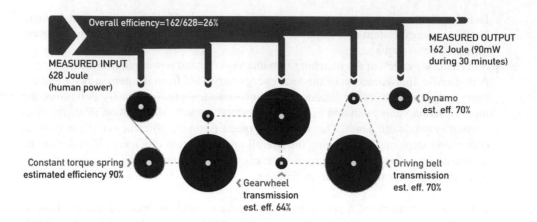

Figure 3.6.8   A Sankey diagram of the BayGen Freeplay radio energy system (the freewheel between the spring and first gear is not depicted here). The input consists of cranking 30 s at 21 W

overloading) the input by inserting an intermediate storage step between the input and conversion device (e.g., a spring). The energy stored in this intermediate storage device can be converted into electricity during its release. The combined spring–conversion system efficiency can easily be optimized due to the invariable force–displacement relation.

2. Adapt the product functionality (i.e., power requirement) to the human-powered energy source (human power) by means of direct power conversion and by omitting stand-by modes. This reduces the need for energy storage and increases the total conversion efficiency.

3. Batteries are relatively cheap, are widely available, have a high energy density, and can be easily replaced or recharged without specific user knowledge. So do not implement human power as a battery alternative *only*; identify additional user benefits. Design or redesign the entire product; simply replacing batteries by a human-powered energy system leads to a suboptimal design.

4. Motion damping from the human body can effectively be used in energy harvesting *without* additional discomfort for the user. So far, negative work like motion damping was applied only in converting the users' heel drop into electricity. This can be effectively applied in other ways as well; damping of motion always implies generation of heat, and identifying motion damping therefore means identifying opportunities for energy harvesting.

Please be aware, though, that only following the steps mentioned in this section unfortunately does not guarantee a successful human-powered product. A sparkling mix of curiosity, physics, design knowledge, and a touch of creativity are at least as important to develop a successful product proposal.

3.6.5   Environmental Aspects of Human-Powered Energy Systems

Both in sales descriptions and in scientific publications, human-powered products are positioned as being environmentally beneficial. This is mainly due to the fact that not using batteries is interpreted as an environmental benefit. In this subsection, we will discuss this assumption.

On one hand, the use of a human-powered energy system reduces a product's environmental impact due to the fact that no electrical power from batteries or the grid is consumed. On the other hand, the additional components used in the human-powered energy system cause an additional environmental impact. Balancing both these effects is the key to answering the question of whether a human-powered product is really environmentally beneficial. This is best illustrated by using the Lifetime Design Strategies (LiDS) wheel (Brezet and van Hemel, 1997; see Figure 3.6.9), which provides a qualitative view on the environmental effects of various design strategies.

Introducing a human-powered energy system means additional use of materials like copper, magnets, gears, and so on, and thus lowers the score on *reduction of materials* and *low eco-impact production*, the second and third categories of the LiDS wheel. The environmental benefit is depicted in the fifth category, *low eco-impact use*.

Besides the straightforward LiDS approach, the answer to the question of whether human power is beneficial to the environment also has a second dimension, namely, where one should draw the system boundaries? In conventional life cycle assessment (LCA) studies based on the ISO 14000 standard, the environmental impact of human power is set to zero. One could argue whether this agreement is still valid when human power is used as a battery alternative. The energy required for growing, transporting, cooling, and displaying food is considerable, and a conversion efficiency of food to muscular power of maximum 20% makes the discussion even more difficult. Including the environmental impact of the supply chain for human food into the system boundaries significantly changes the outcomes of the

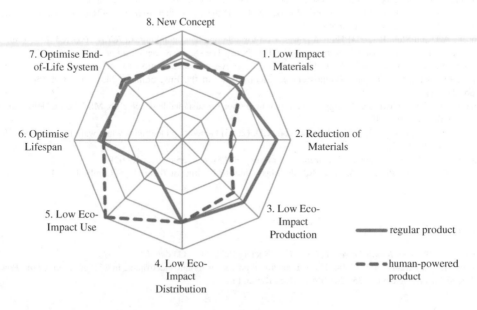

Figure 3.6.9 The Lifetime Design Strategies (LiDS) (Brezet and van Hemel, 1997) applied to human-powered products with respect to regular products

environmental analysis. However, expanding the system boundaries beyond the product might lead to the discovery that calculating the environmental advantages is worthwhile. This can best be illustrated by the example of a smart door-closer system in which the closing motion of the door is damped by means of an electromagnetic system. This system converts the closing motion of the door into electricity, which is then stored in a set of batteries. These batteries can power a wireless communication system, saving the costs of installing wiring to fire doors in larger buildings (Biermann, 2001).

Even in the case of a product with significantly lower environmental performance, still the spin-off may be beneficial for the overall environmental performance. This can be seen in the design of the energy-converting dance floor. This floor was originally designed by Anouk Randag for the Sustainable Dance Club based in Rotterdam (Randag, 2007). The dance floor played a significant role in communicating the objectives of the Sustainable Dance Club by introducing the concept of sustainability to a broader audience; although the environmental performance of the product itself shows no improvement, the product had a noticeable function in communicating sustainability on a different level.

References

Biermann, E. (2001). *Design of an intelligent door-spring system*. MSc Thesis, Faculty of IDE, Delft University of Technology, Delft

Brezet, J.C. and van Hemel, C.G. (1997) *EcoDesign: A Promising Approach to Sustainable Production and Consumption*, UNEP, Paris.

Jansen, A.J. (2004) Advances in human-powered energy systems in consumer products, in D. Marjanovic (ed.), International Design Conference, Design 2004, pp. 1539–1544, University of Zagreb, Zagreb.

Jansen, A.J. (2011) Human power empirically explored. PhD thesis, Product Engineering Research Group Publication 3, Delft University of Technology, Delft.

Jansen, A.J. and Slob, P. (2003) Human power: Comfortable one-hand cranking. In *ICED 2003* (ed. M. Norell, S. Andersson, H. Johannesson, L. Karlsson, and J. Palmberg), The Design Society, Linkoping, pp. 661–670.

Jansen, A.J. and Stevels, A.L.N. (1999) Human power: A sustainable option for electronics. In Proceedings of 1999 IEEE International Symposium on Electronics and the Environment, May 11–13, 1999, Boston, pp. 215–218.

Randag, A. (2007) FluxFloor: Energy-converting dance floor for sustainable dance club. MSc thesis, Delft University of Technology, Delft.

Rome, L.C., Flynn, L., Goldman, E.M., and Yoo, T.D. (2005) Generating electricity while walking with loads. *Science*, **309**, 1725–1728.

Starner, T. (1996) Human-powered wearable computing. *IBM Syst. J.*, **35** (3–4), 618–629.

Wilson, D.G. (2004) *Bicycling Science*, 3rd edn, Massachusetts Institute of Technology, Cambridge, MA.

Further Reading

Dean, T. (2008) *The Human-Powered Home*, New Society Publishers, Golden, CO.

Starner, T. and Paradiso, J.A. (2004) Human generated power for mobile electronics, in C. Piguet (ed.), *Low-Power Electronics Design*, pp. 1–35, CRC Press, Boca Raton, FL.

3.7 Energy-Saving Lighting

Arjan de Winter

Philips Lighting, Mathildelaan 1, Eindhoven, The Netherlands

3.7.1 Energy-Saving Lighting

The invention of the incandescent bulb over a century ago by Edison (in 1879) and the consequent introduction of electrical lighting have meant a huge change in people's life. It not only provides safe illumination, as there is no more open fire of the candles and gas lighting, but also provides an increased convenience, as it can be switched on by a wall switch instead of lighting each individual light point. With the use of electric light, electricity is consumed. With high penetration rates for both indoor and outdoor applications, this consumption has become very considerable. For example, according to the Next Generation Lighting Industry Alliance (NGLIA), lighting accounts for approximately 20% of current North American electricity consumption. Consequently, there has been an ongoing attempt to reduce power consumption for decades.

The attempt for energy reduction started to gain momentum with the higher penetration of electrified light after the Second World War, with fluorescent sources replacing incandescent sources. For higher power ranges, metal halide sources have taken up market share in street lighting and outdoor architecture. More recently the light quality of metal halide has improved to such a level that it has found solid ground in retail lighting. As energy consumption is such a dominant cost factor, most professional light sources are sold based on a total cost of ownership (TCO) calculation, off-setting the higher lamp and fixture costs of energy-efficient sources against the benefits of lower energy consumption and therefore lower operational cost. Next to that, lamp lifetime is an important cost of operation as some lamps, like incandescent, only live for 1000 hours. Though current light emitting diode LED light sources may life up to 50,000 hours, the payback is typically calculated on a 1- to 3-year basis.

In this section, we will go in more detail into the opportunities that are explored to reduce the energy consumption related to lighting. The figures in Table 3.7.1 give an estimate of the cost associated with lighting energy consumption and an indication of the saving potential of a switch to more efficient light sources.

For a typical fashion shop with typically 15 W/m^2 energy consumption and a surface of 200 m^2, one would require 60 of these 50 W incandescent fixtures, resulting in an average annual energy consumption cost of approximately €1,200. An energy consumption reduction of 20%, would lead to a €720 savings based on a 3-year payback. The switch to LED bulbs, as indicated in the table, would reduce the €1,200 energy bill to a mere €310 per year, allowing an initial investment of €2,670, or €45 per fixture. Alternatively, one could choose to use the efficient light source to obtain much higher light levels at the same installed power of 15 W/m^2.

However, as always in energy usage reduction, we have to balance the benefit of energy saving with consequent drawbacks in the application. Therefore in this section we'll focus in the basics of lighting in a number of typical application areas and typical light sources.

Table 3.7.1 Typical energy consumption and operating cost for a 500 lm incandescent and LED fixture for a 3-year payback period at an energy price level of €0.11/KWh (industrial tariff)

		Annual operating hours	Annual energy consumption	Energy cost per year	Energy cost over payback period
50 W/500 lm incandescent bulb	Light usage	1500 h	75 KWh	€8.2	€25
	Average usage	3500 h	175 KWh	€19.3	€58
	24/7 usage	8760 h	438 KWh	€48.2	€145
10 W/500 lm LED bulb	Light usage	1500 h	15 KWh	€1.7	€5
	Average usage	3500 h	35 KWh	€3.9	€15
	24/7 usage	8760 h	88 KWh	€9.7	€29

3.7.2   Lighting Applications

The light output of the fixture itself is typically measured in lumen (lm). Lumen is the total energy that a light source emits across the visible wavelength (luminous flux), corrected for the eye sensitivity, and depicted in the human photopic sensitivity curve (Figure 3.7.1). As the eye is more sensitive to green light, this part of the visible spectrum is taken more into account than, for example, blue or red light. The luminous flux is part of the total energy that is emitted by the light source over all wave lengths (radiant flux), which includes, for example, UV and IR wavelength energy. For example, incandescent sources radiate approximately 8–10 times more IR than visible light.

The majority of lighting fixtures emits light that reaches the eye through reflection on other surfaces. This is called *illuminance*. The light intensity is characterized by the light falling on a surface area, and is measured in lux.

Next to illumination, some fixtures are used for "light to look into", or direct-view fixture like large screens. Their light output is referred to as *luminance* and is measured in "candelas per square meter," or nits. As these are often used for nonfunctional or decorative applications, these are further excluded in this section.

In office environments (see Figure 7.3.2), the light is provided as part of the building, generally with limited individual possibilities to adapt to personal taste or liking. Therefore offices are in many countries subject to norms and regulations, with regard to among others minimal lux levels and glare (blinding effect of light directly reaching the eye). Typical light levels for an open-plan office are 300 lux, with 500 lux at the desk.

In hospitality environments (see Figure 7.3.2), lighting is a dominant factor in creating an atmosphere or moods, with dimmed warm light in bars or restaurants or chandelier-like fixtures in lobbies. Flexibility of lighting is important to allow meeting and banquet rooms, for example, to be used as meeting rooms during the day and as wedding venues at night.

In retail environments (see Figure 7.3.2), light not only is needed to see the environment but also should "sell the products" (e.g., the nice red meat of the butcher) or represent daylight conditions (e.g., for a dress in the latest fashion). Next to that, the lighting scheme

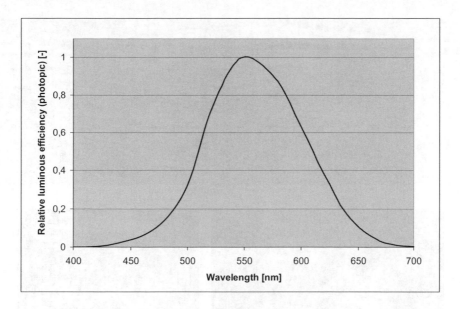

Figure 3.7.1   Relative luminous efficiency (photopic) as a function of wavelength

might be part of the "brand identity," whether "cheap" with many cool white linear fluorescent tubes, or "high-end" with high color-rendering accent lighting.

In outdoor lighting (see Figure 7.3.2), functional lighting of the road is a main application area, but in city centers, many lighting applications are also part of the city beautification scheme and contribute to the city identity with designer poles and illumination of historic and public buildings.

In the consumer environment (see Figure 7.3.2), typically other buying criteria are used. With very little legislation in the application itself, initial price is an even bigger criterion than in professional segments. Though over the last decade the penetration in homes of power-saving or compact self-ballasted fluorescent lamps (CFL-i) has increased due to substantial cost price reductions, the drawbacks of CFL-i (with regard to dimming and the diffuse light and less-than-perfect color-rendering properties) continue to give space to (eco-) halogen and incandescent lamps. In some countries, like the European Union and the United

Figure 3.7.2   Examples of lighting environments suitable for LED technology

Figure 3.7.2   (*Continued*)

States, a recent ban on higher power incandescent lighting has been implemented giving a further boost to power-saving lamps, but for many consumers these energy-efficient alternatives are not acceptable alternatives.

In conclusion, different applications pose very different requirements toward lighting. Therefore, the optimization of energy usage requires different trade-offs for different applications. As these trade-offs have to be made at an application level, the system boundaries should not be limited to the lamp or light-emitting module itself, but should also include the luminaire and even the space or building(s) around it.

In the (LED) lighting industry, these different systems levels (Figure 3.7.3) are referred to as *level 1* for the LED diode, *level 2* for the LED module, and *level 3* for the LED module and

Level 1	Level 2	Level 3	Level 4	Level 5	Level 6

Figure 3.7.3   System levels as defined for LED lighting systems (See Plate 22 for the colour figure)

electronics. *Level 4* represents the luminaire including optics, and *level 5* represents the application.

In some case, *level 6* is referred to as services that can be provided to end users and not so much a physical representation of the system. As such it relates more to the (business) model in which light is sold, maintained, and upgraded. Level 6 is outside the scope of this section.

In practice the situation is even more complex, as for one application different decision-making units (DMUs) decide on their part of the value chain. For example the building owner decides on the initial requirements, but the architect does the design and eventually the installer decides on the actual products purchased. The light designer creates the light plan, but the space user eventually decides on the actual usage. Therefore the systems should be optimized for the different applications and stakeholders over the value chain, leading to multiple and often conflicting requirements.

As LEDs are expected to be the dominant light sources of the future, this section will mainly focus on LED sources, with reference to conventional sources.

3.7.3   Light Source Design: Efficacy

In this section the basics design parameters impacting energy consumption of LED modules are discussed. LED modules (level 2) consist of the light-emitting diode (level 1) and primary optics, and are assembled on a board that provides some electronic functions and electrical and mechanical interconnects (see Figure 3.7.3). On these levels, *efficacy* is used to indicate the amount of light generated per entity of consumed energy. This is expressed as lm/W. Typical efficacies of conventional light sources are 10 lm/W for incandescent, 25 lm/W for halogen lamps, 60–80 lm/W for fluorescent, and 90–100 lm/W for metal halide. These figures are purely indicative, as they can vary considerably per brand, type, power level, and so on.

The efficacy of LEDs has made enormous improvements over the last decades and is expected to further increase in the years to come, eventually becoming superior to other light sources (see Figure 3.7.4). The figure represents an average level of efficacy; as in the design of the LED module, many trade-offs can be made not just between efficacy and cost, but also with respect to color consistency, color rendering, reliability and lifetime, quality of light, and so on.

For example, reducing the current through the LEDs can lead to a 20–30% higher efficacy, but requires more LEDs, resulting in a more expensive device (see Figure 3.7.5).

Another technology option to improve efficacy in white-light applications, like down lights, is the usage of so-called remote phosphor. In this technology, the phosphors used to convert the blue light of the LEDs into white light are not applied directly on the LEDs itself, but on a plate on some distance from the LEDs. This reduces the thermal load on the LEDs itself and so keeps the LEDs in a more efficient low-temperature regime.

A third example of design choices impacting the efficacy is the use of red, blue, and off-white or lime-like LEDs, instead of phosphor-converted LEDs. Each of these LEDs has a high efficacy and as the three together make white light, there is no need for phosphor with associated stoke shift losses. A disadvantage is that additional electronics are needed for

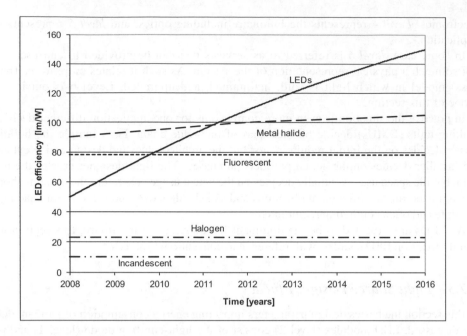

Figure 3.7.4 (Predicted) generic efficacy development of LEDs over time and a reverence to conventional light sources

such a three-channel system and also for a temperature feedback loop to compensate for the temperature dependency of the red LEDs.

The improved efficacy will lead to a massive change-over to LED luminaires. The LED penetration is further accelerated by the rapidly lowering cost of LEDs, reducing the initial cost threshold. The US Department of Energy predicts the cost level of LED luminaires in 2015 to be at 20% of the cost level of 2010.

As LEDs are, unlike the conventional light sources, electronic components, this change-over is also the key enabler for the digitalization of lighting. This digitalization not only changes the design of LED drivers and lighting controls, but also is enabled by the changes of the light source itself.

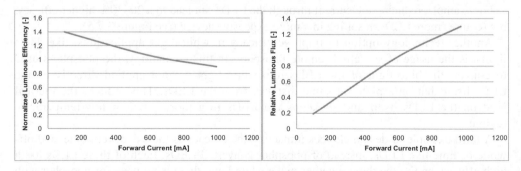

Figure 3.7.5 Typical luminous efficacy and flux characteristics as functions of the forward current of a LED package

3.7.4 Luminaire Design: Optical and Electrical Efficiency

On the next higher system levels, one or more LED modules are connected to a driver (level 3) and assembled into a luminaire or fixture (level 4). See Figure 3.7.3. Apart from the housing and mounting to the outside world, the luminaire also contains secondary optics. For example, in a spot, the secondary optic can be a reflector that creates the beam with a specified beam angle (e.g., a 10° narrow beam or a 60° wide beam). A faceted reflector structure also compensates for inhomogeneities of the LED source. Both driver and optics introduce energy losses impacting the overall efficiency. As with all systems, the overall system efficiency is a multiplication of efficiencies of individual stages in the system. Figure 3.7.6 shows an example of the efficiency buildup of a typical LED luminaire system.

LEDs are current-driven devices; therefore, they typically need a driver like fluorescent and metal halide light sources. In LED luminaires, this typically is a separate driver that converts mains voltage in a current of 350 mA, 700 mA, or 1A depending on how hard the LEDs are driven. Efficiencies of these drivers range between 80% and 95%, with 85% being typical. In some case, for example with multistring RGB–white configurations, the power conversion is separated from the driver itself, with a power supply unit (PSU) and smaller (buck) converters on the LED PCB or close to the LEDs. This for example, allows for pulse width modulation (PWM) dimming of the LEDs, color compensation, and temperature feedback loops on the LED boards. This power configuration, however, introduces an additional conversion step and therewith an additional 5–10% efficiency loss.

In some cases the driver is integrated with the LED board. This increases the thermal load of the driver, thus affecting the lifetime of the device. This, however, allows for a specific optimization of LED configuration and driver. Similar optimizations take place for LED lamps, though in this area size limitations play an important role to keep them within the given mechanical boundaries of the conventional lamps. Next to an optimization toward efficiency, the electronic design has to fulfill requirements with regard to power factor correction and dim-ability.

Not all the light that is generated by the lamp or LED module exits the luminaire. For example, some fluorescent down lights have reported efficiencies in the order of 50%, as

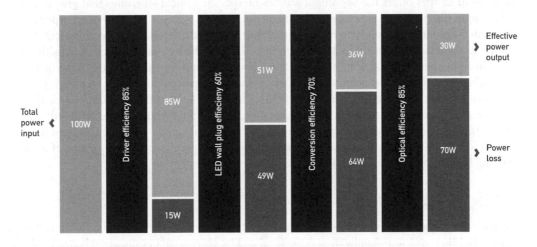

Figure 3.7.6 Efficiency buildup for a typical LED luminaire

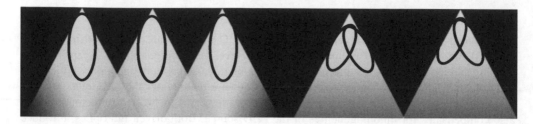

Figure 3.7.7 LEDs allow for further optical design optimization due to a small-sized light source, as in this example the use of "bat wing"–like optics (right side) to get more even light distribution at street level and a consequent reduction of pole numbers to fulfill requirements

only 430 lm exits the fixture of the 860 lm generated by the lamps. The losses occur as part of the light is blocked by the luminaire or is reabsorbed by the lamp. As the LED module design is often closely linked to the design of the LED luminaires, there are more opportunities for optimization. LED luminaire efficiencies can be in the order of 80–95%.

Though the optical efficiency of the luminaire can be calculated based on the lumen exiting the lamp or module compare to the lumen output of the luminaire, the optical efficiency of the luminaire also depends on the application and how efficient the (set of) luminaires is able to illuminate the environment.

Good examples are cove lighting applications, which are typically done based on linear fluorescent lamps. These lamps give an omnidirectional light pattern of which only one-quarter to one-third is effectively used to light up the cove walls. LED cove lights are typically designed to give directional light, only lighting the required area. This can lead to substantial efficiency improvements in the application.

Another example is street lighting. Traditional street poles have a light concentration directly under the pole, with the light intensity rapidly decreasing for distances further away from the pole. Consequently, poles have to be placed close to each other to reach the required minimal light flux at the ground surface (see Figure 3.7.7, left). The LEDs characteristic of a small light emitter offers the opportunity to design specific optical structures that give a "bat wing" light distribution, spreading the light equally over a larger surface, and exactly putting the light where it is needed, also reducing the light spill to the street sides (see Figure 3.7.7, right). In some applications also greenish light is used, as the eye has the highest sensitivity for that light spectrum, allowing the lowest energy consumption. However, this kind of far-stretching minimization of energy consumption also has reported downsides. As the light is only projected at the street itself, people miss the "spilled" light of the omnidirectional conventional light source that conveniently lights the entrance door to the house, or that illuminates strangers passing by and thus gives a general feeling of safety. The greenish light has a very low color rendering. Consequently, people can no longer distinguish the color of a car passing by or get scared by the color of one's skin.

3.7.5   Application Design: Effectiveness

At the next higher system level, numbered 5 in Figure 3.7.3, different luminaires together form a lighting system or application. This level can be seen as an optimization based on the effective use of light, when and where it is needed. This is done by adding intelligence to the system. Lighting controls with functionality beyond *on* or *off* can be installed for different reasons. Though energy saving is a main driver across applications, ambience setting and monitoring system performance are other common motivations.

On a system level, "lighting controls" are added to the systems. In spaces like an open office space, hallway, or living room, there is typically a "line of sight" between luminaires mutually and between users and luminaires. In these environments typically switches, dimmers, sensors, and so on are added to the lighting fixtures. In many cases this is done in with application-specific knowledge captured in the control devices. Examples include algorithms that translate daylight and presence information back to dim levels of lighting fixture close to windows (light) and those more internal (dark).

Typical office implementations, for example, are presence sensors added to the lighting system,to make sure that the lights are switched off when a space is vacated. Though simple in principle, it is a challenge to get this robust at low-cost levels as simple detectors typically detect movement and not presence, leading to arm waving to keep the light switched on. Another disadvantage is that people do not like to sit in a dark environment, even if their own desk is lighted, although these measures can lead to up to 15–25% energy reduction. To stretch savings further, daylight sensing can be added to compensate for daylight entering the space from the outside, in some cases even combined with blinds controls.

These control devices can also be linked between each other and a central control point for situations with a combination of multiple application spaces like a (number of) building(s) or shop(s). This kind of applications comes close to building or home automation systems, as lighting is one of the parameters that are to be controlled next to heating, HVAC, fire safety, telecommunications, blinds, and so on. However, lighting systems are often a separate subsystem, as specific requirements are posed due to dimming or color controls.

These larger control systems allow energy consumption monitoring, benchmarking, maintenance, and scene-setting control. In retail environments, for example, there are many chains that own hundreds or even thousands of shops across the world. Their lighting systems not only are responsible for part of their operational cost, but also lighting "shows and sells their merchandise," and it is a recognizable element of their brand identity and consistent brand presentation to the outside world. Therefore, there is a large need to control the light and make sure it is in line with the lighting design across multiple locations. Moreover, most of these locations are run by very junior and often changing staff that is not familiar with control systems. In these applications, more central-driven systems are used. These systems, for example, allow scheduling with half-light levels before and after opening, when the staff is cleaning and filling the shelves, and full operation during opening times. Other functionality at this level is, for example, benchmarking different locations and energy performance monitoring over time.

Adding lighting controls, beyond simple switches and dimmers, substantially adds cost to the total system, not primarily as a consequence of the hardware cost, as those can be limited to 10–20% of the total lighting installation, but mainly due to the additional cost of installation and commissioning. Especially for larger systems, commissioning can be a substantial effort. This has to be taken into account when calculating total cost of ownership (TCO).

Lighting control systems are currently in rapid development due to the fast increase in calculating power in lower devices like digital drivers and digital LED modules. Another development is the introduction of IP-based networks for lighting control. IP-based communication to lighting systems will stepwise reach down from system-level controls to room controllers and eventually even to the luminaire level. This will be the next generation of lighting controls and a key enabler for the integration of intelligence and flexibility in lighting systems, and a further enabler for, for example, load shedding.

3.7.6   Conclusions and Looking Forward

Lighting makes a substantial contribution to the world's energy consumption and presents an enormous potential for energy saving. In the design of lighting systems, design choices can be made on different system levels that impact the energy consumption of the overall system.

At the lower system level, efficacy (lm/W) is the key measure. At the luminaire level, the focus is on the efficiency of electronic drivers and optics. At systems level, the effective use of the light is crucial, applying light only where and when it is needed.

For all design choices, the benefits of the reduction of energy consumption have to be offset against the trade-offs that have to be made in the application. These will vary depending on the situation and region.

Looking forward, the ongoing LEDification, digitalization, and further penetration of IP into lighting systems offer many new opportunities to be explored in the years to come.

Further Reading

Light-emitting diodes. www.lightemittingdiodes.org (accessed April 20, 2012).
LED lighting explained. www.ledlightingexplained.com (accessed April 20, 2012).
Future lighting solutions: Making LED lighting solutions simple. www.futurelightingsolutions.com (accessed April 20, 2012).
Philips Lumileds LED technology. www.philipslumileds.com/technology (accessed April 20, 2012).
Schubert, E.F. (2006) *Light Emitting Diodes*, Cambridge University Press, Cambridge.

3.8   Energy-Saving Technologies in the Built Environment

Bram Entrop
University of Twente, Faculty of Engineering Technology, Department of Construction Management and Engineering, The Netherlands

3.8.1   Design and Energy Use in the Built Environment

In the European Union, the energy use of the built environment is believed to be more than 40% of the total energy use (EC, 2002). Due to multiple negative environmental effects of this energy use, reducing energy use and making use of renewable energy sources receive a lot of attention. In many countries the energy use of buildings is being reduced through construction regulations. As well as that, the energy performance of buildings is made transparent by energy labels.

It is in this context that this section gives an overview of energy-saving technologies that have been, are being, or presumably will be adopted in the built environment. First the role of energy consumption will be elaborated on by giving a framework for particular energy-saving technologies.

The science involved in the energy use of buildings, ranging from houses to offices and from libraries to hospitals, is building physics. The provision of a certain level of human comfort forms the basis for building physics and for building-related energy use. Users of buildings – for example, residents and employees – want to feel comfortable in a building that gives them shelter from the outdoor climate. Therefore, it is necessary to provide them light, ventilation, low levels of noise, a certain indoor temperature, and humidity. In this section, the focus will be on the constructional and system technologies that help to provide and maintain a comfortable indoor temperature against relatively low environmental costs. Although also much electric energy is used in buildings, the focus in this section is in that regard on thermal energy. For specific information about electrical energy, we refer to the other sections of this chapter and the other chapters of this book.

To come to the adoption of energy-saving technologies in the built environment, a three-step design principle, called *Trias Energetica*, can be used (Entrop and Brouwers, 2010). The Trias Energetica states the following:

Step 1: One first needs to reduce the need for energy as much as possible.
Step 2: When energy is needed, one needs to make use of renewable sources as much as possible.
Step 3: When energy is still needed, one can turn to fossil fuels, using them as efficiently as possible.

This three-step principle has already been known for many years among many Dutch architects, principals, project developers, and construction engineers. Application of the principle is indirectly enforced by the Building Code, which demands that buildings have a minimum level of insulation, a minimum ventilation rate, a maximum infiltration rate, and a minimum energy performance (Entrop, Brouwers, and Reinders, 2010). A dedicated collection of construction technologies and system technologies needs to be applied to a building design in order to meet this energy performance. In this sense, in this section various technologies are explained.

In Section 3.8.2 different construction technologies – often reflecting on the structure and building envelope of buildings – will be explained. In Section 3.8.2 system technologies – often reflected in the interior of buildings – are presented. Besides these construction and system technologies, also some other technologies exist to reduce the energy use of the built environment. These technologies exceeding the envelope of a building are mentioned in Section 3.8.2 and are, in this section, addressed as *transcendental technologies*.

3.8.2 Construction Technologies

Construction technologies form a category of energy-saving technologies that is strongly related to the building envelope. These technologies reflect the structure and the shell of buildings. When there is no match between outdoor temperatures and indoor temperatures that is regarded as comfortable, the structure of this shell needs preferably to offer some resistance to heat transfer. The basic principles by which heat can be transferred are as follows:

Conduction: The transfer of thermal energy through a material. In general, metals are good
thermal conductors.

Convection: The transfer of thermal energy by means of moving molecules in fluids.

Radiation: The transfer of thermal energy by means of electromagnetic waves. All materials
with a temperature above absolute zero emit thermal radiation.

Mass transfer: The transfer of thermal energy by means of a physical transfer of solid matter.

Heat Transmission

Buildings are constructed to provide shelter. Land plots in different areas are subject to different climates. When solar irradiation is high and outdoor temperatures are high, one
expects a building to offer shading and relatively low temperatures. When wind speeds are
high and outdoor temperatures are low, one expects a building to offer relatively high temperatures and protection from the wind. The need to convert fossil fuels (or other sources) in
heat depends on how well a building can keep the heat inside. In a warm climate, it is important to keep the heat outside. The heat transmission through the building shell depends on its
heat resistance, R (m² K/W). This heat resistance is the reciprocal of the overall thermal
transmittance through a particular part of a construction, U (W/m² K).

Often the shell consists of multiple layers of materials. A wall, for example (see
Table 3.8.1 and Figure 3.8.1), can consist of sand–lime bricks (100 mm), insulation

Table 3.8.1 The cross-section of a wall and its thermal resistance

	Thickness	Thermal transmittance or resistivity
R_{si} internal surface	—	0.13 m²·K/W
R_1 plaster	2 mm	±0.45 W/m·K
R_2 sand–lime blocks	100 mm	±1.15 W/m·K
R_3 insulation	120 mm	±0.04 W/m·K
R_{cavity} cavity	40 mm	0.18 m²·K/W
R_4 brickwork	100 mm	±0.85 W/m·K
R_{se} external surface	—	0.04 m²·K/W

		Thickness	Thermal transmittance/ resistivity
R_{si}	Internal surface	-	0.13 m²·K/W
R_1	Plaster	2 mm	± 0.45 W/m·K
R_2	Sand-lime blocks	100 mm	± 1.15 W/m·K
R_3	Insulation	120 mm	± 0.04 W/m·K
R_{cavity}	Cavity	40 mm	0.18 m²·K/W
R_4	Brickwork	100 mm	± 0.85 W/m·K
R_{se}	External surface	-	0.04 m²·K/W

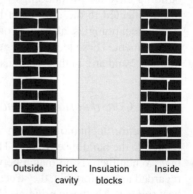

Outside   Brick   Insulation   Inside
          cavity   blocks

Figure 3.8.1   The cross-section of a wall and its thermal resistance

(120 mm), cavity (40 mm), and bricks (100 mm). The total heat resistance R_{total} is determined by the thermal conductivities represented by λ in (W/m K) over the thickness of these individual materials, the heat resistance at the surfaces of the materials, and the heat resistance of the airspace.

$$R_{total} = \sum R_{individual\ construction\ elements} = R_{si} + R_1 + R_2 + R_3 + R_{cavity} + R_4 + R_x + R_{se}$$
$$R_{individual\ construction\ element} = l/\lambda \tag{3.8.1}$$

Where:

$R_{1\ ...\ x}$ = heat resistance of a certain uniform construction material (m^2 K/W);
R_{si} = heat resistance at the internal surface of a construction (m^2 K/W);[1]
R_{cavity} = heat resistance inside an air cavity in a construction (m^2 K/W);[2]
R_{se} = heat resistance at the external surface of a construction (m^2 K/W);[3]
l = length or thickness of the material (m); and
λ = thermal conductivity of the uniform material (W/m K).

In the Netherlands, a minimum heat resistance R_{total}–R_{si}–R_{se} of the construction materials (including air cavity) of 2.5 m^2 K/W is compulsory, but in new dwellings heat resistances of 3.5–5.0 m^2 K/W are already often achieved. When constructing passive houses (with a low annual energy use of 15 kWh/m^2 at maximum) in a moderate climate, a heat resistance R_{total} of 10 m^2 K/W or more is needed.

Increasing the thickness of insulation materials (see Figure 3.8.2) is one way to improve thermal resistance. Passive houses make use of insulation packages with a thickness of up to 400 mm. However, the distance between internal and external brickwork needs to be carefully considered in these cases. When the distance is that large, losses occur regarding strength and stability. New insulation materials, based on the aerospace industry, have been introduced that focus on not only blocking thermal conduction but also blocking radiation. These insulation materials consist of multiple layers of synthetic or metal foils. Although some debate exists regarding the effectiveness of these products, one can imagine that it is important to reflect on blocking both thermal conduction and radiation.

Air Tightness

A well-insulated building cannot reach maximum energy efficiency without addressing its air tightness. Different construction components – for example, walls, roofs, doors, and window frames – need a tight fit that disables heat transfer by air through gaps and cracks. The proper term for this aspect in designing construction is *infiltration*. Current infiltration rates are significantly reduced by paying much attention to construction details on the connections between different building components. Heat losses due to infiltration are being reduced by applying door frames and windows frames with double rubber sealing. The connections

---

[1] According to Dutch standards, $R_{si} = 0.13\ m^2$ K/W for a vertical wall being part of the thermal shell (NNI, 2001).
[2] According to Dutch standards, $R_{cavity} = 0.18\ m^2$ K/W for a heat flow in an (almost) unventilated cavity (NNI, 2002).
[3] According to Dutch standards, $R_{se} = 0.04\ m^2$ K/W for a vertical wall being part of the thermal shell (NNI, 2001).

Figure 3.8.2 Roof construction of a house built in 1982 with only 30 mm of insulation under the roof tiles

between walls and these frames or between the walls and roof are made airtight by using, for example, expanded polyurethane and sealant. The infiltration is often expressed in $m^3/h\ m^2$, often measured at a standard pressure difference of 50 Pa between inside and outside the building. Of course, the final heat losses due to infiltration also depend on the temperature difference between the air inside and outside the building.

Solar Transmittance

Transparent surfaces in the building shell make it possible for solar irradiation to directly enter the building. In this case not only is the mentioned thermal transmittance U (W/m^2 K) or heat resistance R (m^2 K/W) an important aspect, but so also is the level of solar transmittance, expressed by a so-called *g-value* (−), and visible light transmittance, expressed by a *VLT value* (−). All three variables can be used to design windows that fit the future use of a building (see Table 3.8.2). In an office building in a moderate climate, the internal heat load requests windows that keep as much solar irradiation out as possible during daytime. In houses one can benefit from heat due to solar irradiation during approximately three seasons per year. On hot summer days, shading helps preventing internal overheating. The

Table 3.8.2 Overview of some standard values of different types of glazing for dwellings (NEN, 2004)

Type of glazing	U value (W/m^2 K)	g value (−)	VLT value (−)
Single glazing	±5.8	±80%	±90%
Double glazing	±2.8	±70%	±80%
High-efficiency double glazing	±1.2	±60%	±70%
Triple glazing	±0.6	±50%	±60%

application of shadings responding automatically to the weather conditions is especially relevant when buildings are well insulated.

Heat Capacity

Due to the use of concrete and brickwork, buildings are in general heavyweight constructions offering large heat capacities. Temporary buildings are often lightweight constructions. Well known in this category are caravans and so-called portacabins, in which indoor temperatures can rise significantly during hot summer days. A high sensible heat capacity can be used to store heat due to solar irradiation during daytime and to lose heat during relatively cold nights.

Phase Change Materials
The latent heat capacity of a material is in general much larger than the sensible heat capacity. In the case of water, the sensible heat capacity is $4.2 \, kJ/(kg \cdot K)$. The latent heat capacity is $334 \, kJ/kg$ in the case of solidification and $2260 \, kJ/kg$ in the case of vaporization (Verkerk et al., 1992). The liquefaction of ice or evaporation of water for cooling purposes both form ancient techniques, but phase change materials (PCMs) with transition temperatures for specific purposes were introduced in the nineteenth century (Kürklü, 1998).

Multiple materials are offered that already have a melting temperature of around $18-23°C$ (Zalba et al., 2003; Baetens, Jelle, and Gustavsen, 2010), which can be considered to be around comfortable indoor temperatures. It is shown that PCMs can help reduce energy use for maintaining a comfortable indoor temperature (Entrop, Brouwers, and Reinders, 2011; Huang, Eames, and Hewitt, 2006; Sharma et al., 2009; Zalba et al., 2003), since they have the ability to store and release both sensible and latent heat. An ambient temperature below the transition temperature will ascertain that the latent heat in PCMs will be or is emitted to the surroundings. When the ambient temperature rises, the thermal energy necessary for liquefaction will be provided by the surroundings. PCMs will act as a thermal battery storing thermal energy and reducing the need for fossil fuels. Inside a construction they can reduce maximum peak temperatures and increase minimum temperatures by storing in a passive way thermal energy derived from solar irradiation (Entrop, Brouwers, and Reinders, 2011).

Green Roofs
The fact that the latent heat capacity of water is high is also a big advantage in using green roofs. Green roofs consist of a thin earth package or another substrate with vegetation (often sedum; see Figure 3.8.3). Rainwater in these roofs will partially evaporate during hot summer days, which can result in lower indoor temperatures. The substrate and vegetation will also improve the roof's heat resistance in general. Another big advantage is that after precipitation, the ru off to water streams or water treatment facilities will show smaller peaks when large surfaces of green roofs are applied.

3.8.3 System Technologies

The multiple varieties of components making up the building shell have been addressed. In this section, the systems are explained that can provide energy to and withdraw energy from

Figure 3.8.3   Sedums in a green roof

buildings. Often these systems are referred to by the term *heating, ventilating, and air conditioning* (HVAC). The need for thermal energy depends on climate conditions during a specific season, building characteristics, and personal behavior. The need for ventilation is related to the activities conducted in a room or building, the number of people present, building characteristics, and personal behavior. Electric energy is needed for light, mechanical work, or electronic systems.

Ventilation Systems

Humans need oxygen, which is provided by means of fresh air preferably with a humidity of 40–70%. The task of a ventilation system is to provide enough fresh air to a building so the people within it can do their activities (e.g., working, sporting, cooking, or sleeping). Two basic principles can be distinguished in ventilating buildings, namely, natural and mechanical ventilation. These two principles are used in generating an incoming (fresh) and outgoing (stale) airstream. During winter in a moderate climate, the outgoing stale air means a loss of thermal energy. By applying heat recovery systems in the ventilation system, thermal energy from the outgoing stale air is passed over (by means of conduction) to the incoming fresh airstream, without mixing the two airstreams. The advantage of mechanical ventilation compared to natural ventilation systems is the former's better controllability and reliability, and the possibility to apply heat recovery. The disadvantage is, of course, the use of electric energy for the fan and control system.

Heating Systems

Thermal energy needed to offer comfortable living or working space in buildings is often provided by burning fossil fuels. In the Netherlands, natural gas is commonly used for heating buildings, cooking, and heating tap water. In the largest part of the world biomass, in the form of wood, is used for heating and cooking. Heating technologies focusing on energy saving try to make more efficient use of fossil fuels or try to make use of renewable sources. Highly efficient boiler systems, micro-cogeneration heat and power systems, heat pumps, and integral hybrid forms of heat generation will be addressed.

Highly efficient boiler systems, based on the principle of burning fossil fuels, are able to use the heat directly provided by incinerating a specific fossil fuel and the heat indirectly provided by flue gasses originating from the incineration. Based on the lower caloric value of the specific fossil fuel, such as natural gas, the efficiency of these systems can exceed 100%. To a lay observer, this suggests that energy is being created. Nevertheless, when the efficiency is calculated making use of the upper caloric value of the fossil fuel, the efficiency will be below 100%.

The consideration that burning fossil fuels on high temperatures to come to a low indoor temperature of only 20°C, resulted in a search for better usage of these scarce resources. The concept of *exergy*, expressing the potential to perform mechanical work, gave the insight that it would be wishful to integrate the production of electric energy. On a large scale, electric power plants, which make use of natural gas, coal, or nuclear power, can distribute their so-called waste heat to nearby buildings. In many cities district heating, using natural gas or biomass for example, provides electric energy and thermal energy to the neighborhood (i.e., at the meso scale). The smallest in line forms the micro-cogeneration heat and power system (μ-CHP). These systems, using natural gas, can produce some electric energy while providing heat to a single household. For small offices or industrial buildings, mini-CHP systems exist that are also able to make use of wood as a fuel.

By applying the same principle of a refrigerator, a heat pump is able to transport heat from the surroundings into a building or vice versa (see Figure 3.8.4). The thermal energy can be provided by ground, groundwater, open water, or air. Their low-temperature heat is able to evaporate the low-pressurized working fluid. After compressing the gas, it becomes fluid in the condenser, where the thermal energy is used for the building's heating system. The working fluid will pass a valve before entering the evaporator again. In this cycle, mechanical work is executed by the compressor that needs electric energy or natural gas.

Depending on the regional climate and the direct surroundings around a building, certain heating systems can have advantages and disadvantages. Therefore, there exists a large group of hybrid systems, in which multiple systems collaborate in providing heat efficiently for space heating and even tap water. For space heating, a water temperature above 25°C can already be sufficient, but heating tap water demands a higher temperature of at least 60°C at

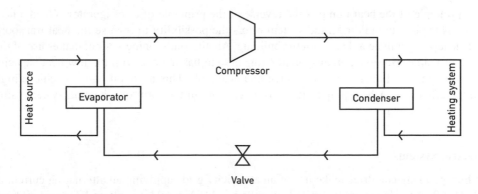

Figure 3.8.4 Representation of the heat pump or refrigeration cycle in which a working fluid is being evaporated and condensed to enable the transport of heat

the water taps. A hybrid system can, for example, be a combination of a solar collector, air source heat pump, and natural gas high-efficiency boiler. In that case, the solar collector and heat pump can be used for space heating and to heat tap water. When outdoor temperatures and solar irradiation decrease, the high-efficiency boiler can be used to increase the esyrt temperature for space heating or tap water. Section 3.8.3.3 will elaborate on more technologies to save energy when heating tap water.

Tap Water–Heating Systems

Compared to space heating, relatively high temperatures are needed for warm tap water to prevent bacteria of the species *Legionella pneumophila* in water pipes. In combination boilers, a temperature of 60°C or more can easily be met. A combination boiler can make use of renewable energy sources by means of a solar collector. This solar connector provides heat via a heat exchanger to a reservoir. The heat reservoir will store heat until the moment heat is needed for space heating or tap water. One speaks of a PVT panel, when a combined photovoltaic panel and thermal solar collector provide electric and thermal energy.

When heated tap water is used for showering, the waste water is 30°C, which is relatively warm. The thermal heat of this waste water can be used to directly heat cold water. A heat exchanger in the drain of the shower or directly on the floor of the shower will heat the cold water running toward the shower. When making use of a thermostatically controlled tap, some part of the heated water will be automatically replaced by the (now) slightly heated cold water. This will reduce the energy use of the boiler system.

Another way to reduce the energy use of a tap water–heating system is to minimize the distance between the heating system and the tap. Less unnecessarily heated water will stay behind in the water pipes when the tap is closed. When new houses are designed, the distance from the water boiler to bathroom and kitchen is kept as short as possible. The same accounts for the distance between the boiler system and kitchenette in new office buildings. In existing buildings, it is quite hard to reduce these pipe lengths.

Cooling Systems

The principle of the heat pump is the reverse of the principle of a refrigerator. When a heat pump is applied in buildings, the system offers the possibility to reverse the heat transport. The heat pump will be able to cool the building. An air-conditioning system makes use of the same principle. However, an air-conditioning system has to cope with the low heat capacity per cubic meter of air at the heat source and heat sink. This makes it hard to achieve high energy efficiency. A heat pump often has a heat sink in the form of a heating system using water as a medium.

Electric Systems

In Europe many countries make use of an electricity grid supplying an alternating current at 230 V at 50 Hz to households. In the United States, 115 V at 60 Hz is offered. Office buildings and small-scale industrial building often make use of high-voltage alternating current, which offers more power than the regular voltage of 230 V. In Europe, the high voltage is 400 V.

However, many appliances in households need only little power, direct current, and low voltages. The number of transformers that are used for mobile telephones, computer screens, computers, game computers, and so on is increasing every year.

Comparable to the situation in which natural gas is being incinerated to offer high-temperature heat for only low-temperature tap water heating and space heating, it also seems strange to generate high-voltage electric energy for low-voltage applications. Photovoltaic systems can provide electric energy with enough power for many of the low-voltage electric appliances in use today. A promising energy-saving measure would be the introduction of a low-voltage (DC) network in houses to which these electric appliances can directly connect. The electric energy for this network can be provided by batteries charged by photovoltaic panels, windmills, or small-scale hydropower.

3.8.4   Transcendental Technologies

The built environment encompasses many different buildings with all sorts of functions. A power plant belongs to a particular category of buildings that is specifically constructed to make an energy transition possible. These power plants offer other buildings the possibility to use (high-voltage) electric energy. In analyzing energy-(saving) technologies, it is very important to conduct system analysis. What forms of energy enter and leave at which quantities? A particular system is, in this case, the object of study. The system boundaries need to be specified. Although elsewhere in this section the boundaries were laid directly around the building shell, this subsection reflects on technologies outside these boundaries. Therefore, the term *transcendental technologies* is used as the title here.

Landscaping

The surroundings, both natural and artificial, have a large impact on the energy use of buildings. Although it will be hard to change the local climate in favor of a lower energy use, multiple elements in the garden can work in favor of the building's internal climate. Trees offer shade in the summer and offer resistance to wind in all seasons. A pond in front of windows can act as a mirror reflecting irradiation toward the building's ceiling. A large garden or estate can provide biomass to heat living space and tap water.

District Heating

The production of electric energy and heat can take place in 1000 MW large power plants, in small 25 kW micro CHP systems, and in medium-sized CHP systems of around 50–250 MW. The last group of CHP systems often supplies heat to a city district. Therefore, the term *district heating* is often applied. Getting rid of heat is a necessary evil in producing electric energy. In the case of micro CHP systems and district heating, this "waste" is used in, respectively, a building and buildings to heat living space and tap water. In the case of district heating, a network of insulated pipes connects the medium-sized power plant to the buildings (see Figure 3.8.5). Offices and industrial buildings have a connection with a bigger capacity

Figure 3.8.5 New insulated metal pipes waiting to replace the existing pipes of a district heating system in a neighborhood in the municipality of Enschede, the Netherlands

than a connection to a house. Inside the buildings only a small heat exchanger is necessary for heating tap water. The radiators can directly be fed by the heated pressurized water of the power plant or one of the intermediate heating stations. In a large network, additional boilers are needed to increase the water temperature.

Figure 3.8.6 At the University of Twente in the Netherlands, a pond is being used as a thermal reservoir to cool surrounding buildings and at the same time enable fire brigades to extinguish fires

Thermal Storage

Buildings cope with two different cycles regarding ambient temperature: firstly, a short cycle of approximately 24 hours covering day and night and, secondly, a long cycle of approximately one year covering four seasons. When it is possible to store heat in a warm period during daytime and/or during summer, it can be used during a cold period in night and/or winter. A ground source heat pump is an example of a system that meks it possible to use natural seasonal thermal storage offered by the ground underneath or close to buildings.

The high thermal capacity of water makes it possible to overcome the day–night cycle. At the University of Twente, for example, an open cooling system is applied (see Figure 3.8.6). An artificial pond emits thermal energy to the atmosphere during night. In summertime the temperature of the water in the pond is even further reduced by making use of cooling overnight. During nighttime, energy prices are significantly lower than during daytime. During the day, the water of the pond is used to cool the buildings around the pond. Gradually the temperature of the water in the pond will increase before the cycle starts all over again.

References

Baetens, R., Jelle, B.P., and Gustavsen, A. (2010) Phase change materials for building applications: A state-of-the-art review. *Energ. Buildings*, **42**, 1361–1368.

Entrop, A.G. and Brouwers, H.J.H. (2010) Assessing the sustainability of buildings using a framework of triad approaches. *J. Building Appraisal*, **5.4**, 293–310.

Entrop, A.G., Brouwers, H.J.H., and Reinders, A.H.M.E. (2010) Evaluation of energy performance indicators and financial aspects of energy saving techniques in residential real estate. *Energ. Buildings*, **42**, 618–629.

Entrop, A.G., Brouwers, H.J.H., and Reinders, A.H.M.E. (2011) Experimental research on the use of micro-encapsulated phase change materials to store solar energy in concrete floors and to save energy in Dutch houses. *Sol. Energy*, **85**, 1007–1020.

European Commission (EC) (2002) Directive 2002/91/EC of the European Parliament and of the Council of 16 December 2002 on the energy performance of buildings (EPBD). http://eur-lex.europa.eu/LexUriServ/ LexUriServ.do?uri=OJ:L:2003:001:0065:0065:EN:PDF (accessed April 12, 2012).

Huang, M.J., Eames, P.C., and Hewitt, N.J. (2006) The application of a validated numerical model to predict the energy conservation potential of using phase change materials in the fabric of a building. *Sol. Energ. Mat. Sol. C.*, **90**, 1951–1960.

Kürklü, A. (1998) Energy storage applications in greenhouses by means of phase change materials (PCMs): A review. *Renew Energ.*, **13**, 89–103.

Nederlands Normalisatie-instituut (NEN) (2004) *Energy Performance of Residential Functions and Residential Buildings; Determination Method*, ICS 91-120-10, NEN, Delft.

Nederlands Normalisatie-instituut (NNI) (2001) *Thermal insulation of buildings: Calculation methods*, NEN 1068, ICS 91.120.10, NEN, Delft.

Nederlands Normalisatie-instituut (NNI) (2002) *Thermal insulation of buildings: Simplified calculation methods*, NPR 2068, ICS 91.120.10, NEN, Delft.

Sharma, A., Tyagi, V.V., Chen, C.R., and Buddhi, D. (2009) Review on thermal energy storage with phase change materials and applications. *Renew Sust. Energ. Rev.*, **13**, 318–345.

Verkerk, G., Broens, J.B., Kranendonk, W. *et al.* (1992) *Binas*, 3rd edn, Wolters Noordhoff, Groningen.

Zalba, B., Marín, J.M., Cabez, L.F., and Mehling, H. (2003) Review on thermal energy storage with phase change materials, heat transfer analysis and applications. *Appl. Therm. Eng.*, **23**, 251–283.

Further Reading

McMullan, R. (2007) *Environmental science in Building*, 6th edn, Palgrave MacMillan, New York.

3.9 Piezoelectric Energy Conversions

Alexandre Paternoster, Pieter de Jong, André de Boer

University of Twente, Faculty of Engineering Technology, Department of Applied Mechanics, P.O. Box 217, The Netherlands

3.9.1 Introduction

Power harvesting using piezoelectric material is a technique that can be useful to power remote or inaccessible sensors. For any location where only a relatively short cable must be drawn, power harvesting is not an option because the engineering and material cost required to develop a power harvester may be significant. This technique is strictly limited to remote use for sensor systems and needs special electronic circuits for generating useful power efficiently.

Applications will range from the nano-Watt to the Watt range, with the latter requiring in the order of 20–30 cubic cm of piezo material. Piezo materials do not possess the energy-density potential to compete with more traditional large-scale means of generating electricity. Some applications are depicted in Figure 3.9.1.

3.9.2 Piezoelectric Material

Introduction

Since its discovery by the Curie brothers in 1880, piezoelectric materials have held the interest of the scientific community for their electro-mechanical properties. Researchers have shown that this class of material has intrinsic properties suitable for actuation as well as power-harvesting applications. The actuation is especially suitable for applications that require a fast response and a large bandwidth (Chopra, 2002). Nevertheless, some characteristics need to be overcome by a thoughtful design of piezoelectric components to achieve the desired force and displacement for actuation.

Piezoelectric materials have also the property to convert mechanical energy into electrical energy. When such a material is subjected to a strain, an electrical charge is created inside the material. This property is called the *direct piezoelectric effect*. Additionally, when the

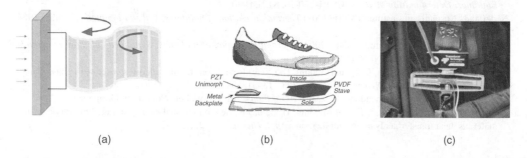

(a)   (b)   (c)

Figure 3.9.1 Applications of power harvesting using piezoelectric material. (a) Vortex shedding in a water flow (Taylor, 2001), (b) impress of an insole in a shoe (Kymissis *et al.*, 1998), and (c) motion of a backpack (Feenstra, Granstrom, and Sodano, 2008) (See Plate 23 for the colour figure)

material is subjected to an electrical field, it deforms according to the electrical field magnitude. This is called the *converse piezoelectric effect*.

Piezoelectric Constants

A piezoelectric material is characterized by the piezoelectric strain constant d_{ij}, which relates the strain to the electrical field. The subscript i indicates the direction of the applied electrical field, and the subscript j indicates the direction of the deformation. Prior to use, the piezoelectric material is poled. The material is heated up to its *Curie temperature* while an electrical field is applied. Above the Curie temperature, electrical dipoles present in the material get aligned with the electrical field. This causes the piezoelectric effect. Consequently, if a material is heated back to its Curie temperature, it will lose its piezoelectric ability as the dipoles will revert back to their initial random distribution. Conventionally the poling direction is along the vertical axis (3-axis) (Figure 3.9.2). When an electrical field (E) is applied in the poling direction, the material is contracting in that direction and extending in other directions (1- and 2-axis). Changing the direction of the electrical field will result in a contraction along the 1- and 2-axis and extension along the vertical axis. To quantify those piezoelectric effects, the direct and shear strains relate to the electrical field in the following way:

$$
\begin{Bmatrix} \varepsilon_1 \\ \varepsilon_2 \\ \varepsilon_3 \\ \gamma_{23} \\ \gamma_{13} \\ \gamma_{12} \end{Bmatrix} = \begin{bmatrix} 0 & 0 & d_{31} \\ 0 & 0 & d_{32} \\ 0 & 0 & d_{33} \\ 0 & d_{24} & 0 \\ d_{15} & 0 & 0 \\ 0 & 0 & 0 \end{bmatrix} \times \begin{Bmatrix} E_1 \\ E_2 \\ E_3 \end{Bmatrix} \qquad (3.9.1)
$$

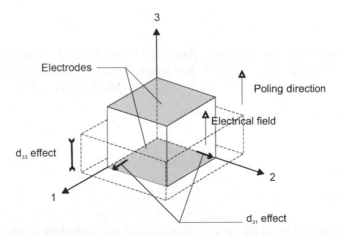

Figure 3.9.2   Axis reference system for piezoceramic components and piezoelectric effects for an electrical field applied in the poling direction

where E_1, E_2, and E_3 are the components of the electrical field; and ε_1, ε_2, and ε_3 and γ_{23} and γ_{13} are, respectively, the strain and shear components. A deformation along the 1- and 2-axis that is characterized by the d_{31} coefficient is called the d_{31} *effect*, and a deformation along the 3-axis is called the d_{33} *effect* as the d_{33} coefficient characterizes this deformation (Figure 3.9.2).

Power Consumption

The power consumed by a piezoelectric actuator to achieve a sinusoidal drive of the actuator depends on the capacitance of the device, the operating frequency, and the driving voltage signal. The average power consumed is equal to the average current that flows through the piezoactuator multiplied by the applied voltage (Physik Instrumente, 2009). This is an approximation as it does not take into consideration any loss in the electromechanical system.

Average current passing through the actuator (A):

$$i_{avg} \approx 2 \cdot f \cdot C_{piezo} \cdot V_P \tag{3.9.2}$$

Average power consumption (W):

$$P_{avg} \approx 2 \cdot f \cdot C_{piezo} \cdot V_p^2 \tag{3.9.3}$$

Peak power consumption (W):

$$P_{avg} \approx 2 \cdot \pi \cdot f \cdot C_{piezo} \cdot V_p^2 \tag{3.9.4}$$

where f is the driving signal frequency (Hz), C_{piezo} is the piezoactuator capacitance (F), and V_p is the driving signal peak voltage (V).

Losses

The power required does not take into account the losses inside the material. Piezoelectric materials are converting electrical energy into mechanical energy and many phenomena during that process can account for their loss in efficiency (Mezheritsky, 2004):

- Induced shear actuator.
- Electric loss.
- Dielectric loss.
- Mechanical loss.
- Elastic loss.
- Piezoelectric loss.

These losses strongly depend on the operation mode of the piezoelectric actuators. Once the actuator is charged, the energy required to keep the obtained deformation is small. The losses are much more important when the material is subjected to a harmonic voltage. These losses come from the hysteresis of the piezoelectric material (Damjanovic, 2006). The energy losses are dissipated with heat. Their importance depends on various factors. Physik Instrumente (2009) states that in large signal condition, the losses can reach up to 12% of the power used to actuate the piezoelectric actuator.

Bandwidth and Actuation Speed

Piezoelectric materials are known to have a large bandwidth that makes them suitable for ultrasonic transducers. The time required for a piezoelectric actuator to reach its nominal displacement depends on the resonance frequency. According to Physik Instrumente (2009), the minimum rise time on a piezoelectric actuator is a third of the period corresponding to the resonance frequency given that the controller can deliver the necessary current in the actuator.

Piezoceramics

Piezoceramics have been widely studied and have been used since the Second World War to manufacture ultrasonic transducers. This material is polycrystalline and needs to be subjected to high electrical fields in order to be poled and to demonstrate electromechanical properties. During this process, the various crystalline cells inside the materials orientate themselves according to the direction of the electrical field. Piezoceramics achieve high force with low strain. When integrating piezoceramic components into a structure, the stresses subjected by the material must be carefully investigated as piezoceramic material will fail for very low uniaxial tensile stresses. Table 3.9.1 lists some piezoceramic materials with their properties. For piezoceramic materials $d_{31} = d_{32}$ and $d_{24} = d_{15}$.

Piezo Polymers

Polyvinylidene fluoride (PVDF) is the main polymer that exhibits a significant piezoelectric effect. Like any piezoceramic material, the constants d_{33} and d_{31} characterize the

Table 3.9.1 Table of some piezoceramic materials. PZT-SP4 and PZT-5A1 are from *Smart-Material Company*. PZT-5H is taken from Chopra (2002)

Name	d_{31} (m/V)	d_{33} (m/V)	e_{33}	T_c (°C)	ϱ (kg/m^3)	S_{33} (m^2/N)
PZT-SP4	−1.23e-10	3.1e-10	1300	325	7500	1.81e-11
PZT-5A1	−1.85e-10	4.4e-10	1850	335	7500	2.07e-11
PZT-5H	−2.74e-10	5.93e-10	3400	193	7500	2.083e-11
PZT-PSt-HD	−1.9e-10	4.5e-10	1900	345	7500	2.1e-11
PZT-PSt-HPSt	−2.9e-10	6.4e-10	5400	155	8000	1.8e-11
PZT-PIC-255	−1.8e-10	4.0e-10	1750	350	7800	2.07e-11
PZT-PIC-151	−2.1e-10	5.0e-10	2400	250	7800	1.9e-11

Note: PZT-PSt-HD and PZT-PSt-HPSt are from *Piezomechanik Company*. PZT-PIC-255 and PZT-PIC -151 are *from Physic Instrumente Company*. T_c is the Curie Temperature of the piezoelectric material, S_{33} is the component of the compliance matrix in the poling direction, ϱ is the material's density, and e_{33} is the component of the relative permittivity of the material in the poling direction.

Table 3.9.2 Comparison of the piezoelectric coefficients of PZT-5H piezoceramic and PVDF piezopolymer. The figures are taken from Chopra (2004)

Name	d_{31} (m/V)	d_{32} (m/V)	d_{33} (m/V)	d_{15} (m/V)
PZT-5H	−274e-12	−274e-12	593e-12	741e-12
PVDF	20e-12	3e-12	−33e-12	0

electromechanical properties of the material. Unlike piezoceramics, piezopolymer does not have shear electromechanical coupling and d_{31} is not equal to d_{32}. Prior to the poling process, the polymer is stretched in order to align the carbon chains. The dipole on the chains is pointing in the direction normal to the applied stress. The polymer is then subjected to heat and an large electrical field (Atkinson, 2006). At room temperature the amorphous part of the polymer is over its glass transition temperature (32°C), therefore the polymer is flexible at room temperature. It is possible to find flexible transducers among the commercial products made with PVDF polymer.

Unfortunately, the performance of the polymer is far behind that of piezoceramic material (Table 3.9.2). The force and displacement are hence lower for the same applied voltage.

3.9.3 Power Harvesting

Introduction

As explained in Section 3.9.2, piezo plates or stacks deform when a voltage is applied. This mode is called the *actuator mode* as it is used to generate a displacement in the mechanical domain. The opposite is also possible: Mechanical straining of the material will lead to the generation of electrical charge or voltage, called the *sensing mode*. This is done by connecting a very high electrical resistance across the piezo element in order to prevent much electrical current from flowing. The developed voltage is then a measure for the mechanical deformation of the piezo material.

It is also possible for the piezo element to actually generate electricity from a mechanical input, current as well as voltage. By affixing a sheet of piezo material to a deforming/or vibrating structure, it will deform with the structure. The piezo will then convert the mechanical vibration energy to electrical current. By conducting this current through a resistor, a voltage difference will be created due to Ohm's law: $U = IR$. Then, through the equation $P = UI$, the power can be calculated. Machines provide the mechanical excitation in the form of a harmonic load, unbalance of rotating parts, or straining of the frame. Examples of power-harvesting research using periodic excitations are impacts, fluid structure interaction (Barrera-Gil, Alonso, and Sanz-Andres, 2010), walking humans (Kymissis *et al.*, 1998), and (Feenstra, Granstrom, and Sodano, 2008) wind or current around blunt bodies (Taylor *et al.*, 2001).

The example of connecting only a resistor across the electrodes of figure 3.9.3 is to demonstrate the power-harvesting concept. To actually generate useful power for sensors and other electronics will require more advanced methods such as synchronous electric charge extraction, synchronized switch harvesting on inductor, pre-biasing, and so on. These will be addressed briefly in this section.

Basic Concepts for a Single Degree of Freedom Piezo Electric Model

Let us consider a 1 degree of freedom (DOF) system consisting of a mass M, mechanical stiffness k, and damping c. There is also a piezo element with properties k_p and Θ, which represent the material stiffness, and a piezo electric coupling value (unit: N/V). Consider the fundamental piezo equations (Priya and Inman, 2009):

$$\varepsilon = S^E \sigma + dE \tag{3.9.5a}$$

$$D = d^T \sigma + \varepsilon^\sigma E \tag{3.9.5b}$$

In this equation ε, S^E, σ, d, E, D, and ε^σ represent the mechanical strain, the compliance matrix (the inverse of the stiffness matrix), the mechanical stress, the piezo electric constant, the electrical field, the charge displacement (the integral of current density), and the permittivity of the material, respectively. If we are to assume a one-dimensional stress state (1D loading) and an actual piece of piezo material with finite dimensions, we can rewrite these equations (here we assume use of the 31 mode) to

$$k_p u + \Theta V = F \tag{3.9.6a}$$

$$\Theta \dot{u} - C_p \dot{V} = I \tag{3.9.6b}$$

In this equation, k_p, u, Θ, V, F, C_p, and I represent the mechanical piezo stiffness, displacement, electromechanical coupling, voltage, applied force, capacitance, and current flowing from the piezo, respectively. Lastly, we want to include the dynamics of this system by introducing mass (M) and damping (C):

$$M\ddot{u} + C\dot{u} + Ku + \Theta V = F \tag{3.9.7a}$$

$$\Theta \dot{u} - C_p \dot{V} = I \tag{3.9.7b}$$

The ideal physical model of this system is given in Figure 3.9.3.

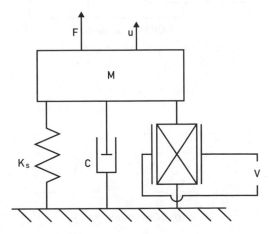

Figure 3.9.3 Ideal physical model of a 1DOF system

Equation 3.9.7a represents the mechanical force balance. We recognize the traditional damped oscillator consisting of the mass M, damping C, and stiffness K. The latter consists of the piezo stiffness and, depending on the model, also some structural stiffness ($K = k_p + k_s$). Through the piezoelectric effect, the piezo element will also generate a force of ΘV. Equation 3.9.7b represents the electrical domain, more specifically a balance of electrical currents at the piezo contacts. Here we see that the current generated by the piezoelectric effect is Θu. Some current is lost to the internal capacitance of the piezo element, which adheres to the equation of a capacitor $I = CdU$.

For a conceptual discussion, we will consider the system with a resistor R, rectifier, and buffer capacitor C_s, known as direct current impedance matching (DCIM). The rectifier is an electronic device consisting of four diodes that is capable of converting the alternating current from the piezo element to a direct current that is more suitable for sensors and electronics. Together with the buffer capacitor C_s and properly tuned electric load R, a harmonic excitation of the system will lead to a nearly constant rectified voltage V_{dc} allowing for a constant dissipation of energy in the resistor. Figure 3.9.4a shows the circuit layout, and Figure 3.9.4b conceptually shows the alternating piezo voltage V_p under a given displacement u. Ideally the resulting constant voltage V_{dc} is equal to the maximum V_p. Two modes of operation can be seen in the voltage curve. Between t_1 and t_i, the piezo element is transitioning from $-V_{dc}$ to $+V_{dc}$ and no current is flowing from the piezo element. Between time t_i and t_2, the voltage remains constant and current is flowing from the patch. Shu and Lien (2006) provide a detailed analytical

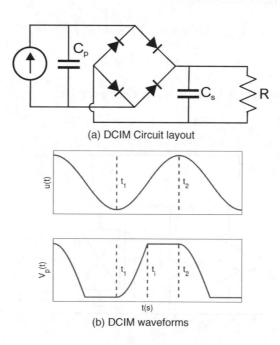

(a) DCIM Circuit layout

(b) DCIM waveforms

Figure 3.9.4   DCIM circuit layout and typical waveforms

description of the circuit. For most, cases this circuit possesses a single optimal resistance, written as

$$R_{opt} = \frac{\pi}{2C_p\omega}$$ (3.9.8)

With a lower resistance, more current will flow but at a lower voltage. With a higher resistance, less current will flow but at a higher voltage. Recalling $P = UI$, it is not either the voltage or current that must be optimized, but the product. Let us also consider some other properties of the circuit. The electromechanical coupling Θ tells us how much force is generated for 1 V in the piezo material. The important question is how this "apparent stiffness" relates to the mechanical stiffness of the system. This is indicated with the variable k_e (Shu and Lien, 2006):

$$k_e^2 = \theta^2/(KC_P)$$ (3.9.9)

Looking ahead, note that the equation for k_e^2 is valid only at the optimum resistance. The coupling k_e^2, in combination with the damping ratio $\zeta = 2C/2\sqrt{KM}$, is the most critical in the analysis of a power-harvesting system (Shu and Lien, 2006). For low values ($k_e^2/\zeta < 0.01$), the influence of the generated piezo voltage on the mechanical domain is negligible. Analyzing the behavior of the 1DOF system under a harmonic excitation is then straightforward: First the mechanical domain is solved without taking the piezo material into account. The resulting displacement is then inserted into equation (3.9.7b) and solved for V.

This leads to the first important property for efficient power-harvesting systems. Maximizing the displacement of the harvester leads to more generation of charge from the piezo element. For ambient vibrations, it is therefore important to tune the harvester's natural frequency to the ambient vibration frequency. This maximizes the amplitude of the harvester. As the coupling k_e^2 and the ratio k_e^2/ζ increase, the piezo material becomes more dominant in the system. The dynamic behavior of the mechanical domain becomes dependent on the amount of piezo material, the electrical load resistance, and so on. This requires that both equations must be solved simultaneously.

The natural frequency will, for instance, shift as well. Assuming an optimally matched load, the change in frequency for a DC impedance matched system is written as follows (Shu et al., 2006):

$$\omega_{em} = \sqrt{k(1 + k_e^2)/m}$$ (3.9.10)

with ω_{em} denoting the (now coupled) electromechanical natural frequency of the system. Another issue is that one may install so much piezo material that the total power output of the harvester will actually decrease (Lefeuvre et al., 2005). There is clearly an optimum amount of material. A qualitative description of this phenomenon is analogous to that of the optimal load resistance: No piezo material will result in zero harvested power but a large vibration. As the amount of material is increased, the power output will increase as well. When too much piezo material is added, it will provide so much damping that the harvester will not achieve a large enough vibration level.

Resonant versus Nonresonant Systems

In the previous section, it was stated that resonant systems are most efficient. At resonance a small input force will lead to a large displacement because energy is efficiently transferred from the input force to the harvester. There are, however, some applications in which this may be impossible, such as very high-force applications. Although not as prevalent, there are some applications (De Jong, 2012) that generate enough useful power simply due to the large force or strain present.

These applications are typically subresonant or displacement driven, as opposed to force driven. For displacement-driven systems, the dynamics clearly become irrelevant with respect to the power-harvesting calculation. For subresonant systems, the frequency of actuation is far lower than the natural frequency of the component. This also greatly simplifies analysis because a quasi-static approach can be used when calculating the deformation of the system, again ignoring the dynamics of the mechanical domain. Damping and natural frequencies can be ignored as well.

In subresonant systems, most of the energy from the input is transferred to strain deformation of the host structure and released again as the force or deformation decreases. This is why these systems are very inefficient from an energy standpoint. This does not mean that subresonant power-harvesting systems do not generate useful energy, but that they convert only a smaller amount of the available mechanical energy present to electrical energy.

A final critical difference for resonant systems is that, regardless of which circuit is used, the maximum achievable output remains the same. The main difference is that the amount of piezo material that is required varies with the more efficient circuits requiring less material. In displacement-driven systems, the more efficient circuits increase the output significantly for the same amount of piezo material (Lefeuvre et al., 2006).

Alternative Electrical Circuits

The impedance matching circuit is the first circuit that was developed for power harvesting but it is comparatively inefficient. Considering Figure 3.9.4 the material is not conducting any energy out of the piezo element for a significant portion of one oscillation. For an optimally designed circuit, it transfers current for only 50% of the time, between t_i and t_2. Much time is lost while the piezo element must change voltage between $+V_{dc}$ and $-V_{dc}$ to again conduct energy through the rectifier. A number of nonlinear methods have been developed to greatly increase the output of harvesting systems.

These methods require active circuits to control switches within the circuit in order to manipulate the piezo voltage. These circuits will, of course, need to be powered by the piezo element as well. This means that for micro-scale harvesters, the output may be so small that it is not wise to use active circuits: the added output of the circuit may be fully consumed by the control electronics. The active circuits become viable options when the output of the DCIM circuit is in the order of mW or more. An example is the synchronous electric charge extraction (SECE) circuit proposed by Lefeuvre et al. (2005). This circuit (see Figure 3.9.5) functions by breaking the contact between the piezo element and the rest of the circuit. Contact is made only when the circuit detects a maximum in the piezo voltage after which the energy stored in the piezo element is rapidly discharged through the remainder of the circuit. For the same displacement excitation of the mechanical domain, this circuit will generate approximately four times more power than the DCIM circuit. For resonant systems, the SECE circuit requires only a quarter of the amount of material that DCIM requires. The advantage is the lack of an optimum resistance.

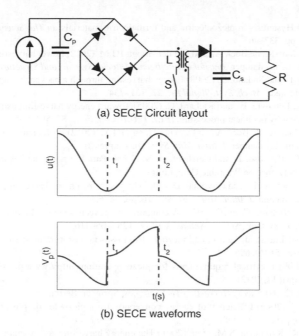

(a) SECE Circuit layout

(b) SECE waveforms

Figure 3.9.5 SECE circuit layout and typical waveforms

Another circuit is the synchronized switch harvesting on inductor (SSHI) circuit. For more information on this circuit, see Guyomar *et al.* (2005).

3.9.4 Conclusions

Power harvesting using piezoelectric material is a technique that can be useful to power remote or inaccessible sensors. For any location where only a relatively short cable must be drawn, power harvesting is not an option because the engineering and material cost required to develop a power harvester may be significant. This technique is strictly limited to remote use for sensor systems and needs special electronic circuits for generating useful power efficiently.

The energy balance must also be considered. In some cases, adding a power harvester somewhere in a system (e.g., a drive train, vehicle, or rotating equipment) may increase the load on the motor and as a consequence the energy consumption of the system. No energy is then actually harvested but more conversions take place, greatly reducing efficiency. Power harvesting must not be seen as a technique to generate a significant amount of power. Applications will range from the nano-Watt to the Watt range, with the latter requiring in the order of 20–30 cubic cm of piezo material.

References

Atkinson, G.M. (2006) *MEMS: Design and Fabrication*, vol. **2**, CRC Press Taylor & Francis Group, Boca Raton, FL.

Barrera-Gil, A., Alonso, G., and Sanz-Andres, A. (2010) Energy harvesting from transverse galloping. *J. Sound Vib.*, **329**, 2873–2883.

Chopra, D. (2002) Review of state of art of smart structures and integrated systems. *AIAA J.*, **40** (11), 2145–2187.

Damjanovic, D. (2006) Hysteresis in piezoelectric and ferroelectric materials, in *The Science of Hysteresis*, Academic Press, Oxford, pp. 337–465.

De Jong, P.H., Loendersloot, R., de Boer, A., and van der Hoogt, P.J.M. (2012) Power harvesting in a helicopter rotor using a piezo stack in the lag damper. *Journal of Intelligent Material Systems and Structures* (in press).

Feenstra, J., Granstrom, J., and Sodano, H. (2008) Energy harvesting through a backpack employing a mechanically amplified piezoelectric stack. *Mech. Syst. Signal Pr.*, **22**, 721–734.

Guyomar, D., Badel, A., Lefeuvre, E., and Richard, C. (2005) Toward energy harvesting using active materials and conversion improvement by nonlinear processing. *IEEE T. Ultrason. Ferr.*, **52**, 584–595.

de Jong, P.H., Loendersloot, R., de Boer, A., and van der Hoogt, P.J.M. (2012) Power harvesting in a helicopter rotor using a piezo stack in the lag damper. *J. Intel. Mat. Syst. Str.* (in press).

Kymissis, J., Kendall, C., Paradiso, J., and Gershenfeld, N. (1998) Parasitic power harvesting in shoes. In Second IEEE International Conference on Wearable Computing.

Lefeuvre, E., Badel, A., Richard, E., and Guyomar, D. (2005) Piezoelectric energy harvesting device optimization by synchronous charge extraction. *J. Intel. Mat. Syst. Str.*, **16**, 865–876.

Lefeuvre, E., Badel, A., Richard, C. *et al.* (2006) A comparison between several vibration-powered piezoelectric generators for standalone systems. *Sensor. Actuat. A-Phys.*, **126**, 405–416.

Mezheritsky, A.V. (2004) Elastic, dielectric, and piezoelectric losses in piezoceramics: How it works all together. *IEEE T. Ultrason. Ferr.*, **51** (6), 695.

Physik Instrumente (2009) Electrical requirements for piezo operation. http://www.physikinstrumente.com/en/products/ (accessed April 12, 2012).

Priya, S., and Inman, D. (2009) *Energy Harvesting Technologies*, Springer, Berlin.

Shu, Y.C., and Lien, I.C. (2006) Efficiency of energy conversion for a piezoelectric power harvesting system. *J. Micromech. Microeng.*, **16**, 2429–2438.

Taylor, G.W., Burns, J.R., Kammann, S.M. *et al.* (2001) The energy harvesting eel: A small subsurface ocean/river power generator. *IEEE J. Oceanic Eng.*, **26**, 539–547.

4

Using Energy: Beyond Individual Approaches to Influencing Energy Behavior

Daphne Geelen[1] and David Keyson[2]

[1]Delft University of Technology, Faculty of Industrial Design Engineering, Design for Sustainability, Landbergstraat 15, The Netherlands
[2]Delft University of Technology, Faculty of Industrial Design Engineering, Department of Industrial Design, Landbergstraat 15, The Netherlands

4.1 Introduction

While many studies in the past have focused on energy consumers in relation to providing feedback and stimulating energy-efficient behavior, this chapter deals with the upcoming role of end users as consumers and producers of energy.

Calls to reduce greenhouse gas emissions and current and anticipated constraints in energy resources continue to increase the pressure to improve energy efficiency. This also applies to the energy efficiency of households. Energy-consuming products play an important role in households and thus in household resource consumption. Technical innovations may increase the efficiency of the products used, but ultimately the decisions and habits of their users will have a determining effect on total energy consumption (e.g., Elias, Dekoninck, and Culley, 2007; Gardner and Stern, 1996; Groot-Marcus et al., 2006). Studies have shown that 26–36% of in-home energy consumption is due to residents' behavior (Wood and Newborough, 2003).

In light of upcoming changes in the energy system toward a more distributed system of energy provision, the role of the energy user is also expected to change. Technology

The Power of Design: Product Innovation in Sustainable Energy Technologies, First Edition.
Edited by Angèle Reinders, Jan Carel Diehl and Han Brezet.
© 2013 John Wiley & Sons, Ltd. Published 2013 by John Wiley & Sons, Ltd.

development is currently underway, but the social acceptance and social changes needed for successful implementation hardly receive attention (Wolsink, 2011).

In this chapter we argue for the potential of social approaches to support behavior change, and we relate it to the current practices in design of influencing energy consumption behavior in terms of interventions as well as the design of interactions between a house and its users.

4.2 The Changing Roles of End Users and Residents in the Energy Provision System

The traditionally centralized electricity provision system is becoming increasingly distributed. *Distributed generation* means that there are multiple, small generating units that are situated close to where energy is consumed (Ackermann, Andersson, and Söder, 2001; Alanne and Saari, 2006).

The integration of renewable energy generation into the electricity supply system contributes to a lower dependency on fossil fuels, as well as lower CO_2-equivalent emissions related to fossil fuel consumption. Additionally, distributed generation avoids transport losses because long-distance transport can be minimized.

One characteristic of the electricity system is that there has to be a balance between supply and demand. Variability in supply caused by the intermittent nature of renewable energy sources like wind and solar poses a challenge to the reliability of the power system. Energy loads from end users as well as fine-tuning of the energy supply have to be taken into account for regulation of the power system. Both issues require upgrading the grid toward a "smarter grid" (Charles, 2009). A *smart grid* makes use of information and communications technology (ICT) to manage the balance between supply and demand, and as such to accommodate the integration of renewable energy sources into the power system. Development in this area has focused mainly on technology development. As mentioned, social acceptance and shaping household behavior have received scant attention (Wolsink, 2011).

Currently the balancing of supply and demand is invisible for household end users. The energy system transition, however, expects end users of electricity, from industry to consumers, to play an active role in the management of the system (European Technology Platform SmartGrids, 2011; International Energy Agency, 2011).

A restructuring of utilities infrastructures is already taking place, as Van Vliet, Chappells, and Shove (2005) described. It is expected that a variety of large-scale and small-scale systems will coexist (Wolsink, 2011; Marris, 2008). The big issue is to establish a sustainable system of energy provision, in which local energy networks with co-providing end users operate in cooperation with larger scale utility companies. This implies a change in the technologies mediating between provision and consumption, a change in the roles that consumers play in the energy provision system, and as a consequence a change in energy-related behavior.

The shift to decentralized energy generation allows consumers to play an active role in energy provision. Apart from being a "normal consumer" who buys energy from an energy provider, consumers can choose to become producers of energy and thus participate in the energy market. A concept from literature to capture this is *co-provision* (Van Vliet, Chappells, and Shove, 2005). Van Vliet and coauthors argue that the ongoing restructuring of utilities infrastructure leads utilities and users to act as "co-managers" in order to establish environmentally sustainable infrastructures for water, electricity, and waste services. They argue that the role of consumers is becoming differentiated, and as a consequence so are the ways in which services are provided and demand is managed.

They define four consumer roles:

1. *Captive consumer*, who is normally associated with monopolistic modes of provision.
2. *Customer-consumer*, who the service providers have a stake in keeping satisfied due to competition.
3. *Citizen-consumer*, who is also concerned with societal issues.
4. *Co-provider*, the consumer who participates in the provision of utility services.

Given the idea that the energy provision system would develop into a system in which microgrids with distributed generation are connected to the larger grid, households in local communities would become participants in the local microgrid. As Wolsink argued, when users become the managers of production in a microgrid, distributed generation is physically close and also at a closer "social distance." When these users decide to cooperate, for example by integrating their "distributed generation units" in a cooperative microgrid, they constitute a community (Wolsink, 2011).

Whereas Wolsink looked at smart grid developments on the macro level in terms of social acceptance and governance, Section 4.3 of this chapter will address the micro level with energy-related behavior in households.

Changes in the system of energy provision predict that households will add an extra dimension to their energy consumption behavior. In addition to the need to facilitate energy efficiency, co-management of supply and demand in the local grid has to be facilitated. This means not only being energy efficient (energy saving) but also taking advantage of energy that is available from distributed generation in the local community.

4.3   Stimulating Energy Behavior Change in Current Design Practice

In the design discipline, there is growing interest to stimulate sustainable behavior through product and service design (e.g., Ehrenfeld, 2008; Kuijer and De Jong, 2009; Manzini, 2008). The design discipline operates at the intersection between technology and human behavior. The challenge for designers is thus to come up with products and services that enable sustainable lifestyles. A quick glance at design handbooks reveals that traditional design approaches consider individual users. This is also reflected in design strategies aimed at facilitating sustainable behavior in both product and interaction design.

4.3.1   Design Strategies to Stimulate Behavior

Strategies can be applied on a wide range of products, from making use of physical affordances (e.g., not allowing people to sleep on a public bench) to creating awareness (Lilley, 2009; Lockton, Harrison, and Stanton, 2010; Wever, Van Kuijk, and Boks, 2008). While these strategies provide handles on how to stimulate or constrain certain behaviors through product design, the design is always aimed at the behavior of individual users rather than groups of users.

4.3.2   Interaction Design

Many of the usability issues of common home energy management systems were recently highlighted by van Dam *et al.* (2012). One of the most important issues for the design of such a system is whether the user feels in control over the product, or feels the product is in control. The issue of control should be seen along a continuum from a product taking no action, giving suggestions and collaborating on tasks, to, at the far end, taking action completely autonomously. A great deal of criticism has been expressed toward smart products, such as cars equipped with adaptive cruise control that take action without due regard to the user context and needs, for example speeding up when the car is exiting the freeway simply because no car in front is detected (Norman, 2007). Ideally, a smart system should build a sense of trust over time, such that the user will begin to delegate mundane or repetitive tasks to the product. This implies that the system has some notion of the user's current energy consumption–related activities and a model of costs versus benefits in taking a particular form of action.

An important usability issue is the degree of transparency afforded by a home energy controller, that is, to what degree can the user understand, follow, and possibly change decisions or suggestions made by the system? Too much insight may become confusing, creating information overload, whereas too little may create a *black box syndrome*, whereby the user is less likely to trust and engage with the product. In the case of products with rich embedded functionality, a collaborative and narrative approach may be useful. In terms of tangible products, communicating what the product can or cannot do can be conveyed through the product's physicality. Work by Djajadiningrat and colleagues (2004) has focused on the role of design in creating a balance between a product's physical appearance and action potential, as communicated by perceived affordances. In short, at the cognitive level the user should be able to understand what the smart product does and how it can be of help in performing a certain task or fulfilling a certain goal.

In considering how user–product communication is currently supported, one is confronted with the problem of many of today's products: They interact with the user at a low hardware–software function level. Product interaction should support interaction with the product at a higher goal level. Many home control systems provide the user with a button or controller for every hardware function. One should consider user product interaction in terms of the need to support a range or hierarchy of control acts from communicating rudimentary settings to performing acts related to higher goals (Rasmussen, 1986).

4.3.3   Collaborating on Energy Management

Principles of collaboration can be used to enable the user to discuss with a home energy controller options for saving energy while maintaining or improving comfort. Depending upon the relative knowledge of the user and system, the product agent may transition back and forth between a tutoring or more assistive role. In a more assertive role, the product agent may even attempt to persuade the user to follow a certain goal that otherwise may have been neglected. Smart products, such as an Ibo robot, are commonly perceived as complex due to the amount of hidden functionality and their range of functions (Rijsdijk, 2006). However, studies conducted by Rich, Sidner, and Lesh (2001) demonstrated that the perceived complexity of a product may be reduced by designing a more goal-based collaborative system rather than one based on controlling low-level functions. Goals can be

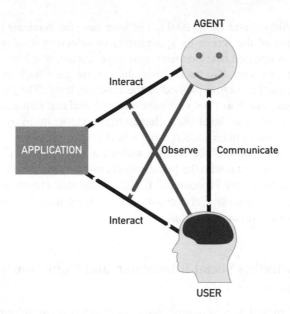

Figure 4.3.1   The collaborative paradigm (Rich, Sidner, and Lesh, 2001)

communicated between the user and product such that the product can help the user meet a goal (Figure 4.3.1). The advantage of a fully integrated collaborative interface is that the user is still encouraged to explore and learn about the system, rather than become dependent upon a help or wizard agent. For example, the intelligent thermostat concept (Keyson *et al.*, 2001) shown in Figure 4.3.2 is based on models of collaboration stemming from the Collagen agent

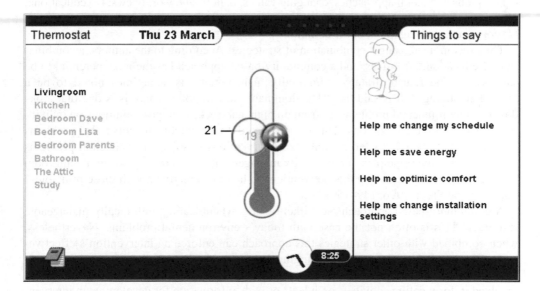

Figure 4.3.2   The intelligent thermostat prototype

architecture (Rich, Sidner, and Lesh, 2001). The user can communicate task-level information via the right side of the screen by speaking in or selecting displayed sentences. The "Things to say" list is updated based on user actions in relation to a hierarchical task model. For example, if the user says "Go back", then the system goes back to the previous task, rather than undoing the last button pressed at the function level. The user can request help with a higher level goal, such as "Help me save energy," and can also communicate with the product at the function–feature level (e.g., change room temperature) by selecting items via the touch screen or speaking in terms. Knowing what the user can say to a product is critical. For example, DeKoven (2004) observed that without a "Things to say" window, one user asked the intelligent thermostat who the prime minister of Canada was during a trial session. Usability studies conducted by Freudenthal (2002) found that elderly people in particular appreciated being able to converse at a conversational level with the interface, rather than having to recall menu names and functions.

4.4   Toward Including Social Interaction and Community-Based Approaches

In this chapter we argue that community-based approaches to stimulating energy-efficient behavior need further exploration. Human behavior is shaped by the social context in which it takes place. Review studies have shown, however, that the influences of social interaction between individuals have received little attention in comparison to studies addressing individual energy-related behavior (see, e.g., Breukers *et al.*, 2011; Jackson, 2005; Lutzenhiser, 1993; Wilson and Dowlatabadi, 2007).

In their work about the relationship between environmental problems and human behavior, Gardner and Stern (1996) described four strategies for changing consumer behavior: (1) religious and moral approaches: changing values, beliefs, and worldviews; (2) educational interventions: changing attitudes and providing information; (3) changing the incentives; and (4) community management of the commons.

They recommend using a combination of strategies. According to the authors, a combination of education, incentives, and a community-based approach has the most potential to be successful. The fourth strategy, community management, is sometimes noted to be a "forgotten strategy." It is said that "The dominant view in policy analysis is one that looks down on environmental problems, as from on high, and seeks to impose solutions on individuals, groups, or organizations that are presumed to be unable to solve the problems themselves." As noted, "A large number of interventions to solve environmental problems have neglected the principles of community management . . . there may be significant room for improvement by remaking the interventions to be more congruent with those principles" (Gardner and Stern, 1996, 149–150).

A community-based approach most likely works when dealing with locally manageable resources. This is often not the case with today's environmental problems. Nevertheless, when combined with other strategies this approach can enforce an intervention's effectiveness. Community involvement in changing behavior is effective due to word-of-mouth publicity, active citizen groups or leaders, and a participatory approach in which citizens are involved in formulating activities, or at least provide information for developing an intervention program.

Considering that energy provision is expected to be organized more into microgrids connected to a larger grid, community management of the commons is applicable (Wolsink, 2011). According to Wolsink, such an approach would additionally improve the technologies' social acceptance.

Similarly, Heiskanen *et al.* (2010) argued that many behavior change programs do not address the socially grounded nature of human behavior. They elaborated barriers to behavior change that call for solutions at the community level rather than individual level, and investigate how a number of low-carbon communities can achieve lasting behavior change. They considered a community-based approach to be a solution for four persistent problems in energy demand-side management: social dilemmas, social conventions, sociotechnical infrastructures, and the helplessness of individuals when faced with the enormity of the climate challenge.

A complementary perspective on the influence of social context on behavior change is social innovation. Jégou and Manzini (2008) described it as follows:

The term social innovation refers to changes in the way individuals or communities act to solve a problem or to generate new opportunities. These innovations are driven more by changes in behavior than by changes in technology or the market and they typically emerge from bottom-up rather than top-down processes.

Collaborative communities, as they call these groups of people, come up with ways to fulfill their daily life needs by organizing themselves differently. For instance, in one case a lack of safe roads and proper public transport leads to a "walking bus": Parents take turns in walking a group of children to school. The role of design in these initiatives is to facilitate the adoption of these solutions by a broader public. In order to establish a sustainable society, the authors argue that designers should, rather than translate new technology to end users, learn from end users for new directions of technology development. Design could fulfill a role in fostering collaborative communities. This approach is similar to the recommendation of Gardner and Stern (1996) of involving the community in the development of interventions.

Facilitating interaction between community members is key to the success of interventions and activities. The question is, now, if and in what ways design can provide "enabling solutions."

4.5   Approaches to Using Social Interaction in Relation to Energy-Related Behavior

To illustrate how social interaction and community-based approaches can be used to influence energy-related behavior. The following sections provide some examples.

4.5.1   Interventions Using Interactions between Participants

EcoTeams

The EcoTeam Program was devised in the 1990s. EcoTeams were small groups of people (e.g., neighbors, friends, and family) who came together once every month to exchange information about energy-saving options. Additionally they received information about pro-environmental behavior and feedback about their own energy savings, as well as the savings of other EcoTeams. The approach proved to be successful in reducing energy consumption in several domains, both shortly after the program and during a follow-up of 2 years.

In terms of energy consumption, the program resulted in 20.5% savings on gas use and 4.6% on electricity use. Two years later, the savings were 16.9% for gas use and 7.6% for electricity use (Staats, Harland, and Wilke, 2004). The program thus resulted in significant energy saving. Since the program consisted of a combination of interventions (information, feedback, and social support), it was difficult to attribute the role of individual elements to its success. Potential contributions to the intervention's success can, however, be supported by several studies investigating the impact of social interaction on behavior (Hopper and Nielsen, 1991; Lewin, 1947; Weenig and Midden, 1991).

Perspective Project

The Perspective project, executed in the years 1995–1998, aimed to investigate whether households were able to develop lower energy consumption patterns, in terms of both indirect consumption through the consumption of goods and services, as well as direct consumption of gas and electricity. Twelve households participated for 2 years. The goal of the households was to spend a certain amount of money and at the same time lower their energy consumption. In addition to continuously monitoring household consumption and giving the households rules to adhere to, the project assigned a personal coach to each household and regular meetings among all participants were held. The households indicated that personal contact with the coach was very supportive in helping them fulfill the project goals. The social interaction between participants was found to be motivating, for they could exchange experiences and tips and motivate each other to continue.

Bottom-Up Initiatives for Collaborative Communities

Whereas Eco-Teams and the Perspective project were top-down approaches that were set up as research projects, bottom-up development takes place in the formation of collaborative communities such as Transition Towns. The top-down approaches generally require a lot of financial resources for their execution, because of the involvement of organizers, researchers, and fees for participation. The bottom-up initiatives, in contrast, are driven by the enthusiasm of their participants and are executed on a completely voluntary basis. Obviously they are not set up as research projects and are thus harder to use for sound research; however, these experiences provide valuable information about social processes. Breukers *et al.*, (2011) have approached this issue via action research that involves active involvement of the researchers in communities.

Transition Towns is a successful example of a collaborative community that started locally in a UK town and whose principles and organizational structure were adopted worldwide, either as new activities or as a means to reinforce existing activities in communities. The goals of the Transition Towns movement are to deal with peak oil and to mitigate the effects of climate change on the local community. One of the "instruments" that is used is the Energy Descent Plan, which is developed in the local community to help it become independent of fossil fuels (Hopkins, 2010). Naturally, the Transition Towns movement addresses the sociotechnical context of energy consumption and the Energy Descent Plan is one of the means to address changes in individual behavior as well as in social and technical contexts.

As Jegou and Manzini (2008) pointed out, design could contribute to the establishment and growth of collaborative communities with enabling solutions. In the case of Transition Towns, one notices the extensive use of social media and wikis for communication and knowledge sharing among participants, in addition to regular meetings. Furthermore, since the community focuses on local production and consumption of food and energy, designers can play a role in finding solutions for local problems (e.g., organic waste management) or monitoring systems to evaluate progress of the local Energy Descent Plan. Such systems could in turn be connected to the social media used by the communities.

4.5.2   Games as Means for Social Interaction in a Community

In the examples given in Section 4.4.2, social interaction takes place through meetings and Internet-based media for interaction. Games with multiple players have also been studied as a means to enhance community awareness and influence behavior (e.g., Ritterfeld, Cody, and Vorderer, 2009).

Playing a game allows people to step outside of the ordinary (Caillois, 1962; Huizinga, 1949). In the "as-if" situations of a game, people can try out different behaviors safely. For example, a serious game called Energy Battle served as a means to influence energy-related behavior in households. Energy Battle was designed to engage home occupants in a fun way toward saving energy via a competition with other households. The challenge enabled home occupants to gain insight regarding their energy consumption and actively involve them in reducing energy consumption. The energy reduction was registered online, and information and tips on the online platform were given to help participants reduce energy consumption. The game was tested in 20 student households in the city of Rotterdam. Energy Battle resulted in an average reduction of 24% in electricity consumption. In an evaluation 8 months later, however, the participants indicated that although the participants had become more aware of energy consumption, only a few of the behaviors developed in the game were maintained. The teamwork and competition between households had been strong motivating factors for energy saving during the game (Geelen et al., 2010).

Another type of game is the simulation game, a type of serious game that simulates situations, often in an abstracted way. An example is Sasér, which is about smart grids (Costa, 2011). The goal of this paper-based game is to provide end users insight regarding the mind shift that is expected in a smart grid situation. A game session takes about 2 hours. The players have to complete a personal challenge; to complete this challenge, they needs to win elements by matching energy consumption with the available energy in the system. As the game moves into a "smart grid era" with distributed energy production, this matching of supply and demand becomes more important. Evaluation of the game prototype showed that players gained insight in the upcoming changes and what would be expected of them as end users in local microgrids. This game does not yet connect to real-life situations of energy consumption like Energy Battle does. Consider a combination of the Energy Battle and Sasér, in which the (simulated) smart grid situation is played in connection to real energy consumption and production data. The game setting would motivate players to try out new behaviors in their home regarding balancing supply and demand in the community of players. When, at the same time, smart grid technology is available in households, the behaviors developed in the game could (partly) be continued and a community of active co-providers developed.

4.5.3   Social Interaction in Interaction Design

As pointed out in this chapter, the user can be motivated to respond to goals put forward by a home energy system providing the rational is clear and the user is able to act upon them. Similarly, the user could be triggered to respond to social-goals. For example, users could be motivated to consume at or below the energy consumption levels of similar households in a local neighborhood.

The interface between the house and its user-residents can be considered a means of communication. Given the development of local microgrids in which not only the supply and demand of individual houses but also those of the whole community are relevant, energy consumption and production on a community level have to be provided. Using social interaction to facilitate energy-efficient behavior of the community is a logical step.

Concerning interaction design, there is little work still relating to social interaction and community-based approaches to stimulate energy-efficient behavior. A recent literature review of "sustainable HCI" publications (Disalvo, Sengers, and Brynarsdóttir, 2010) pointed out that the area is still very new for the HCI research community. They argue that there is a lack of common understanding or views on sustainable HCI research and that many studies are overlapping. They observe that, in line with the dominant research literature on behavior change, individual consumers have frequently been addressed. The authors argue that it is also necessary to extend knowledge about stimulating sustainable behavior by addressing groups of people.

An example that focuses more on social interaction is Energy Mentor. Energy Mentor, complementary to an energy meter, can create great potential to save energy as it transforms the sterile indications of the energy meter into rich suggestions and tangible actions. The Energy Mentor creates a network of experts that can be called upon at a moment's notice to answer questions about energy and sustainable living. By integrating the Energy Mentor in people's existing workflow of using a cell phone or computer, one can easily and quickly ask questions and get personal, precise answers. Listening to real stories of people that have already taken on the challenge to minimize their energy consumption can be more engaging and fascinating than browsing endless lists on the Web. People can text ideas and solutions to each other or even send a video of themselves, for instance, making home energy improvements. Last but not least, the Energy Mentor offers the opportunity to find other local users who are trying to live a more sustainable lifestyle. This can create great dynamics on a local level, fostering more chances to grow grassroots movements and fostering the success of centers for sustainable living (Papantoniou, 2009; Papantoniou et al., 2009; Figures 4.4.1, 4.4.2, and 4.4.3).

Figure 4.4.1   Asking and receiving answers with Energy Mentor (Papantoniou, 2009)

Figure 4.4.2   Visualising the location of Energy Mentor community members (Papantoniou, 2009)

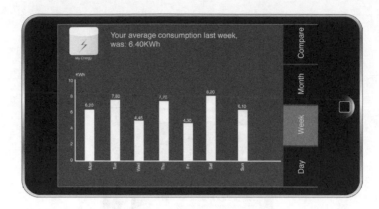

Figure 4.4.3   Energy meter as part of the Energy Mentor service (Papantoniou, 2009)

4.6   Conclusions

This chapter elaborated on the role of social interaction as a means to stimulate energy-efficient behavior. In light of the ongoing developments toward smart grids, people are supposed to become active participants in the management of (local) energy provision system. Whereas most past interventions aimed to stimulate energy-efficient behavior, these developments can particularly benefit from a social, community-based approach to stimulating behavior change so that residents are not only physically but also socially connected in the electricity provision grid. Community-based approaches have already shown their potential to leverage behavior change, but research about energy consumption behavior is dominated by individual approaches. This is reflected as well in design strategies aimed at stimulating pro-environmental behaviors.

The potential of knowing the consumption levels of neighbors, as a means of social influence, is highlighted in the design of OPOWER, which has delivered about 400 GWh of electricity in the United States (Lasky and Kavazovic, 2011) and provides descriptive social-normative messages as a central aspect of home energy reports. Further research about how design can contribute to social approaches toward facilitating energy-efficient behavior is required to support developments toward a smarter energy system and allow end users to be truly active participants in the management of supply and demand.

References

Ackermann, T., Andersson, G., and Söder, L. (2001) Distributed energy generation and sustainable development. *Renew. Sustainable Energy Reviews*, **57**, 195–204.

Alanne, K. and Saari, A. (2006) Distributed energy generation and sustainable development. *Renew. Sustainable Energy Reviews*, **10**, 539–558.

Breukers, S.C., Heiskanen, E., Brohmann, B. *et al.* (2011) Connecting research to practice to improve energy demand-side management (DSM). *Energy*, **36**, 2176–2185.

Caillois, R. (1962) *Man, Play and Games*, Thames and Hudson., London.

Charles, D. (2009) Renewables test IQ of the grid. *Science*, **324**, 172–175.

Costa, E. (2011) SASÉR: Educative serious game about Smart Grids and renewable resources. Master's thesis, Delft University of Technology, Faculty Industrial Design Engineering, Delft.

DeKoven, E. (2004) Help me help you: Designing support for person-product collaboration. PhD thesis, Delft University of Technology, Delft.

Disalvo, C., Sengers, P., and Brynarsdóttir, H. (2010) Mapping the landscape of sustainable HCI. CHI 2010 Conference, Atlanta.

Djajadiningrat, J.P., Wensveen, S.A.G., Frens, J.W., and Overbeeke, C.J. (2004) Tangible products: Redressing the balance. *Personal and Ubiquit. Computing*, **8**, 294–309.

Ehrenfeld, J.R. (2008) *Sustainability by Design: A Subversive Strategy for Transforming Our Consumer Culture*, Yale University Press, New Haven, CT.

Elias, E.W.A., Dekoninck, E.A., and Culley, S.J.C. (2007) The potential for domestic energy savings through assessing user behavior and changes in design. EcoDesign 2007 Conference, Tokyo.

European Technology Platform SmartGrids (ETPS) (2011) Smart Grids European Technology Platform. http://www.smartgrids.eu/ (accessed April 14, 2012).

Freudenthal, A. (2002) Designing transgenerational usability in an intelligent thermostat by following an empirical model of domestic appliance usage. 21st European Annual Conference on Human Decision Making and Control, Glasgow July 15–16 (ed. C. Johnson), pp. 134–142.

Gardner, G.T. and Stern, P.C. (1996) *Environmental Problems and Human Behavior*, Pearson Custom Publishing, Boston, MA.

Geelen, D., Brezet, H., Keyson, D., and Boess, S. (2010) Gaming for energy conservation in households. ERSCP-EMSU Conference, Delft, The Netherlands (ed. R. Wever, J. Quist, A. Tukker, J. Woudstra, F. Boons, and N. Beute).

Groot-Marcus, J., Terpstra, P., Steenbekkers, L., and Butijn, C. (2006) Technology and household activities, in *User Behavior and Technology Development* (ed. P. Verbeek and A. Slob), Springer, Berlin.

Heiskanen, F.E., M., Robinson, S. *et al.* (2010) Low-carbon communities as a context for individual behavioral change. *Energ. Policy*, **38**, 7586–7595.

Hopkins, R. (2010) *Transition Towns Handbook*, Chelsea Green Publishing, White River Junction, VT.

Hopper, J.R. and Nielsen, J.M. (1991) Recycling as altruistic behavior: Normative and behavioral strategies to expand participation in a community recycling program. *Environ. Behav.*, **23**, 195–220.

Huizinga, J. (1949) *Homo Ludens: A Study of the Play-Element in Culture*, Routledge and Kegan Paul Ltd, London.

International Energy Agency (2011) *Technology Roadmaps – Smart Grids*, OECD/IEA, Paris.

Jackson, T. (2006) *Motivating Sustainable Consumption – A Review of Models of Consumer Behaviours and Behavioural Change: A Report to the Sustainable Development Research Network*, Sustainable Development Research Network, London.

Jégou, F. and Manzini, E. (eds) (2008) *Collaborative Services – Social Innovation and Design for Sustainability*, Edizioni POLI.design, Milan.

Keyson, D.V., Freudenthal, A., Dekoven, E., and de Hoogh, M.P.A.J. (2001) European patent no. 1014792. Intelligent Thermostat. TU Delft.

Kuijer, L. and De Jong, A. (2009) A practice oriented approach to user centered sustainable design. Ecodesign 2009 Conference, Sapporo.

Lasky, A. and Kavazovic., O. (2011) OPOWER: Energy efficiency through behavioral science and technology. XRDS Magazine, June 1.

Lewin, K. (1947) Group decision and social change, in *Readings in Social Psychology* (ed. T.M. Newcomb and E.L. Hartley), Holt, New York, pp. 459–473.

Lilley, F.D. (2009) Design for sustainable behavior: Strategies and perceptions. *Design Studies*, **30**, 704–720.

Lockton, D., Harrison, D., and Stanton, N.A. (2010) The design with intent method: A design tool for influencing user behavior. *Appl. Ergon.*, **41**, 382–392.

Lutzenhiser, L. (1993) Social and behavioural aspects of energy use: Annual reviews. *Energy Environ.*, **18**, 247–289.

Manzini, E. (2008) New design knowledge, introduction to the changing the change conference. Changing the Change Conference, 2008, Torino.

Marris, E. (2008) Upgrading the grid. *Nature*, **454**, 570–573.

Norman, D.A. (2007) *Design of Future Things*, Basic Books, New York.

Papantoniou, L. (2009) Making the leap from awareness to action: The Energy Mentor. Master's thesis, Delft University of Technology, Delft. http://repository.tudelft.nl/view/ir/uuid%3Af6df30d5-658a-491a-a144-82c60ce2235a/ (accessed April 20, 2012).

Papantoniou, L., De Jong, A., Crul, M.R.M., and Geelen, D. (2009) Energy Mentor. ERSCP-EMSU, Delft.

Rasmussen, J. (1986) *Information Processing and Human–Machine Interaction*, North Holland, New York.

Rich, C., Sidner, C., and Lesh, N. (2001) Collagen: Applying collaborative discourse theory to human–computer interaction. *AI Magazine*, **22** (4), 15–25.

Rijsdijk, S.A. (2006) *Smart products: Consumer evaluations of a new product class*. PhD dissertation, Delft University of Technology, Delft.

Ritterfeld, U., Cody, M., and Vorderer, P. (Eds.). (2009) *Serious Games: Mechanisms and Effects*. Routledge, New York.

Staats, H., Harland, P., and Wilke, H.A.M. (2004) Effecting Durable Change: A Team Approach to Improve Environmental Behavior in the Household. *Environment and Behav.*, **36**, 341–367.

Van Dam, S., Bakker, C.A., and Van Hal, J.D.M. (2012) Insights into the design, use and implementation of home energy management systems. *J. Design Research*, **10** (1–2), 86–101.

Van Vliet, B., Chappells, H., and Shove, E. (2005) *Infrastructures of Consumption: Environmental Innovation in the Utility Industries*, Earthscan, London.

Weenig, W.H. and Midden, C.J.H. (1991) Communication network influences on information diffusion and persuasion. *J. Pers. Soc. Psychol.*, **61**, 734–742.

Wever, R., Van Kuijk, J., and Boks, C. (2008) User-centred design for sustainable behavior. *Int. J. Sustainable Engineering*, **1**, 9–20.

Wilson, C. and Dowlatabadi, H. (2007) Models of decision making and residential energy use. *Annual Review of Environment and Resources*, **32**, 169–203.

Wolsink, M. (2011) The research agenda on social acceptance of distributed generation in smart grids: Renewable as common pool resources. *Renew. Sustainable Energy Reviews*, **16**, 822–835.

Wood, G. and Newborough, M. (2003) Dynamic energy-consumption indicators for domestic appliances: Environment, behavior and design. *Energ. Buildings*, **35**, 821–841.

Case A

SolarBear: Refrigeration for the Base of the Pyramid through Adsorptive Cooling

Leonard Schürg, Jonas Martens, Roos van Genuchten and Marcel Crul

Delft University of Technology, Faculty of Industrial Design Engineering, Landbergstraat 15, Delft The Netherlands

A.1 Introduction

Big challenges exist for developing off-grid, reliable, and affordable solar refrigeration for businesses at the economic base of the pyramid (BoP). SolarBear, an initiative started in 2010 by DUT MSc students of the Faculty of Industrial Design Engineering, is looking for solutions to this challenge by applying adsorption-cooling technology to consumer products for the BoP.

A.1.1 The Need for Off-Grid Refrigeration in BoP Small-Scale Businesses

Major problems are linked to absent or unreliable refrigeration in emerging economies. The lack of preservation and cooling capacity for the crop and livestock of farmers and fishers in BoP markets causes high price volatility. In addition, cooling in BoP "hot zones" requires a lot of power. Grid-connected electricity is still not available in many locations, and off-grid energy production is costly. In India alone, hundreds of millions of people live without access to electricity, and therefore do not have access to reliable refrigeration.

The Power of Design: Product Innovation in Sustainable Energy Technologies, First Edition.
Edited by Angèle Reinders, Jan Carel Diehl and Han Brezet.
© 2013 John Wiley & Sons, Ltd. Published 2013 by John Wiley & Sons, Ltd.

A.1.2   Existing Cooling Solutions

Off-grid refrigeration solutions exist for many different applications and markets, from gas-powered medical vaccination cooling over cool boxes with ice to generator-powered (household) refrigerators. All entail high costs of use. Evaporation coolers can have low initial costs and costs of use, but they work only in very dry climates. Photovoltaic (PV) solar-powered coolers usually have high initial costs. However, there are several solutions for low-cost refrigeration for the BoP: Based on a thorough technical exploration, the SolarBear initiative has chosen adsorption cooling as one of the most cost-effective, relatively accessible, and possibly modular technologies suitable for the main demands of BoP markets.

The decisive argument for using adsorption technology in combination with a simple flat-plate heat collector was the competitive low-price promise. In case of a high-production volume of an integral modular system, very low prices can be reached (see Figures A.1.1 and A.1.2).

A.2   The SolarBear Approach

To come to an integral solution, a parallel approach was chosen in the SolarBear initiative. This means that while technical research into practical application of adsorption cooling starts with assumptions about the market, the product design (prototyping cycles) will be driven by predictions about the technology. A third activity is market development in a selected initial market – in this case, India. This leads to the approach discussed throughout the remainder of this section.

A.2.1   Market Opportunities Research

Overlay maps of solar insulation, Gross Domestic Product (GDP), and grid connection rate were produced in order to identify areas of large interest. Possible markets were identified together with experts, and the size of the market was roughly calculated. Based on this exploratory research, teams were sent to verify the market expectations in various regions and identify local needs; field studies were performed in Surinam, Ghana, and India.

A.2.2   Technical Opportunities Research and Prototyping

First, a literature review was performed that reviewed existing solutions to similar problems and evaluations, zooming in on solar adsorption cooling. An in-depth review of that technology's possibilities followed, serving as the starting point for the first prototype.

With the results of this first effort, a second prototype system was developed that satisfied identified market needs with the given technological possibilities. Further improvement of the system was made, in dialog with the next prototyping cycle.

A.2.3   Market Development in India

As a first step, possible partners in India were identified, followed by local research on the cooling needs of business chains in the BoP. Market opportunities, established as combinations of needs and technical possibilities, were defined and elaborated.

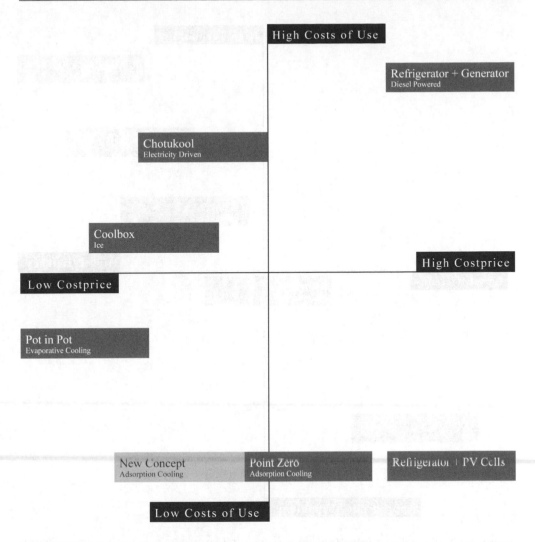

Figure A.1.1 Cost price and cost of use position of the new SolarBear concept, compared with the competition

A.3 Results of the First Cycle of Product Development: Proof of Concept and Market

An overview of the wide range of available cooling technologies was made, and the possibilities were assessed. The same was done for solar collection technologies. Adsorption cooling was chosen as the preferred technology, as it can be provided with solar power and off the grid, with relatively simple components that can be produced and/or assembled locally, a low-cost price, and low costs of use.

Adsorption cooling is performed in an intermittent cycle, collecting energy in phase 1 of the cycle during the day (in the sun), and cooling the system in phase 2 during the night (out of the sun). When ice is produced during this cooling phase, this can be used to cool products

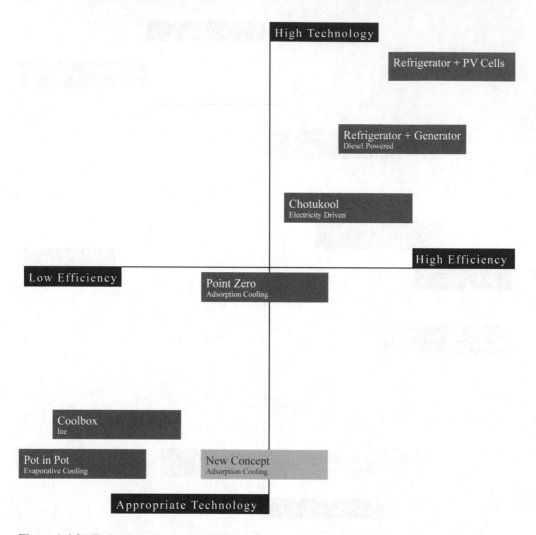

Figure A.1.2 Technology level and efficiency position of the new SolarBear concept, compared with the competition

the next day. The system works basically at two different pressure–temperature equilibrium conditions: the cooling phase (phase 2) at low pressure and temperature, and the generating phase (phase 1) at high pressure and temperature. The speed and efficiency of the process are dependent on the temperature and pressure differences achievable between the three main parts of the system: collector, condenser, and evaporator (Figure A.3.1).

A.3.1 First Prototype

As a first result, the basic functionality of adsorption cooling was proved with a first prototype (Figure A.3.2). During the prototyping phase, gas tightness of the system appeared to be the most severe problem. This should be solved in the production prototype.

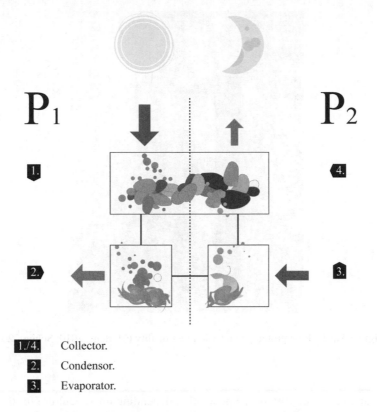

Figure A.3.1　Intermittent adsorption cycle simplified (Martens, 2010)

A.3.2　Second Prototype

Next, a second prototype was developed in which market requirement choices were made and connected to the technology. The prototype was made as a small system for a low-income BoP market, to be used by an entrepreneur selling cooled items during daytime, outside in the sun. The very basic system is provided (with or without frame), and it is adjustable to one's own needs and liking, which means that it can be built with locally available materials within the context of one's local culture (Figure A.3.3a–c).

A.3.3　Product-Service System Development for SolarBear in Lakshmikantapur, India

Having developed the prototypes, a thorough market entry strategy and business model were developed. A focus on rural West Bengal in India was chosen for field research into what would be the best market in which to further develop the technology. India has 400 million inhabitants who do not have access to the electricity grid. The northeastern states, one of which is West Bengal with its capital city of Kolkata (Calcutta), are the least developed in

Figure A.3.2   First prototype on basic functionality (Martens, 2010; Schürg, 2010)

India, with an average of 70% of the population depending on agriculture for their incomes. The culture is one of long-term orientation, more so than the short-term-oriented African countries, but the uncertainty avoidance of the region is not as high as in other Asian countries (Hofstede, 1991), which makes them adopt changes more easily. The humid and hot climate makes crops perish quickly and off-grid forms of evaporation cooling unsuccessful. The specific choice of West Bengal also depended on locating a viable partner for the project, which was found in ONergy in Kolkata.

A product service system (see Figure A.3.4) was needed to effectively implement the product in the market. This was because low-income people who have little to spend might need personal economic value creation by buying the product in order to afford it. However, affordability and desirability are different things: A study into what people actually wanted to spend their scarce money on was also needed.

Two high-potential target groups were chosen for further concept development. Different from the original assumption, one direct, attractive market was for domestic fridges, whose consumers did not require additional value creation. The other target group was small farmers, who typically owned 0.5–1.5 hectares of land and produced various kinds of crops. Because of the lack of cooling, at that time 30% of harvests perished. Furthermore, these small farmers are dependent on the prices offered by middlemen on the day of harvest, which makes them vulnerable to low prices and lack a good bargaining position.

A modular SolarBear system was defined in which parts can be produced in a local factory and installed locally by micro entrepreneurs who will be trained by ONergy, the local distributor (see Figure A.3.5). The modular system consists of an on-roof vacuum collector, a pipe

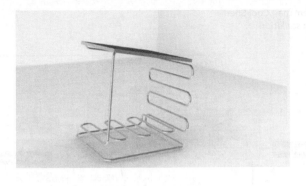

Figures A.3.3 (a–c): (Top) Bare system; (middle) system with frame; and (bottom) system with example exterior made from local materials

to connect the collector to the rest of the system, and the cooling system itself. For the domestic fridge, the refrigerator room consists of an existing fridge housing with a bigger condenser and evaporator built in. For the cold storage room, the same collector and condenser and evaporators are used, but the amount will vary and the room will be built according to the required volume. A local installer will build it with locally used clay walls but will

Product service system map Cool Shed
> *system of sale and installation*

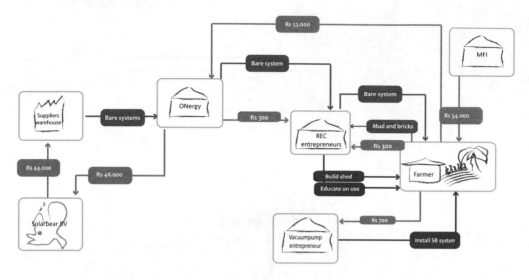

For the actual sale and installation of one cold storage shed, the depicted service, product and money streams are active:

A) ONergy pays Solarbear, who pays the manufacturer and orders the products to be sent
B) All parts of the bare system (the cooling technology) are being shipped to ONergy
C) The farmer gets a loan at a micro finance institute (MFI) and pays ONergy the purchase price
D) The farmer assembles the necessary mud and bricks for the new shed and prepares the tush and husk into building material
E) The farmer pays the REC entrepreneur to build the shed and deliver and install the technology
F) The farmer pays the vacuum entrepreneur for installation

Figure A.3.4   Part of the product service system (PSS) cool shed

also build an extra isolated cold chamber within with Styrofoam boards. The investment needed for this installation will come from a microloan lent by a micro finance partner, Vive-kananda Sevakendra O Sishu Uddyan (VSSU).

When the system is successfully developed and built, a pilot can be set up in Lakshmi-kantapur using VSSU facilities for warehousing and repairing. After evaluation, sales can start with promotion campaigns and by training micro entrepreneurs together with ONergy.

A.4   Future Work: A Working Prototype and Further Development by Enviu

The second cycle of product development starts with the design of a new system that is a combination of the successful factors of previous designs. The new prototype (Figure A.4.1)

Cool shed assembly

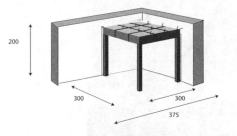

1. build the four shed walls (of which only two are depicted here)
2. build frame with the 9 evaporator boxes
3. place styrofoam walls to outerwall and inside for room division

4. connect condensors to the evaporators
5. Bring the systems on the right pressure per system

6. connect collectors with the condensors
7. open the valves to connect the systems

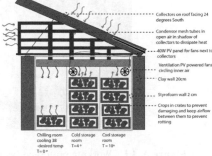

8. bring all systems to their desired pressure, depending on junctions:
a. chilling: evaporation tem -8 C for forming ice
b. keep cold: evaporation temp 4 C
c. keep cool: evaporation temp 10 C

Figure A.3.5   Assembly steps of rural cold shed (van Genuchten, 2011)

includes a modular connection system, which enables the exchange of every part and measures the changed performance. Advanced sensor technology logs every change of temperature and pressure, so that correct understanding of all situations is possible. By a series of tests, the system will be improved bit by bit.

Since 2011, the concept of affordable cooling for BoP entrepreneurs, as generated in this SolarBear series of MSc graduation projects, has been adopted by the Dutch sustainable innovation organization Enviu. Enviu is now taking up the challenge to improve the cold chain in India, and is working on several concepts that will help Indian farmers to store and transport harvests in a more sustainable and profitable way, and reduce postharvest losses.

The Enviu project will further investigate the technical and social solutions of improving the cold chain, of which adsorption technology as developed in SolarBear is one of the options.

Figure A.4.1   Working prototype (van Genuchten, 2011) (See Plate 24 for the colour figure)

References

Enviu (2012) Enviu Wow! Project Cold Chain. http://enviu.nl/?ac=wow%21projects-29-1 (accessed April 10, 2012).

Hofstede, G. (1991). Cultures and Organisations: Software of the Mind. London, McGraw-Hill.

Martens, J.R.M. (2010) Design of a user-centered Solar Refrigerator. TU Delft MSc Thesis.

Schürg, L. (2010) A low-cost Refrigeration System for Developing Countries. TU Delft MSc Thesis.

van Genuchten, R.P. (2011) Solar Refrigeration in Rural West Bengal – Design of a Product-Service System. TU Delft MSc Thesis.

Case B

Environmental Impact of Photovoltaic Lighting

Bart Durlinger

RMIT University, Centre for Design, Melbourne, Australia

B.1 Introduction

Photovoltaic (PV)–lighting products are increasingly used to provide rural households in developing countries with clean and safe lighting. These households commonly have no access to electricity from a power grid (Legros, Havet, Bruce, & Bonjour, 2009). They currently rely on either kerosene lamps or electricity from car batteries. PV-lighting products deliver a higher quality of service than conventional lighting sources; they are safer and more reliable (Figure B.1.1).

The PV-lighting products that are studied in this case study are designed for use in rural areas in Cambodia, where no electricity grid is currently available. The products are typically smaller than solar home systems in terms of both physical dimensions and energy capacity. There are indications that PV-lighting products are environmentally beneficial when compared to conventional lighting solutions. A method to qualify and quantify such an environmental benefit is through undertaking a life cycle assessment (LCA). This case study uses the LCA framework to assess the environmental impacts of the lighting products considered.

LCA is the process of evaluating the potential effects that a product, process, or service has on the environment over the entire period of its life cycle.

Figure B.1.2 illustrates the life cycle system concept of natural resources and energy entering the system while products, waste, and emissions leave the system.

The Power of Design: Product Innovation in Sustainable Energy Technologies, First Edition.
Edited by Angèle Reinders, Jan Carel Diehl and Han Brezet.
© 2013 John Wiley & Sons, Ltd. Published 2013 by John Wiley & Sons, Ltd.

Figure B.1.1 Impressions of solar lighting, in this case the Moonlight manufactured for the Cambodian market (See Plate 25 for the colour figure)

The International Standards Organization (ISO) has defined LCA as follows:

A compilation and evaluation of the inputs, outputs and the potential environmental impacts of a product system throughout its lifecycle.

(International Organization for Standardization, 2006)

An LCA consists of four main components of equal importance:

- Goal and scope, that is, definitions of the systems considered and what needs to be resourced.
- A life cycle inventory, a collation of all relevant process data including material use, energy inputs, and emissions.
- A life cycle impact assessment, relating the emissions to a discrete number of environmental impacts.
- An interpretation of these stages.

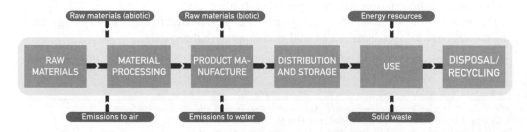

Figure B.1.2   Life cycle system concept (International Organization for Standardization, 2006)

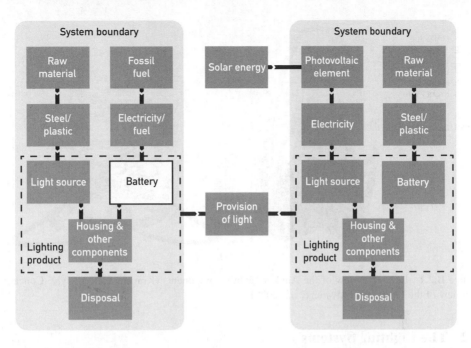

Figure B.1.3   Simplified diagrams of (left) a product system relying on an external power source, and (right) a product with integrated photovoltaics (PV)

A system boundary describes the processes that are taken into consideration. For this particular case study, a system boundary diagram is shown in Figure B.1.3.

The PV-lighting products considered in this study are sold to rural households in Cambodia to substitute kerosene lights, battery-operated torches, or candles. Some products have enough capacity to power more than just lights. However, for this case study these other functions are not taken into consideration (i.e., it is assumed that all energy is used for lighting). Because the lighting systems vary in size and capacity, it would be unfair to compare products one on one. Instead, the systems are compared by their delivery of a certain amount of service, in this case the provision of light. Therefore the following unit of service (or *functional unit*) is used in this study:

> Functional unit: The provision of light, with strength of 100 lumens, for 3 hours a day, over a period of 1 year.

This functional unit represents typical lighting conditions for the range of products that are considered in this study. The systems can be compared by scaling them to this size (scaling the operation time and lighting strength). A lighting strength of 100 lumens is comparable to the amount of light that is emitted by an 8 W incandescent light bulb, or a 2 W compact fluorescent light bulb (CFL). Lights are typically used for 3 hours daily. The period of 1 year is included for practical reasons; it eliminates possible effects from seasonal differences (e.g., irradiation and operating time), so yearly averages are used. Also lifetimes of investigated products are expressed in years and can therefore more easily be compared to the functional unit.

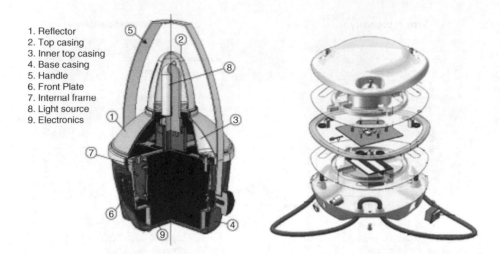

1. Reflector
2. Top casing
3. Inner top casing
4. Base casing
5. Handle
6. Front Plate
7. Internal frame
8. Light source
9. Electronics

Figure B.2.1   Left: Overview of the Angkor Light's components (Kranen, 2007). Right: Conceptual overview of the Moonlight (Alvarez *et al.*, 2008)

B.2   The Lighting Systems

B.2.1   System 1: Angkor Light

The Angkor Light (see Figure B.2.1, left) is a solar lantern fitted with a 3 W CFL and therefore able to deliver a considerable amount of light. The light can be used as a focused light (using the reflector) or as a general-purpose light (for lighting a room). The lantern contains a 4.5 Ah sealed lead–acid battery and comes with a 4.5 Wp solar panel.

B.2.2   System 2: Moonlight

The Moonlight is a smaller scale solar-lighting product. The lantern (pictured in Figure B.2.1, right) is powered by a 0.7 Wp solar panel. The energy that is produced in the daytime is stored in two AA NiCd batteries. This allows the Moonlight to produce light in the evening and night, using six light-emitting diodes (LEDs). The Moonlight can produce 29.5 lumens for 4 hours, or dimmed light for a longer period.

B.2.3   System 3: Solar Home System

The solar home system (SHS) that is used in this study is the Kamworks Solar Home System "medium" (see Figure B.2.2). This system is equipped with a 40 Wp Kyocera solar panel and a 48 Ah lead–acid battery. Also included are the home box (which contains the battery and charging electronics) and three 7 W compact fluorescent lights.

   The system can deliver a larger daily amount of energy than is needed to power the light bulbs; this additional energy can be used to power a radio or television, for example. However, to compare the solar home system to the other product systems, it is assumed that all

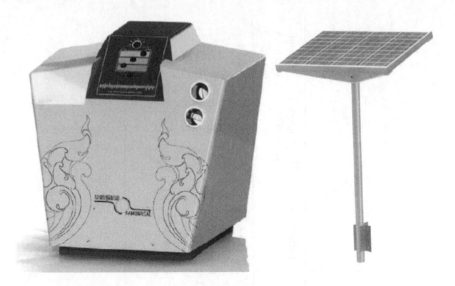

Figure B.2.2   Solar home system medium, home box, and solar panel with mounting structure (Diessen, 2008)

available energy will be used for lighting (by using more light bulbs, e.g., to light multiple homes).

Technical details of the solar-lighting products are summarized in Table B.2.1.

B.2.4   System 4: Light Delivered by Battery (Charged at Charging Station, Using Diesel)

The most common form of electricity use in Cambodia is the charging of lead–acid batteries at a charging station, and using the charged battery at home to power lights, radio, and/or television. The size of the batteries varies, as does the charging regime. There are many operators who offer their services to villagers, with varying knowledge and capabilities. Generally speaking, batteries are charged using electricity from a diesel-powered generator. The battery has to be transported to and from the charging station by the customer (often by bike or on foot; Figure B.2.3).

Due to poor battery management, battery lifetime and capacity are much lower than what are theoretically possible. Van Diessen (2008), as a student who worked in Cambodia on a lighting project, describes the situation as follows:

> The way of charging is very damaging, the battery capacity decreases rapidly and usually a battery breaks down within a year of use. The batteries are charged without a charge controller, directly by the generator. They are often just tied together by some uncovered copper wires, independent on their state of charge, type or age. Therefore unsafe situations easily occur and overcharging is impossible to prevent. For example, during the field research an operator just determined the moment to stop charging, when a battery in the line started to boil.
>
> (Diessen, 2008, p. 20)

Table B.2.1 Technical details of PV-lighting products

System	Lighting strength (lumens)	Light source	Solar panel size (Wp)	Solar cell type	Battery capacity	Battery type	Daily energy input (Wh)	System efficiency (estimated)
System 1: Moonlight	29.5	6 LEDs	0.7	a-Si	2000 mAh	NiCd (AA type) (1.5 V)	3.15	63%
System 2: Angkor Light	150	One 3 W CFL	4.5	a-Si	4.5 Ah	Lead–acid (12 V)	20.25	44%
System 3: Solar home system	1050	Three 7 W CFL	40	Multicrystalline Si	48 Ah	Lead–acid (12 V)	180	65%

Figure B.2.3   Impressions of battery charging at a charging station (Diessen, 2008)

To model this situation, it is assumed that a battery has a lifetime of 1 year, and during its lifetime has an average capacity of 70% of its rated capacity (thus, a 100 Ah lead–acid battery will deliver only 70 Ah on average). The charging efficiency is highly dependent on the state of charge (SOC) (Stevens and Corey, 1996); the higher the SOC, the lower the efficiency is. At a very low SOC the charging efficiency can be as high as 95%, whereas the efficiency drops to around 50% at a very high SOC. An average of 70% was taken as a typical charging efficiency.

B.2.5   System 5: CFLs and Electricity from the Grid

To determine if PV-lighting solutions are beneficial when there is also access to grid electricity, a system of a CFL connected to grid electricity is also considered. A readily available model for Chinese electricity (largely coal-fired power plants) is used as a proxy. The system comprises a CFL, electrical wiring, and emissions associated with electricity production.

B.2.6   System 6: Kerosene Lamp

Another widely used source of light is kerosene lamps (Figure B.2.4). A fuel is burned on a wick to provide light. The fuel consumption rate can be adjusted by moving the wick up or

Figure B.2.4   Example of a kerosene lamp

Table B.2.2   Technical details of conventional lighting systems

System	Lighting strength (lumens)	Light source	Battery capacity	Battery type	Daily energy input (Wh)
System 4: Battery charged at charging station plus fluorescent light tube (CFL 18 W)	900	One 18 W CFL	70 Ah	Lead–acid (12 V)	23.3 Ah, used daily
System 5: CFL plus grid electricity	900	One 18 W CFL	—	—	33 Wh of grid electricity used daily
System 6: Kerosene light	45	Kerosene combustion	—	—	0.035 l kerosene/hour

down, adjusting the size of the flame. According to a report by Alsema (2000), the fuel consumption of a typical kerosene lamp is between 0.02 and 0.05 L per hour, delivering a lighting strength of 45 lumens. Technical details of the "conventional" lighting products are shown in Table B.2.2.

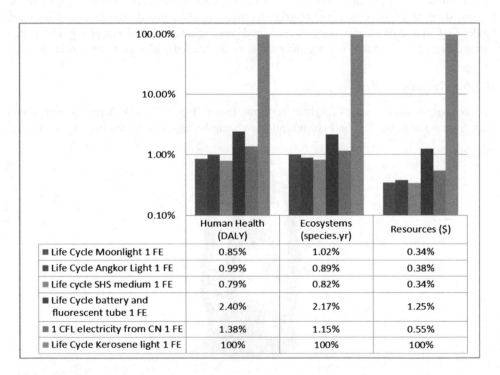

	Human Health (DALY)	Ecosystems (species.yr)	Resources ($)
■ Life Cycle Moonlight 1 FE	0.85%	1.02%	0.34%
■ Life Cycle Angkor Light 1 FE	0.99%	0.89%	0.38%
■ Life cycle SHS medium 1 FE	0.79%	0.82%	0.34%
■ Life Cycle battery and fluorescent tube 1 FE	2.40%	2.17%	1.25%
■ 1 CFL electricity from CN 1 FE	1.38%	1.15%	0.55%
■ Life Cycle Kerosene light 1 FE	100%	100%	100%

Figure B.3.1   Damage assessment using a ReCiPe impact assessment method, logarithmic scale

Table B.3.1   Life cycle impacts for lighting products, kerosene light excluded

Damage category	System 1: Moonlight	System 2: Angkor Light	System 3: Solar home system	System 4: Battery and fluorescent tube	System 5: CFL plus grid electricity
Human Health (DALY)	36%	41%	33%	100%	57%
Ecosystems (species/year)	47%	41%	38%	100%	53%
Resources ($)	27%	30%	27%	100%	44%

B.3   Environmental Impacts and Discussion

Environmental impacts are assessed by using the ReCiPe 2008 impact assessment method's (Goedkoop *et al.*, 2009) end points, which are damage to human health, damage to ecosystems, and depletion of resources.

The results in Figure B.3.1 show that the PV-lighting products generally have a lower environmental impact when compared to the other, "conventional" sources of light. The kerosene lamp is the most harmful to the environment on all categories considered. When the kerosene lamp is removed from the assessment, the relative differences between the other lighting systems become more visible. This is shown in Table B.3.1.

From Table B.3.1, it can be seen that the battery charged with a diesel generator has a larger impact than the other systems considered; however, the impacts from the light plus grid are comparable to those of the "worst" PV-lighting product (although still slightly higher).

It is interesting to see that the drivers of impacts are widely different between the "conventional" lighting solutions and the PV lights. Figure B.3.2 shows that Moonlight's impacts mainly originate from manufacturing processes, and to a lesser extent the disposal phase. However, impacts from the kerosene light are mainly fuel related, with the majority of emissions on site (when the product is used).

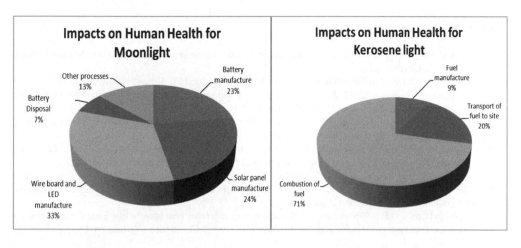

Figure B.3.2   Shares of impacts on human health. Left: The Moonlight. Right: The kerosene light. Note that the total impact of the Moonlight is a factor 100 less than that of a kerosene light

Although the results seem optimistic for the PV lights, there is reasonable uncertainty regarding some aspects of the inventory (Durlinger, Reinders, & Toxopeus, 2012). Some system components have a relatively large impact and large uncertainty (e.g., battery disposal and the printed circuit board), and therefore have to be researched in more detail. Especially worrying in that sense is the relatively high impact that Ni from NiCd batteries has on ecotoxicity. It is shown, in the sensitivity analysis, that a responsible handling of batteries will improve Moonlight's environmental profile. Due to poor data quality on batteries, it is impossible to recommend alternatives to the currently used NiCd batteries.

B.4   Conclusion

The solar-lighting products considered in this case have a lower impact than current lighting solutions, but are comparable to grid power. However, their environmental profile can be improved by adequate battery waste management. Still, one should be careful when interpreting the results of this study, as there are a number of data points with a relatively low data quality. For instance, waste management of PV is not well understood; because of the short existence of PV technology, end-of-life scenarios have not yet been perfected. Also, no good models exist yet to determine the effect of materials degradation directly in nature. LCA data are often based on European conditions, which assume well-managed landfill sites.

Acknowledgments

This case study summarizes part of the work that was done during my graduation project in 2009–2010 at the University of Twente. Many people were involved in this work, particularly Dr. Angèle Reinders and Ir. Marten Toxopeus. Also, many of the data used in this study originated from Kamworks, which develops and sells photovoltaic (PV) lighting systems for the Cambodian market.

References

Alsema, E.A. (2000) *Environmental Life Cycle Assessment of Solar Home Systems*, Department of Science Technology and Society, Utrecht.

Diessen, T.v. (2008) *Design of a Solar Home System for Rural Cambodia*, TU Delft and Kamworks, Delft.

Durlinger, Reinders, Toxopeus, (2012) *A comparative life cycle analysis of low power PV lighting products for rural areas in South East Asia. Renewable Energy*, 41, May 2012, Pages 96–104.

Goedkoop, M., Heijungs, R., Huijbregts, M. *et al.* (2009) *ReCiPe 2008, Part I: Characterisation*, Ministry of Housing, Spatial Planning and the, Environment, The Hague.

International Organization for Standardization (2006) *ISO 14044: Environmental Management – Life Cycle Assessment – Requirements and Guidelines*, International Organization for Standardization, Geneva.

Legros, G.,Havet, I., Bruce, N., & Bonjour, S. (2009). *The energy access situation in developing countries - a review focussing on the least developed countries and sub Saharan Africa. Report*. New York: United Nations Development Programme and World Health Organisation.

Stevens, J.W. and Corey, G.P. (1996) A study of lead-acid battery efficiency near top-of-charge and the impact on PV system design. Paper presented at the 25th Photovoltaic Solar Energy Conference, Washington, DC.

Case C

Restyling Photovoltaic Modules

Michael Thung

Department of Design, Production and Management of University of Twente, Enschede, The Netherlands

C.1 Introduction

At present the fastest-growing renewable energy technology is photovoltaics (PV). This development provides many opportunities for solar cell manufacturers. Manufacturers become more aware of the importance of increasing solar cell efficiency and of the possibilities of designing new types of PV modules with solar cells with an innovative design.

Most of the crystalline PV solar cells have a standard design and appearance. They are blue colored, and the metal contact pattern of these cells is often bounded by an H-shaped pattern (Figure C.1.1). Colors other than blue can be achieved by varying the thickness of the antireflective coating on the top of the cells. This is interesting for PV modules that are implemented primarily in an urban environment, which asks for not only high efficiency but also an aesthetically pleasing appearance.

This case considers the restyling of PV modules by modifying the metal contact pattern on multicrystalline solar cells. The main focus is on two-sided contact solar cells, which are usually screen printed on the front side with an H-shaped contact pattern consisting of two busbars that are connected to many smaller contacts, called *fingers*. Radike and Sumhammer (1999) investigated new possible contact patterns; their paper is the starting point of this case, which was formulated as "the development of a new screen-printed solar cell by designing contact patterns."

The aim was to explore the possibilities in contact pattern designs and to realize functioning prototypes of the redesigned solar cell. During the project, several parties were involved: solar cell producer Solland Solar, the University of Stuttgart, the University of Twente, and PV module producer Ubbink BV.

The Power of Design: Product Innovation in Sustainable Energy Technologies, First Edition.
Edited by Angèle Reinders, Jan Carel Diehl and Han Brezet.
© 2013 John Wiley & Sons, Ltd. Published 2013 by John Wiley & Sons, Ltd.

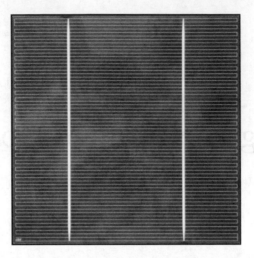

Figure C.1.1   Solar cell with H pattern

In this case description, the design process and results of the research are presented. The design process consists of several steps, which are shown in Figure C.1.2. This case describes two phases: the analysis phase (a and b) and design phase (c). Most of the time spent in the design process is on analysis and pattern design. At the end of this case, conclusions are given that are obtained from the prototype and evaluation phase (d).

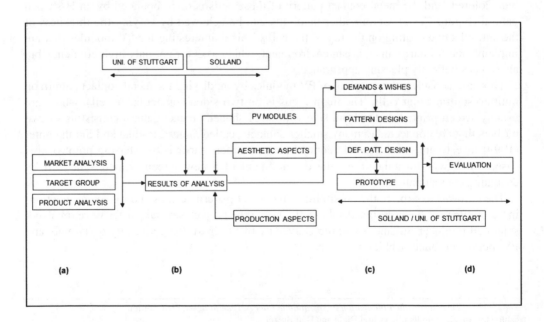

Figure C.1.2   Design process

C.2   Analysis Phase

Before redesigning the solar cells, it is necessary to obtain a better understanding of PV technology. Therefore, PV modules are analyzed because solar cells are generally used in a PV module. Next to this, the PV market and target group are explored. The derived information from this analysis is used to determine the demands and wishes of the new solar cell and PV module.

The characteristics and production of solar cells provide information about the limitations and possibilities in cell design. An important aspect is the production of the cell, because the metal contact pattern is applied by screen printing and a redesigned contact pattern must be printable.

The screen-printing process of a solar cell is similar to that of textiles. The machine squeezes silver paste through a patterned screening mesh onto the cell surface. After the paste is applied, the paste is dried at low temperature and then fired at a higher temperature. This step drives off the remaining organics and allows the silver regions to combine.

The advantages of screen printing are in its simplicity and cost-effectiveness. The disadvantages of screen printing are the relative large line width, the high contact resistance, and the low aspect ratio of the metal lines (broad and flat), which results in large shadowing losses.

The limitations and possibilities of screen printing lead to boundaries in metal line widths and heights. Beside these boundaries, the shape of the lines must be taken into account. Lines with sharp corners are difficult to screen print, because the metal paste is not able to reach the triangular part of the corner. Metal lines in the squeegee direction are more easily printed and are straight lines as opposed to curved lines.

Many patterns can be screen printed; however, the efficiency will be influenced by the pattern design:

1. *Shading:* The metal contacts on the top surface create shadow and prevent part of the incoming light from striking the solar cell. The redesign of the front contact involves metal contacts, and the metal area plays a big role in the efficiency loss.
2. *Series resistance:* The transmission of electric current produced by the solar cell involves ohmic losses. The resistance must be as low as possible, and this is dependent on the metal contacts on the solar cell.

Many possibilities are available in terms of geometries and aesthetics. Radike and Sumhammer (1999) described in their paper that newly designed contact patterns are possible but that efficiency loss is not preventable. The losses are, however, small: 0.71% efficiency loss in the worst-performing cell with respect to an H-patterned cell. This should not hold back the application of appealing front contact patterns. Radike and Sumhammer designed and simulated 15 contact patterns. These patterns can be used in a PV module creating a decorative appearance. Figure C.2.1 shows the standard cell and 15 new contact patterns of their research.

A new pattern can be produced on a screen mesh. The screen mesh in the screen-printing machine can be easily replaced by this new designed screen mesh. The need to change the machine's settings is dependent on the mesh and the opening of the screen.

The appearance of the solar cell is dependent on the metal contact pattern and the surface color. The front surface of the solar cell reflects light while the cell is illuminated. This

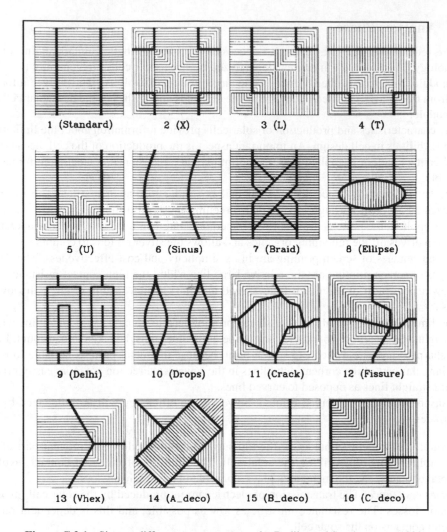

Figure C.2.1   Sixteen different contact patterns by Radike and Sumhammer (1999)

Figure C.2.2   Colored solar cells

reflection has to be reduced as much as possible. Therefore, an antireflection coating or anti-reflective coating (ARC) is applied on the solar cell.

The color of the crystalline silicon cell can be altered by varying the thickness of the ARC on the cell's top surface. A standard solar cell has a blue color, while other colors are available, such as purple and green (Figure C.2.2).

With the gathered information from the analyses, demands and wishes for a multicrystal-line silicon cell with a dimension of 6 inches are formulated. This input is then used in the design phase.

C.3 Design Phase

In the design phase, the solar cell's front contact pattern will be redesigned to make the cell and PV module more attractive. The analyses, demands, and wishes provide functional input, but little direction in terms of aesthetics. Four categories are considered in developing design directions: architecture, nature, art, and textiles. Several collages are made with these categories. Figure C.3.1 depicts a collage with textiles.

Other patterns are examined by scanning pattern books. Some examples are Chinese lattice design, Indian patterns, motifs, panels, folk art, floral patterns, and borders. Figure C.3.2 shows an example of a pattern.

The generation of patterns is executed in three steps:

1. Small sketches of new contact patterns are put on paper.
2. Redesigned cells are Photoshopped.
3. PV modules with the redesigned cells are Photoshopped.

Figure C.3.1 Collage textiles (See Plate 26 for the colour figure)

Figure C.3.2   Example of a pattern

The redesigned solar cell must have an attractive appearance, leading to the following direction: The aesthetical value must increase, while a small efficiency loss due to the redesign is acceptable.

A single solar cell can have a beautiful design with a small contact pattern. However, this pattern is not visible from a distance. The idea is to have single solar cells as puzzle pieces. Each cell has a pattern that can be combined with other, differently patterned cells. These different puzzle pieces are combined to form a big puzzle, the PV module. The PV modules can also be seen as puzzle pieces in a PV array.

The puzzle idea enables the target group to customize their own PV modules. Several new designed solar cells are selected to form a set of puzzle pieces (Figure C.3.3), called the *ABC set*. The inspiration is to design cells that can be combined with standard H cells. Solland Solar produces H cells and is able to use the new solar cells as add-on cells. The advantages are clear: The new designed cells do not compete with the H cells and provide more possibilities in PV module configuration. Architects and designers have the freedom to choose and design their own patterns. Every building is different, so providing that freedom in design is a valuable aspect. While experimenting with the ABC set, a new idea arose to have a less geometric and less strict appearance of the PV module. This approach led to figurative and customizable contact patterns. In this way, a contact pattern of a cell can have different shapes, such as a butterfly (Figure C.3.4), flower, logo, and more.

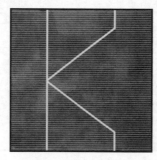

Figure C.3.3   One cell of the ABC set

Figure C.3.4   Figurative cell

C.4   The "Flower Cell"

The ABC set and figurative cell fulfill both the demands and wishes. The ABC set has geometric lines, while the figurative cell has curved lines and is less familiar. Curved lines are difficult to screen print, and it is possible that these lines are incompletely filled with metal paste. To demonstrate that this contact pattern is producible, the figurative cell is chosen for detailing. This figurative cell has an abstract flower-shaped contact pattern and is named the *flower cell*.

A standard multicrystalline Si wafer was used as the foundation for the flower cell. The cell's dimensions were 15.6 × 15.6 cm, and the four final contact points have been located at the same location of the H cell for interconnections. The cells were blue colored and had the same finger pattern. The flower cell was inspired by the poppies of Andy Warhol's *Flowers 1*. The idea was to have a pattern that can be interpreted not only as a flower but also as a different figure. The detailing of the design was executed in several steps.

Firstly, pictures of flowers were gathered. A selection of flower shapes was collected in a collage (Figure C.4.1).

Figure C.4.1   Collage of flower shapes

Figure C.4.2 Definitive flower design

Secondly, many sketches of possible flowers were developed. One flower was selected for the contact pattern, which is shown in Figure C.4.2. This cell can be used in four different configurations due to its asymmetric design. It is possible to place the flower cells in a pattern in a PV module. Next to that, the PV modules can form a pattern in the PV array.

The flower cell has no sharp cornered lines but curved lines. The lines are designed such that they connect with each other, enabling collection of the generated electricity. The design of the fingers was kept the same because of the efficient screen printing. Four busbar lines, starting from the sides of the solar cell, are 2 mm wide; the lines of the flower are 1 mm wide. The height of the metal lines is 20 μm.

C.5 Prototyping

The definitive concept was detailed. A render visualized the solar cell, but it was not physical. Additionally, the performance of the solar cell cannot be tested. A functioning prototype was the solution to enable the evaluation of the product.

Figure C.5.1 Flower cells at Ubbink BV (courtesy of P. de Boer)

Figure C.5.2   Soldering tabs on a flower cell (courtesy of P. de Boer)

The University of Stuttgart offered to order the screen and screen print the new designed cells. The University of Twente generously provided a budget for the prototyping. Solland Solar agreed to offer 200 solar cells for the prototyping. This allowed for a greater number of functioning flower cells. The screen and solar cells were delivered to Stuttgart, and the University of Stuttgart produced the solar cells.

After testing the redesigned solar cells, the cells were sent to Ubbink. Ubbink made three functioning PV modules with these cells in combination with H cells. Figures C.5.1, C.5.2 and C.5.3 depict several production steps. Figure C.5.4 shows two of the three PV modules.

After finishing production of the PV modules, the modules were tested.

Figure C.5.3   PV module production (courtesy of P. de Boer)

Figure C.5.4   Final PV modules (courtesy of E. Pupupin)

C.6   Test Results

C.6.1   H Cell versus Flower Cell

The flower cells were prototyped by University of Stuttgart and tested in the Sunsimulator. Standard H cells and flower cells are tested to provide data. The Sunsimulator used standard test conditions (STCs): 25°C, an irradiance of 1000 W/m^2, with an air mass 1.5 spectrum (AM1.5). During the simulation, the temperature was 25.2°C.

The calibration cell of the H cell and flower cell was the same, providing the possibility to compare the cells with each other.

In PV technology, the performance of solar cells is illustrated in current–voltage (I–V) curves. The best performing H cell and best performing flower cell were selected for the I–V curves. From these curves it was apparent that both cells were performing well. The curve that compares the two cells is illustrated in Figure C.6.1.

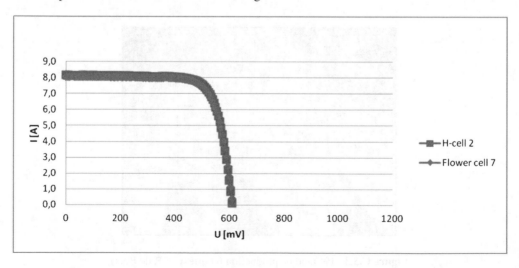

Figure C.6.1   Measured current–voltage (I–V) curves of an H cell and flower cell

The flower cell had an average output of 0.6 V, 7.9 A, and a power output of 4.8 W. In reference to the demands and specifications, the data showed that the demand was fulfilled (Table C.6.1). This result gave a positive outlook for the development of new contact patterns.

Table C.6.1 Demands and specifications

Function	Requirement	Technical specification
The solar cell has to convert light into electricity.	The solar cell produces enough electrical output.	The new solar cell has an output of at least 0.5 V, 7 A, and 3.5 W at standard test conditions (STC).
The solar cell has to be able to transport the generated current.	The produced current is collected by metal contacts on the cell.	The cell has metal lines that are connected, leading the current to the contact points of the cell.
The solar cells have to be connectable in a product (e.g., a PV module).	The cells will be connected to each other by contact points.	The cell must have a minimum of four contact points, divided at the sides of the cell.
	A cell generally has an output of 8 A. This output must be sufficiently distributed; otherwise, the cell can be overheated.	
The solar cell can be screen printed.	The metal pastes can be applied on the cell.	A standard H cell fulfills these requirements. The metal contact-stripped H cell is the base cell for the new solar cell. Silver alloy paste and aluminium alloy paste can be applied on this cell.
	The squeegee must be able to handle the metal paste (no metal paste must stay behind at the squeegee).	The direction of the squeegee is always in the direction of the small structures, like the fingers.
		The angle in the screen, between open structures and the wires, is in general 22.5°.
	The metal lines can be screen printed.	The screen fits in recent screen-printing machines.
		The height of the fingers and busbars can range between 10 and 20 μm.
		The width of the fingers can range between 110 and 120 μm. The minimal width is 60 μm.
		The spacing between fingers can be 1–5 mm.
		The busbar width ranges between 1.6 and 2.0 mm.

(*continued*)

Table C.6.1 (*Continued*)

Function	Requirement	Technical specification
		An achievable minimal width is 110 μm.
The new solar cell has to function efficiently.	The efficiency will decrease by shadowing and resistances of the metallization.	The new solar cell has a minimal efficiency of 12%.
	Efficiencies of standard cells (Solland Solar) range between 14.6% and 16.6%. The new solar cell must have an acceptable efficiency.	The absolute efficiency loss of the cell must not exceed 2%.
The new solar cell's cost price must be acceptable.	The cost price must not be too high. The new solar cell adds aesthetical value and can be sold for a higher selling price.	The cost price may rise a maximum of 5%.

C.6.2   Redesigned PV Modules

The PV modules were prototyped and flashed six times to provide data. The PV modules with flower cells have a lower output and efficiency than standard PV modules (e.g., 175 W). Several factors played a role in causing this result. The prototypes contained lower class cells, which had less output and efficiency. The soldered tabs were not completely running over the cells and led to higher resistances. This resulted in output losses. During production, the flower cells were soldered by hand and not by machine, causing less accuracy. Each hand-soldered cell took 0.5 W off the total output.

Finally, the glass top of a standard PV module is coated, while the prototyped modules did not have this coating. This led to a loss of 8–10 W. It was logical that the prototypes had lower outputs, but for a prototype of a "design" PV module this output of 150 W was acceptable.

C.6.3   Expected Costs

The flower cell has an increased value in aesthetics. This enhancement adds to the selling price of the solar cell and PV module. In comparison to a standard H cell, the flower cell of a 100,000-cell batch has a cost increase of only 0.5%.

The prototyped 50-celled PV module contains 18 flower cells. This module will have a cost increase of 0.95% compared to a standard 50-celled PV module. This result is acceptable, especially for a redesigned PV module.

C.7 Conclusions

The assignment was executed, and the design process delivered satisfying results. The PV market continues to grow, and it is expected that multicrystalline silicon solar cells will have the major market share in future years. The possibilities in redesigning solar cells are found in the metal contact patterns and the use of colored cells. The flower cell proves that the ideas of Radike and Sumhammer can be executed. Their paper described the theoretical possibilities of new front contact pattern designs. By applying their theory in practice, the flower cell was developed. The prototypes of the flower cells and PV modules with these cells give positive results. The efficiency loss of a new contact pattern is indeed low and acceptable. This is promising for the further development of new front contact patterns. The flower cells and cells with new front contact patterns are viable, especially in large batches. Every new contact pattern involves added costs, as such small-production series are expensive. For a design product in large volumes, however, these added costs are acceptable. Therefore, the flower cells can offer a feasible upgrade of PV module aesthetics.

The aesthetics of PV products have been less investigated, but there is awareness about the importance of aesthetical PV products. PV modules and building-integrated photovoltaics (BIPV) are installed in urban environments, and therefore their appearance plays a big role in the acceptance of PV systems. Finally, increasing the aesthetic value of PV systems is not only a matter of cell design: Integration of PV modules in the built environment also plays an important role.

For future research and applications of metal contact patterns of solar cells, this case proves that there are more opportunities for increasing the benefits of front contact solar cells.

Reference

Radike, M. and Sumhammer, J. (1999) Electrical and shading power losses of decorative PV front contact patterns. *Prog. Photovoltaics: Research and Applications*, **7** (5), 399–407.

Case D

Selection of Power Sources for Portable Applications

Bas Flipsen

Delft University of Technology, Faculty of Industrial Design Engineering, Landbergstraat 15, The Netherlands

D.1 Introduction

In general, 10–30% of the total volume in a portable electronic application is used by its power source. This percentage is increasing in newer applications, where the electronics size is decreasing and the quest for more power and energy is increasing. This makes the power source an important component influencing a great deal of the application's form.

Usually a power source is chosen at the very end of the design process (usually called *off the shelf*) instead of at the beginning of the conceptualization phase of the design process. Most industrial design engineers operate in the concept phase, and almost 80% of the product is defined in this phase. Based on first assumptions the industrial design engineer wants to size the application, which leads to important and sometimes irreversible choices. The industrial designer engineer makes use of experience from past projects, knowledge developed at school and university, expert knowledge, or mother wit. When it comes to power sources, no tools are commonly used that support the industrial designer.

Therefore, in this case a review is given of tools and methods that give conceptual designers a first estimate of the issues they have to deal with when designing the power source. Section D.3.1 provides a first approach to a second-order model and method that should produce an optimized volumetric design of the fuel cell (FC) hybrid, being used in the concept phase of design.

Also in this case we will show a tool that has been developed for choosing and sizing the right power source combination for a portable electronic application during the concept

The Power of Design: Product Innovation in Sustainable Energy Technologies, First Edition.
Edited by Angèle Reinders, Jan Carel Diehl and Han Brezet.
© 2013 John Wiley & Sons, Ltd. Published 2013 by John Wiley & Sons, Ltd.

phase of design. A literature review on tools and methods is also executed, described in detail in Section D.2.

In Section D.3 a first approach toward a new algorithm for a power source selection tool is presented. For this tool, different transfer functions have to be developed. In Section D.3.2, a simple analytical model for the direct methanol fuel cell (DMFC) hybrid system is presented. This model is evaluated by designing a FC-powered MP3 player (Flipsen, 2007, 2009) as presented in Section D.3.3. The simple model was found not to be sufficient in the design of an FC hybrid system, and is updated based on the redesign, in Sections D.3.4 and D.3.5.

D.2   An Overview of Selection Strategies

To give the designer more insight into the world of (alternative) power sources, this section will review tools and methods in which different power and energy sources are compared. Different search and selection strategies are applicable during the concept development phase:

1. *Power source selection for a specific application* (Section D.2.1). This is especially interesting for the industrial designer while designing portable electronics with specific performance characteristics.
2. *Application selection* (Section D.2.2). Instead of seeking power sources to fit a new application, how can we identify applications for a new power source?
3. *Power source design tools* (Section D.2.3). These help the designer in dimensioning the power source.
4. *Optimization tools* (Section D.2.4). Here all possible configurations of power sources and energy containers are evaluated and optimized for single or multiple objectives.

In the rest of Section D.2, these strategies will be described more in depth.

D.2.1   Power Source Selection Tools

Four main aspects of power sources for portable electronics are of interest for the designer: capacity and power per unit of (1) size, (2) weight, and (3) costs. Flipsen (2006) compared alternative power sources among each other based on these aspects. A large database of commercially available power and energy sources and combinations thereof is used to compare them as shown in Figure D.2.1. Due to its restriction to commercially available products, opportunities for short-term but especially long-term developments in portable power sources are overlooked. Emerging power sources like FCs are found to be an interesting alternative to lithium-ion batteries. Other power sources described and compared are ethersmog, human power, thermoelectric generators, piezo generators, electromechanical devices, photovoltaic cells, micro fuel cells, and micro combustion engines.

Fu *et al.*'s (2005) "Database and Selection Method for Portable Power Sources" is based on the "free search" selection strategy for materials and processes from Asby *et al.* (2004), and is implemented in the Cambridge Engineering Selector (CES, 2009). The CES selection tool is widely used by industrial designers to screen thousands of materials and production processes for their applicability to a specific product, part, or function. The search engine of

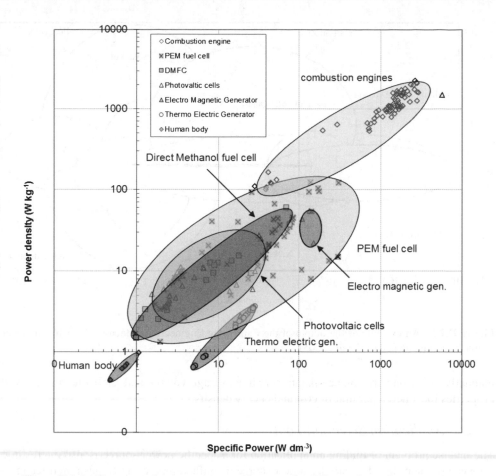

Figure D.2.1 The power density versus specific power for different alternative power sources

CES is a "free search" selection method, one of three possible selection strategies available (Asby *et al.*, 2004):

1. The free search is based on quantitative analysis of performance and normalized characteristics. So-called design indices are used to direct the designers in their search for the optimal choice.
2. The questionnaire-based search relying on expertise and knowledge guides the uninformed user through a more or less structured set of decisions, using built-in expertise to compensate for the user's lack of expertise.
3. The inductive reasoning and analogy search is based on a library of previously solved problems or cases.

The CES power source selector is mainly based on basic characteristics such as the mass m, volume V, and cost c of existing components. In Figure D.2.2, an example is given of the output of the power source selection method. Besides size and mass, a selection can be made

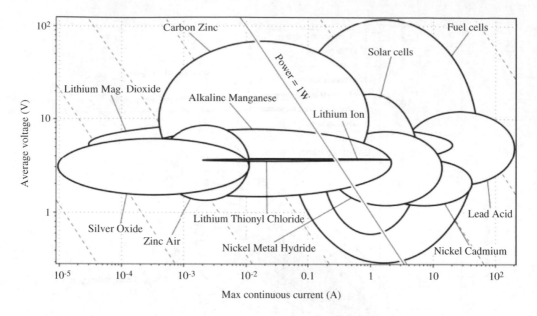

Figure D.2.2 An example of the output of the Cambridge Engineering Selector (CES)–based power source selection method (Fu *et al.*, 2005)

manually based on performance parameters like average voltage and shelf life, and derived properties like energy per unit of cost and energy density.

D.2.2 Application Selection Tools

The inverse problem of finding an application for a specific power source is also interesting. This is a problem faced mostly by power source manufacturers trying to push their technology into the market.

In Flipsen (2007, 2009), a simple method to find applications for DMFC power systems is described. A so-called Ragone plot is set up to screen the various fields of opportunity for the DMFC and the lithium-ion battery (Figure D.2.3).[1] In Figure D.2.3, an overlay is made of different existing applications. Low-power, long-endurance applications seem to be especially interesting for DMFC systems.

D.2.3 Designing Alternative Power Sources

A third problem that designers face is the dimensions of a new power source. Component volume and size are of particular importance when giving form to an application. Batteries are a main functional component accounting for, in general, from 10% (for laptop computers) to 30% (for cell phones) of the total volume.

For primary batteries a limited number standard forms exists, making the choice easier when designing. When the lithium-ion rechargeable battery was introduced, the form was

---

[1] This method specifically compares lithium-ion batteries with DMFCs, but it can also compare other power sources or combinations thereof.

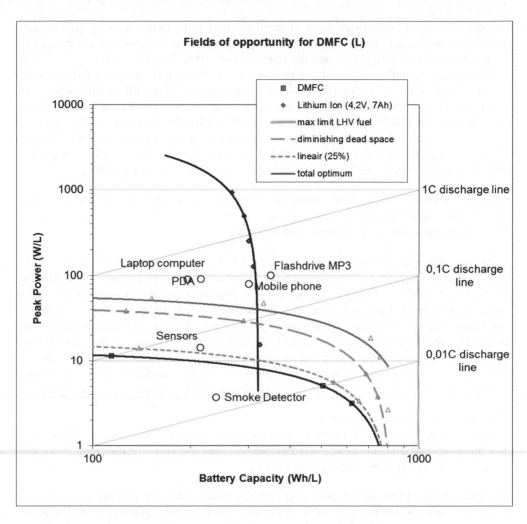

Figure D.2.3 Fields of opportunity for a DMFC power system compared the lithium-ion battery (Flipsen, 2007)

not standardized. Cell phones and other flat electronic devices directed the shape from cylindrical to a thin flat prismatic form, the cheaper pouch cell, or even the less energy-dense flexible cell. In general, however, the battery is available in prismatic form. Dimensioning a battery in the concept phase depends on three important constraints: (1) the energy it has to contain (Wh), (2) charge and discharge currents (C), and (3) the number of cycles (i.e., the cycle life of the battery).

The procedure is almost as easy as dividing the energy the application needs by the energy density of the specific battery $(Wh\,L^{-1})$. The knowledge to make the choice can be derived from an expert, a handbook, or specification sheets from battery manufacturers. This procedure is well implemented in the CES-based power source selection tool described in Section D.2.1.

However, to overcome some of the limitations of electrochemistry, size, and shape, often two or more power supplies are used within the same system. Typically a hybrid power supply combines a high-power dense element for high-power pulses, and a high-energy-dense element for low sustaining power. Different studies have shown that improved performance is possible with hybrid supplies (Drews *et al.*, 1999; Wu *et al.*, 2000).

Giving form to multicomponent power sources like FC hybrids and other hybrid power sources is not as easy as giving form to stand-alone batteries. When minimizing size, mass, or costs in hybrid power source design, one specific optimization tool is of interest: Power Optimization for Wireless Energy Requirement (POWER), developed at the University of Michigan's Wireless Integrated Microsystems Engineering Research Center (WIMS-ERC) (Albano *et al.*, 2007; Cook and Sastry, 2005; Cook *et al.*, 2006). Factors influencing selection of the power supply can be summarized as follows (Cook and Sastry, 2005):

- Electrochemistry: cell potential, discharge–charge profile, capacity, and lifetime.
- Geometry: surface area, volume, and mass.
- Environment: temperature, pressure, and exposure.

Cook and colleagues (Cook and Sastry, 2005; Cook *et al.*, 2006) describe a systematic method for the selection and design of a power system for a WIMS Environmental Monitor Testbed (EMT). The method focuses on selecting and designing power supplies for microelectronic devices that are prescribed after the design is nearly complete. The methodology constraints are operating temperature, energy and power density, and specific energy and power. Further requirements and constraints on rechargeability, mass, volume, and lifetime are allowed in selecting the appropriate battery electrochemistries or configurations like parallel, series, and combinations thereof. The algorithm (implemented in MATLAB code) separately evaluates the results of three strategies (approaches) to system design, specifying:

1. a single aggregate power profile;
2. a power system designed to satisfy several power ranges (from micro-, milli-, and Watts); or
3. a power system designed to be housed within specified spaces within the application, with device constraints on volume and surface area.

The methodology focuses on micro-power systems combining single primary and secondary batteries, often applied in implantable electronics where volume and weight of the power source are of great importance.

D.2.4   Optimizing Tools

In del Real *et al.* (2008) hybrid power systems, combining larger power systems with energy storage systems, are optimized for element sizing. The optimizing is based on costs, not on size. So-called energy hubs are introduced, an optimal hub layout of a power system described as in Figure D.2.4. Converters link inputs and outputs through coupling factors that can be considered to be the converter's steady-state energy efficiency. In general, a "system approach" is applied, and the costs of the system are optimized based on normalized investment cost per hub element (e.g., a wind generator costs approximately $2 per W) and normalized input data (e.g., from available wind power). Other optimization tools use dynamic models to optimize hybrid electric energy generators with two main objectives in mind: the efficiency of the system and the unmet load. These optimization tools are focused on the total net present cost (NPC) throughout the lifespan for both off-grid and grid-connected power systems for remote, stand-alone, and distributed generation (DG)

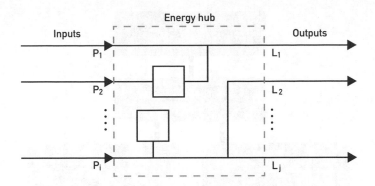

Figure D.2.4   General energy hub diagram (del Real *et al.*, 2008)

applications. Enumerative methods like HOMER (Lilienthal and Lambert, 2009) evaluate all possible solutions requiring excessively high calculation time. To reduce time heuristic techniques, (like) genetic and evolutionary algorithms are applied; an example is the hybrid optimization by genetic algorithm (HOGA) (Bernal-Agustin and Dufo-Lopez, 2009a, 2009b).

D.2.5   Discussion

A few methods and tools have been found that could support the industrial designer in making a choice for a power source in different phases of design. The method described by Cook and colleagues (Cook and Sastry, 2005; Cook *et al.*, 2006) is generally used in the end phase of design, when all constraints (application characteristics) are defined and a power source is chosen "off the shelf." This is common practice, but choosing the power source would ideally take place at the beginning of the concept phase.

The comparison methods described by Fu *et al.* (2005) and Flipsen (2006) are very useful at the start of the concept phase, when only the constraints and requirements are known. The methods indicate the possibilities and opportunities of new power sources and energy storage devices. After "screening" power sources, the next step should be sizing the chosen power system. In the method of Fu *et al.*, the mass and volume parameters are simplified. For instance, the mass of the FC is considered as equivalent to the mass of the fuel, which in general provides a good estimate but may result in problems in the outer application field. As described in Flipsen (2007), the strength of DMFCs seems to be in their low-power, long-endurance applications, where the fuel mass probably exceeds the mass of the FC system.

The tools described in Lilienthal and Lambert (2009) and in Bernal-Agustin and Dufo-Lopez (2009a, 2009b) can do both in one step. The methods described are used in larger scale power systems like wind power in combination with alternative storage systems. All combinations of power sources and energy storage systems are evaluated, and the optimal (i.e., least cost) combination is chosen. This method is quite enumerative and thus time-consuming.

D.3   Power Source Selection Tool Method

D.3.1   First Approach

In Section D.2, different tools and methods were presented. Most of them are either used by the institute itself (e.g., POWER) or not implemented (e.g., the CES database of power

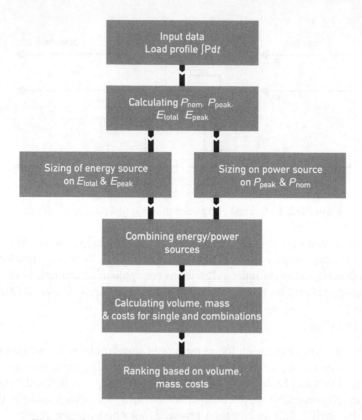

Figure D.3.1   Flow chart of the Hybrid Energy Selection Tool

sources). The methods described by Cook and Sastry (2005) and Fu *et al.* (2005) are both interesting solutions to the problems that industrial design engineers face when making a choice for a power source during the concept phase. In this section, a first approach toward a power source selection tool is proposed, based on both the tools presented.

The general outline of the tool is described in Figure D.3.1. One of the building blocks, in the sizing part of the algorithm, is the analytical model of different power sources and energy containers. For the tool these models have to be developed, making a good estimation of volume, mass, and costs. Because the tool will be used during the concept phase of design, not much is known about the power source's characteristics. Basic characteristics like user profile, number of cycles per period of time, and nominal and peak power are used as inputs for the different transfer functions.

The algorithm will have to make combinations of power and energy sources, single or hybrids, parallel or in series. Besides a single power source, multiple power sources should be evaluated. In the end, the tool has to present the different options of power and energy combinations in an ordered way.

In Section D.3.2, a first-order analytical model for sizing a FC (hybrid) system is presented. To evaluate this model, a design of a hybrid power source powering a MP3 player is described in Section D.3.3. The function is evaluated (Section D.3.4) and modified (D.3.5), and a simple analytical formula is proposed that quickly sizes FC hybrid systems based on basic power and energy specifics of the application.

D.3.2 Analytical Model for Sizing an FC (Hybrid) System

A FC power system consists of not only fuel but also a power converter, the FC, a fuel tank, and auxiliary components such as pumps, tubing, and electronics (balance-of-plant components). In this section an analytical model, the transfer function, for quick volume and weight estimation is introduced.

Xie *et al.* (2004) discussed the design of an FC system designed for powering the Motorola charger unit. The system is divided into three main parts:

1. The FC stack.
2. The fuel tank.
3. The auxiliary units like pumps, electronics, tubing, wiring, and empty space.

The weight and volume of these major parts for the 2 W prototype are 0.69 kg and 0.92 L, respectively. The subdivision is listed in Table D.3.1.

Based on these figures, a first approach to an analytical model is proposed in which the volume and weight of an FC system are sized. It is assumed that dimensions of the FC stack and the balance-of-plant components are based on power specifics (P) and the dimensions of the fuel tank are based only on energy specifics (E). The following analytical model of the volume (L) is derived:

$$V = V_{fc} + V_{fuel} + V_{BoP}$$
$$= \frac{P_{peak}}{(p\rho)_{fc}} + \frac{E}{\eta_{sys} c_{v_{can}} (u\rho)_{fuel}} + \frac{P_{peak}}{(p\rho)_{BoP}} \qquad \text{(D.3.1)}$$

where:

$$(p\rho)_{fc} = 77.3 \, \text{W L}^{-1}$$
$$(u\rho)_{fuel} = 4780 \, \text{Wh L}^{-1}$$
$$(p\rho)_{BoP} = 16.1 \, \text{W L}^{-1}$$
$$\eta_{sys} = 20\% \; (\text{Xie } et \; al., 2004)$$
$$c_{v_{can}} = 0.85^2$$

Table D.3.1 Volume and weight of the three major components of the Motorola DMFC charger (Xie *et al.*, 2004)

	Volume (L)	Weight (g)
Fuel cell stack	0.087	180
Fuel tank	0.413	351
Balance-of-plant components	0.416	155

[2] Volumetric canister coefficient: 15% of the total volume consists of the canister, based on figures from Xie *et al.* (2004).

Balance-of-plant components comprise the largest contributor to the total systems volume ($\approx$45%). Less than 20% of these components' volume is claimed by the electronics, air pump, mixing chamber, fuel pumps, and others. More than 80% of these total components' volume is empty space, electrical interconnects, and plumbing (Bostaph *et al.*, 2003; Xie *et al.*, 2004). This is 34% of the total volume. Thus diminishing empty space is one of the key issues to increase energy and power density.

For weight (kg), the following analytical model is derived:

$$m = m_{fc} + m_{fuel} + m_{BoP}$$

$$= \frac{P_{peak}}{p_{fc}} + \frac{E}{\eta_{sys} c_{m_{can}} u_{fuel}} + \frac{P_{peak}}{p_{BoP}} \tag{D.3.2}$$

where:

$p_{fc} = 37.2 \, \text{W} \, \text{kg}^{-1}$
$u_{fuel} = 6306 \, \text{Wh} \, \text{kg}^{-1}$
$p_{bop} = 43.2 \, \text{W} \, \text{kg}^{-1}$
$\eta_{sys} = 20\%$ (Xie *et al.*, 2004)
$c_{m_{can}} = 0.80^3$

The energy density and specific energy of methanol are based on the lower heating value.

These simple models can now be used to compare different applications with each other in so-called Ragone plots (Figures D.2.3 and D.3.2). The plot for a Sony 383450A8 prismatic lithium-ion battery (4.2 V, 7 Ah) is also included (Sony, 2004).

In general, FCs outperform lithium-ion batteries in areas that need high energy and low peak power. Compared to the FC, the battery is a high-power-dense energy source.

D.3.3   Design of a DMFC Power System for an MP3 Player

To test the method described in Section D.3.2, a design is made of a DMFC in combination with a battery powering a MP3 player (Flipsen, 2007, 2009). Figures D.2.3 and D.3.2 show the potential for improvement when a DMFC hybrid is applied to a flash-drive MP3 player.

The MP3 player has the load characteristics pictured in Figure D.3.3. This coarsened load profile is repeated for 2 h ($\approx$30 cycles) per day. The MP3 player's peak power has to be at least 868 mW during startup, and nominal power is 150 mW; power output of the FC is approximately 185 mW, and the efficiency of the balance-of-plant components (η_{bop}) is 81%. The runtime of the system should be at least 17 h for intensive users, which is equal to 3.1 Wh of energy.

According to the model described in Section D.3.2, the volume and weight of the system will be 69 mL and 47 g, respectively.

Two designs have been made: one making use of standard commercially available components, the *preliminary design* (Flipsen, 2007), and an optimized version improving volume by using scaled-down components, the *optimized redesign* (Flipsen, 2009). Figure D.3.4

---

[3] Mass canister coefficient: 20% of the total mass consists of the canister, based on figures from Xie *et al.* (2004).

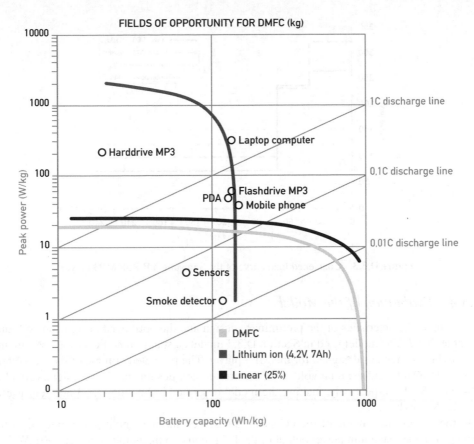

Figure D.3.2 Mass Ragone plot for the Sony Li-ion battery and the model for the DMFC power system

shows the exploded view of the latter design, and Figure D.3.5 and Table D.3.2 show the volume breakdowns of both designs.

The power systems size of the optimized version is $83.2 \times 36.0 \times 6 \, mm^3$ (18.0 mL), and the total volume of the main components alone is 14.0 mL. In the preliminary design, the fuel tanks were designed as a box. In the optimized redesign, the tanks are formed to fit in between the tubing and act as a platform where all components can be attached. This gives two advantages over the preliminary design: (1) More empty space is effectively used; and (2) during assembly the components can be fitted to the tanks, making the subassembly easier to handle. Table D.3.2 compares the lithium-ion battery used in the MP3 player with the preliminary design and the optimized redesign. The redesign has been a great improvement compared to the preliminary design, but still does not fit into the compartment available (8.7 mL). Power and energy density also lags behind that of the lithium-ion battery.

Unfortunately, at the time of this writing it was not yet feasible to draw up the weight characteristics from the computer-aided design (CAD) models.

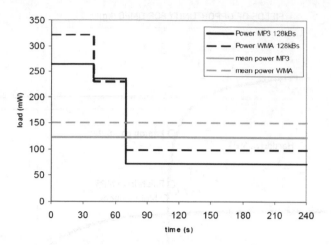

Figure D.3.3   Coarsened load curve of the Samsung YP-Z5F MP3 player

D.3.4   Evaluation of the Model

The volume characteristics of the preliminary design and the optimized design can be found in Table D.3.2. The model used in Section D.3.3 models a stand-alone FC power system and not a hybrid system, as designed for the MP3 player. The predicted volume of the FC system is equal to 69 mL, and the total volume of the preliminary design and the redesign is equal to 24.4 and 18 mL, respectively, making the model described in Section D.3.3 not very useful for a DFMC hybrid.

In the model, the volume of the FC stack is based on the maximum power output. In the design case, the maximum power output of the FC is equal to the nominal power 150 mW, less than the maximum power output of the whole system, 868 mW. Using the nominal power of 150 mW in the model, a volume of 15 mL is derived, lower than the outcome from the design.

The difference between the FC system designed by Motorola and the one described in this case is the extra added accumulator, used to deliver power bursts for short periods of time. The FC stack strongly depends on the nominal power, but this intermediate accumulator depends strongly on peak power. Because no intermediate accumulator is used in the Motorola case, an extra accumulator volume should be added for the FC hybrid model:

$$V_{accu} = \frac{P_{peak}}{(p\rho)_{accu}} \tag{D.3.3}$$

In addition to the power difference, the designed system has two tanks, one methanol fuel tank and one water tank. The Motorola design case has only one tank containing 100% methanol and no extra water. All water produced by the cell is recycled and used to dilute the methanol to a 3% wt. mixture. The effective energy density of the fuel tank decreases in this redesign case by a factor of 2. An extra concentration parameter c_{meoh} is introduced in the model. This parameter indicates the amount of methanol over the total liquid amount. In the Motorola case this parameter is equal to 1.0, and in the MP3 design case this parameter is equal to 0.5.

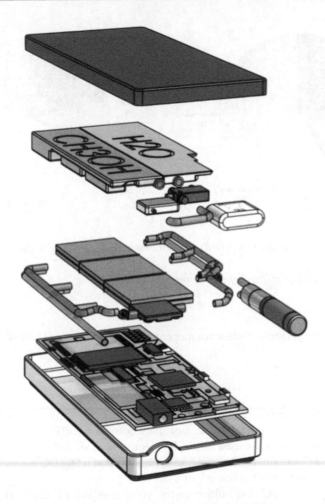

Figure D.3.4 Exploded view of the "optimized" design of the DMFC battery system (Flipsen, 2009)

The volume of the FC *flat pack* in the design case has a power density equal to about one-half, or $51.5\,W\,L^{-1}$, of the Motorola case $(100\,W\,L^{-1})$.[4] The Motorola FC is a four-cell *stacked* version using bipolar plates that are thinner than the end plates. Making use of a flat-pack design, linking several independent cells to each other in series, increases the volume of different cells and also results in an overall thinner power source. An extra parameter is added to the final analytical model, the flat-pack constant c_{fp}, equal to 1 for stacked design and 1.5 for a flat-pack design.

System efficiency η_{sys} is difficult to predict. In contrast, the FC's efficiency (η_{fc}) can be derived from its specification sheets, based on load. In the Motorola case the FC voltage efficiency is 29% and the total systems voltage efficiency is 22%.[5]

---

[4] This is the direct power output over the stacks volume. The power density as used in this case is based on the systems power output over the fuel cell stacks volume $(77.3\,W\,L^{-1})$.

[5] It is assumed that the overall system efficiency equals the systems voltage efficiency. The overall system efficiency (E_{out}/E_{in}) of the Motorola FC system is 20%.

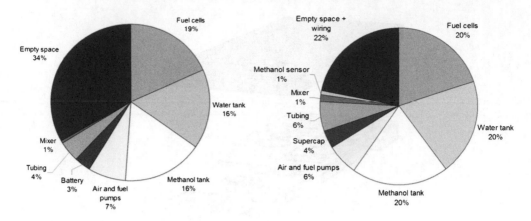

Figure D.3.5 Volume breakdown of (left) the preliminary design and (right) the "optimized" design

The system's voltage efficiency can be modeled as

$$\eta_{sys} = \eta_{fc}\eta_{BoP} \tag{D.3.4}$$

where η_{BoP} is the balance-of-plant component efficiency (Xie *et al.*, 2004):

$$\eta_{BoP} = \frac{\eta_{DC/DC}}{1 + r_{au}} \tag{D.3.5}$$

With r_{au} the ratio of auxiliary power to net power ($\approx 18.75\%$) and $\eta_{DC/DC}$ the efficiency of the DC/DC convertor ($\approx 90\%$), the balance-of-plant component efficiency is equal to 76%. If power conditioning circuitry is used inside the electronic device, the device can be powered directly from the FC without conditioning, meaning without a DC/DC convertor, and achieve a balance-of-plant component efficiency of 84%.

The efficiency of the FC also influences the volume of the FC stack. Higher efficiency means higher power density. For now the power density is set at $(p \cdot \rho)_{fc} = 100\,\text{W L}^{-1}$. Higher power densities up to $500\,\text{W L}^{-1}$ have been reported by PolyFuel (Membrane Technology, 2007).

Table D.3.2 Comparison between the lithium polymer battery, the preliminary design, and the optimized redesign

		Lithium polymer battery	Preliminary design	Optimized design
Volume	(mL)	8.7	24.4	18.0
Component	(mL)	—	16.3	14.0
Dimensions	(mm)	$66 \times 33 \times 4$	$86 \times 35 \times 8$	$83 \times 36 \times 6$
Energy	(Wh)	3.1	3.3	3.17
Energy density	(Wh L^{-1})	348	135/202	176/226
Power density	(W L^{-1})	103	37/55	50/64
Fuel cell efficiency	(−)	—	21%	29%

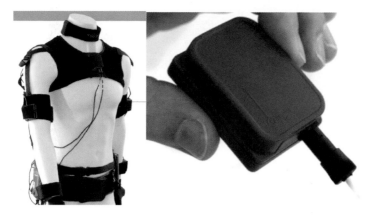

Plate 16 Figure 2.8.2 Xsens' MVN Biotech motion measurement system (left) and MTx motion tracker (right).

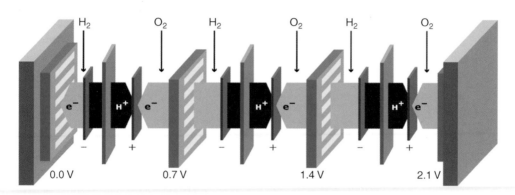

Plate 17 Figure 3.4.1 Schematic drawing of a fuel cell stack containing three repetitive units, using a polymer electrolyte membrane fuel cell (PEMFC) as an example.

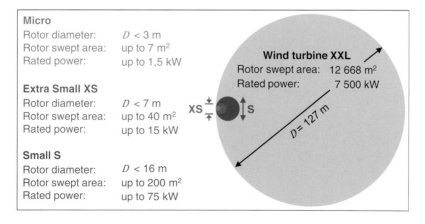

Plate 18 Figure 3.5.1 Small wind turbines divided into three size classes.

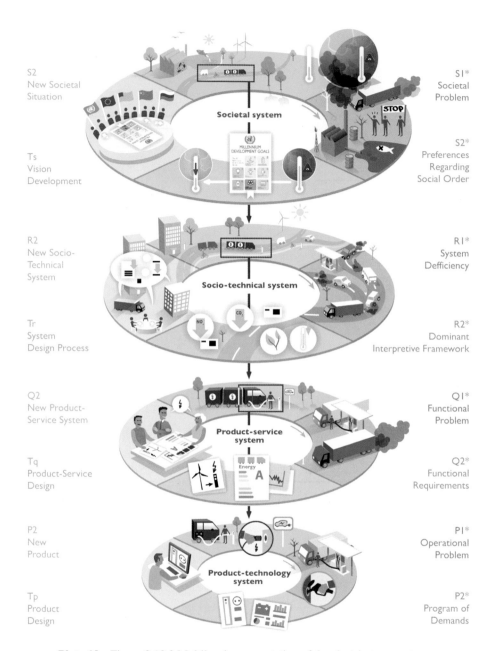

S2
New Societal
Situation

S1*
Societal
Problem

Societal system

Ts
Vision
Development

S2*
Preferences
Regarding
Social Order

R2
New Socio-
Technical
System

R1*
System
Defficiency

Socio-technical system

Tr
System
Design Process

R2*
Dominant
Interpretive Framework

Q2
New Product-
Service System

Q1*
Functional
Problem

**Product-service
system**

Tq
Product-Service
Design

Q2*
Functional
Requirements

P2
New
Product

P1*
Operational
Problem

**Product-technology
system**

Tp
Product
Design

P2*
Program of
Demands

Plate 19   Figure 2.10.2 Multilevel representation of the electric transport case.

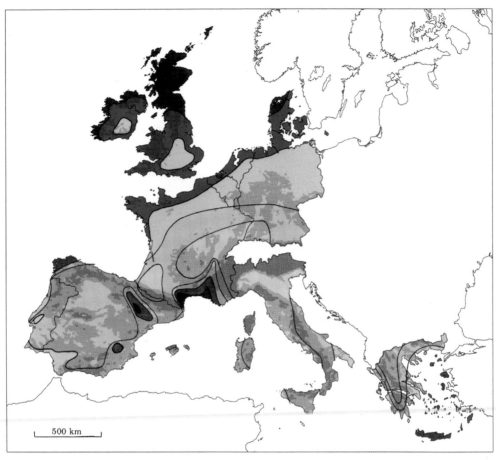

Wind resources[1] at 50 metres above ground level for five different topographic conditions									
Sheltered terrain[2]		Open plain[3]		At a sea coast[4]		Open sea[5]		Hills and ridges[6]	
m s^{-1}	Wm^{-2}	m s^{-1}	Wm^{-2}	m s^{-1}	Wm^{-2}	m s^{-1}	Wm^{-2}	m s^{-1}	Wm^{-2}
> 6.0	> 250	> 7.5	> 500	> 8.5	> 700	> 9.0	> 800	> 11.5	> 1800
5.0-6.0	150-250	6.5-7.5	300-500	7.0-8.5	400-700	8.0-9.0	600-800	10.0-11.5	1200-1800
4.5-5.0	100-150	5.5-6.5	200-300	6.0-7.0	250-400	7.0-8.0	400-600	8.5-10.0	700-1200
3.5-4.5	50-100	4.5-5.5	100-200	5.0-6.0	150-250	5.5-7.0	200-400	7.0- 8.5	400- 700
< 3.5	< 50	< 4.5	< 100	< 5.0	< 150	< 5.5	< 200	< 7.0	< 400

Plate 20 Figure 3.5.3 Wind resources at 50 m above ground level for five different topographic conditions from the *European Wind Atlas* (Troen and Petersen, 1989).

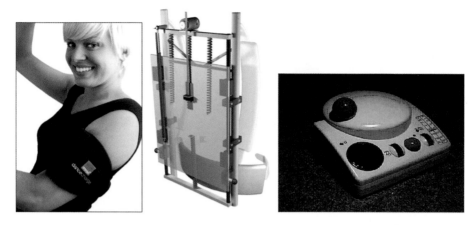

Plate 21 Figure 3.6.3 (Left to right): Orange dance charger, backpack charger (Rome *et al.*, 2005), and My First Wind-Up Sony (DUT student project by Yannic Dekking).

Plate 22 Figure 3.7.3 System levels as defined for LED lighting systems

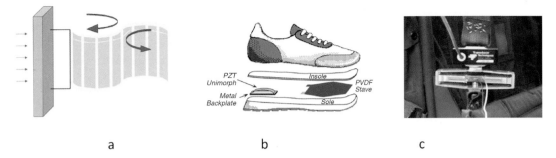

a   b   c

Plate 23 Figure 3.9.1 Applications of power harvesting using piezoelectric material. (a) Vortex shedding in a water flow (Taylor, 2001), (b) impress of an insole in a shoe (Kymissis *et al.*, 1998), and (c) motion of a backpack (Feenstra, Granstrom, and Sodano, 2008).

Plate 24 Figure A.4.1 Working prototype (van Genuchten, 2011).

Plate 25 Figure B.1.1 Impressions of solar lighting, in this case the Moonlight manufactured for the Cambodian market.

Plate 26 Figure C.3.1 Collage textiles.

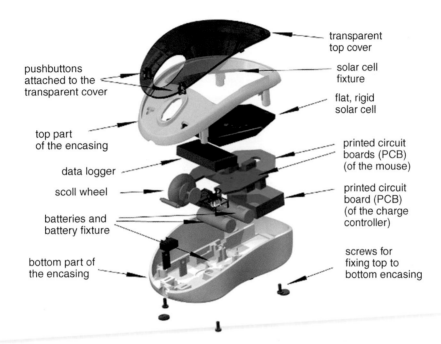

transparent
top cover

pushbuttons
attached to the
transparent cover

solar cell
fixture

flat, rigid
solar cell

top part
of the encasing

printed circuit
boards (PCB)
(of the mouse)

data logger

scoll wheel

printed circuit
board (PCB)
(of the charge
controller)

batteries and
battery fixture

bottom part of
the encasing

screws for
fixing top to
bottom encasing

Plate 27   Figure E.2.3 Exploded view of the SPM.

Plate 28   Figure F.5.1 The UMC prototype and illustrations of its use.

Plate 29   Figure G.2.5 Co-development with local stakeholders.

Table D.3.3 Volume output from the CAD model of the preliminary and optimized design, versus the first-approach model modified in this section

		CAD-optimized design	Modified analytical model	Deviation
Volume fuel cell	(mL)	3.6	3.9	+8%
Volume fuel tank	(mL)	7.2	7.7	+7%
Volume balance-of-plant components	(mL)	2.7	2.7	0%
Volume empty space	(mL)	3.9	3.8	−4%
Volume accumulator	(mL)	0.7	0.7	0%
Total volume	(mL)	18.0	18.8	+4%

The volume of the FC can now be modeled as follows:

$$V_{fc} = \frac{c_{fp}P_{nom}}{\eta_{fc}(p\rho)_{fc}} \tag{D.3.6}$$

The volume for balance-of-plant components is based on the volume of wiring, tubing, air and fuel pumps, other auxiliaries, and empty space. A large chunk of this volume is empty space, and in the standard design case almost 32% of the total volume was taken up by empty space. In the redesign, however, this amount was decreased to almost 20%. For the function proposed in this case, the balance-of-plant component volume depends solely on the nominal power P_{nom} and power density of the components (including empty space) $(p\rho)_{BoP}$. The values for the power density of the balance-of-plant components have to be adjusted, because empty space is diminished. The empty space will be excluded from the balance-of-plant volume and extra volume is added; the empty space volume is V_{empty}:

$$V_{BoP} = \frac{P_{nom}}{(p\rho)^*_{BoP}} \tag{D.3.7}$$

$$V_{empty} = c_{es}V_{total} \tag{D.3.8}$$

where:

$$(p\rho)^*_{BoP} = 70\,\mathrm{W\,L^{-1}}$$
$$c_{es} = 0.2\text{–}0.32$$

D.3.5 Modification of the Model

Taking all the modifications of Section D3.4 into account, the volume can now be modeled as

$$V_{total} = \frac{1}{1 - c_{es}}\left(V_{fc} + V_{fuel} + V_{BoP} + V_{accu}\right) \tag{D.3.9}$$

According to the model described by equation (D.3.9) and Section D.3.4, the estimated volume for this FC hybrid is equal to 18.8 mL. Table D.3.3 compares the predicted volumes by the optimized function with the CAD model. The predicted value deviates +4% from the CAD model, which is within reasonable bandwidth (<10%).

D.4 Conclusion and Discussion

When searching for a power source for a specific application, a first guesstimate during the concept phase can be done very well with the Power Source Selector from Liang Fu and the general overview in Flipsen (2006). In Section D.3, an analytical model, the transfer function, was introduced that produces weight and volume characteristics of a FC hybrid system based on general input parameters like load profile and use time. The model is optimized to predict the volume for a FC hybrid system based on general input parameters.

Besides analytical models for FCs and FC hybrids, other power sources and combinations of power sources have to be developed:

- Power sources of interest are hydrogen FCs, micro internal combustion engines, and photovoltaic cells.
- Energy sources of interest are hydrogen and other fuels.
- Combined power sources like primary and secondary batteries and capacitors are also of interest.

Because the basic parameters, like the power density of FC stacks, will change over time, a database of subcomponents has to be made. The database should be updated regularly, and new characteristics of components should be added. In Figure D.3.1, the flowchart for the power source (hybrid) selection algorithm is shown.

References

Albano, F., Chung, M.D., Blaauw, D. *et al.* (2007) Design of an implantable power supply for an intraocular sensor, using power (power optimization for wireless energy requirements). *J. Power Sources*, **170**(1), 216–224.

Asby, M.F., Brchet, Y.J.M., Cebon, D., and Salvo, L. (2004) Selection strategies for materials and processes. *Mater. Design*, **25**(1), 51–67.

Bernal-Agustin, J.L. and Dufo-Lopez, R. (2009a) Efficient design of hybrid renewable energy systems using evolutionary algorithms. *Energ. Convers Manage.*, **50**(3), 479–489.

Bernal-Agustin, J.L. and Dufo-Lopez, R. (2009b) Multi-objective design and control of hybrid systems minimizing costs and unmet load. *Electr. Pow. Syst. Res.*, **79**(1), 170–180.

Bostaph, J., Xie, C., Pavio, J. *et al.* (2003) 1w direct methanol fuel cell system as a desktop charger, http://www.docstoc.com/docs/38924878/1W-DIRECT-METHANOL-FUEL-CELL-SYSTEM-AS-A-DESKTOP (accessed April 12, 2012).

CES (2009) Cambridge engineering selector software, http://www.grantadesign.com/ (accessed April 12, 2012).

Cook, K.A. and Sastry, A.M. (2005) An algorithm for selection and design of hybrid power supplies for mems with a case study of a micro-gas chromatograph system. *J. Power Sources*, **140**(1), 181–202.

Cook, K.A., Albano, F., Nevius, P.E., and Sastry, A.M. (2006) Power (power optimization for wireless energy requirements): A MATLAB based algorithm for design of hybrid energy systems. *J. Power Sources*, **159**(1), 758–780.

del Real, A.J., Arce, A., and Bordons, C. (2008) Optimization strategy for element sizing in hybrid power systems. *J. Power Sources*, **193**, 315–321.

Drews, J., Wolf, R., Fehrmann, G., and Staub, R. (1999) Development of a hybrid battery system for an implantable biomedical device, especially a defibrillator/cardioverter (ICD). *J. Power Sources*, **80**(1–2), 107–111.

Flipsen, S.F.J. (2006) Power sources compared: The ultimate truth? *J. Power Sources*, **162**(2), 927–934.

Flipsen, S.F.J. (2007) Design challenges for a fuel cell powered MP3 player, in International Power Sources Symposium, Bath, Great Britain.

Flipsen, S.F.J. (June 2009) Designing micro fuel cells for portable products, in Proceedings of FuelCell2009, ASME.

Fu, L., Lu, T.J., and Huber, J.E. (2005) Database and selection method for portable power sources. *Adv. Eng. Mater.*, **7**(8), 755–765.

Lilienthal, P. and Lambert, T. (March 2009) HOMER, the optimization model for distributed power. https://analysis.nrel.gov/homer/ (accessed April 12, 2012.

Membrane Technology (2007) Membranes push boundary of power density for DMFC stacks. *Membrane Technology*, **2007**(11), 8–9.

Sony (2004) Announcement of new lithium ion batteries, realizing industry's highest level of energy density (a8 series, 383450a8), http://www.sony.net/SonyInfo/News/Press_Archive/200412/04-060E/ (accessed April 12, 2012).

Wu, Q., Qiu, Q., and Pedram, M. (2000) An interleaved dual-battery power supply for battery-operated electronics, in Proceedings of Asia and South Pacific Design Automation Conference, pp. 387–390.

Xie, C., Bostaph, J., and Pavio, J. (2004) Development of a 2w direct methanol fuel cell power source. *J. Power Sources*, **136**(1), 55–65.

Case E

Design of a Solar-Powered Wireless Computer Mouse

Wilfried van Sark[1] and Nils Reich[2]

[1]*Utrecht University, Copernicus Institute, Utrecht University, Budapestlaan 6, The Netherlands*
[2]*Fraunhofer Institute for Solar Energy Systems ISE, Heidenhofstr. 2, Germany*

E.1 Introduction

Photovoltaic (PV) cells have been introduced to consumer products since as early as the mid-1980s and have found many applications in consumer products since that time, for example calculators and wristwatches. A large percentage of the PV products presently available on the market, however, do not function properly or have shortcomings that make them non-competitive with their conventionally powered counterparts (Kan, Van Beers, and Brezet, 2004). In addition, the present knowledge base on systems powered by small PV units does not cover the knowledge necessary for suitable consumer product designs (Veefkind, 2004). This may result in poor energetic performance, which users perceive as inappropriate, especially for PV-powered devices.

To investigate industrial design processes with regard to technical engineering solutions, we selected a test product: the solar-powered wireless mouse (SPM) (Alsema *et al.*, 2005). The choice for this SPM fulfilled our aim to tackle the many problems frequently encountered by PV product designers. The key issue – to select a challenging product case – was fulfilled to such an extent that it was initially unclear whether an SPM product could be a feasible product concept at all. Variations of device use patterns and available irradiation lead to uncertain energy balances. In the case of heavy device use and little irradiation, it may not be possible to power an SPM on a solely PV-generated charge. Furthermore, the hand of the user will have to touch the SPM cover with the solar cells directly underneath,

The Power of Design: Product Innovation in Sustainable Energy Technologies, First Edition.
Edited by Angèle Reinders, Jan Carel Diehl and Han Brezet.
© 2013 John Wiley & Sons, Ltd. Published 2013 by John Wiley & Sons, Ltd.

thereby shading the solar cells and reducing the amount of generated charge. Moreover, the uncertainty of the energy balance poses the question of how far users are willing to sunbathe their PV-powered mouse. The energy balance is thus one, if not the most critical, issue for an SPM product type. Integrating solar cells into a wireless computer mouse in an aesthetically pleasing and energetically acceptable way poses yet another challenge.

In this case study, we will present technical engineering solutions in combination with correlated industrial design processes, which are required to build a wireless computer mouse equipped with solar cells (Reich *et al.*, 2009). First, we aim to distinguish different design concepts of computer mice that are to be equipped with PV. Therefore, in Section E.3 we will analyze the product and design criteria of this kind of product. The design process itself also comprises research on and analysis of user requirements and desires, and a preliminary assessment of suspected energy balances. Second, in Section E.4 we aim to identify one optimized design concept that is related to both technical aspects and design aesthetics. Here, the selection and dimensioning of the required secondary battery type and size are discussed together with required charge controller electronics. Also the selection of appropriate solar cell types, especially with respect to appropriate low-light performances of the various PV technologies that might be used, will be presented.

The integrated approach, especially the cooperation of PV and industrial design experts, contributes to improving the product design process of PV-powered devices, which might help to open up further markets for solar cell applications.

E.2 Product Design Process

Exploratory studies were conducted to determine the kind of project and product the SPM idea would develop into. This so-called *quick scan* included market research (existing products, market segmentation, and corridor of price), user research (observational research, interviews, and focus groups), and rough energy balance calculations (charge generation of PV indoors, device charge demands, and typical use times). The quick scan and the preliminary energy balance estimations (discussed further in this case) indicated that the development of a wireless solar mouse would be feasible. At this stage, three obvious product concepts could be distinguished: (1) an SPM that is worry free, (2) an SPM that requires special user care, and (3) an SPM that has a larger surface area than traditional mice.

Ideally, the SPM would be energetically self-supporting, thus not requiring any special user care. This can only be achieved, however, if the SPM is only occasionally used (option 1). To guarantee 100% solar operation for medium- and heavy-use SPMs, they require pro-active "sunbathing" to recharge their batteries (option 2). An alternative is to increase the PV area (option 3). However, this inevitably leads to a mouse pad–sized product, which would not require special user care due to the sheer inconvenience of sunbathing a larger product. We decided to opt for option 2.

E.2.1 Focus Group Research

So-called focus group research, in which several potential users are interviewed qualitatively, allows a closer interpretation of user criteria related to device designs. Focus group research highlighted that the SPM's perceived quality is greatly influenced by the way PV cells are integrated in the encasing. Figure E.2.1 shows three mouse designs that were assessed by

(a)   (b)   (c)

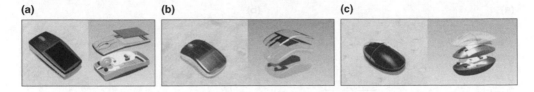

Figure E.2.1 Integration of various PV cell types using different encasing shapes and technologies, with (a) flat panel integration as part of an outer encasing; (b) integration of crystalline silicon cells connected in series; and (c) bent amorphous silicon cell integration

potential users. According to the focus group participants, the PV cell must follow the shape of the encasing. If this is not the case, PV cells are perceived as vulnerable and/or less reliable. This is a key issue, since the same focus group research pointed out that ergonomics is the most important criterion of a user, with organic, rounded shapes being perceived as most attractive. It thus also indicates that rectangular mice, such as that depicted in Figure E.2.1a, are not an option.

E.2.2   Energy Balance Scenarios

To estimate energy balances for the SPM, both the charge demand of the device and the charge generation potential of the solar cells to be incorporated need to be determined. Market research found commercial products' charge demands to vary by a factor of 4–5. The commercially available product with the lowest power demand at the time (end of 2005) was sold by Microsoft as the Intellimouse. Meanwhile, energetically more economic mice operating with a laser instead of a light-emitting diode (LED) have been introduced. The SPM product is based on Intellimouse electronics and consequently LED technology. Charge demand ranges can easily be calculated based on expected device use times of 4–27 hours of mouse motions per week (Percept Technology Labs, 2004). The Microsoft Intellimouse uses three energy management system (EMS) levels to minimize energy consumption. Additionally, battery runtime is increased by adapted LED intensity depending on the underlay material. Furthermore, the DC/DC converter provides the electronic circuitry of the mouse with sufficient voltage (3.5 V) with (primary) battery voltages as low as 0.6 V. However, the product is not equipped with a simple energy-saving on-off switch, but relies on "sleeping mode" activation.

The amount of PV-generated charge depends on the solar cell area, the effective solar cell efficiency, and the amount of available irradiation. Active solar cell area was initially estimated to be between 20 and 30 cm^2, based on typical mouse product dimensions; the final product incorporates 28 cm^2 of PV. As the "effective" PV efficiency depends on several factors (Reich *et al.*, (2010a, 2011)), we assumed overall 10% constant efficiency to reduce complexity (that is, neither irradiance intensity nor temperature-dependent PV performance is considered, and spectral effects are not accounted for). Consequently, the amount of PV-generated charge can be easily estimated based on irradiance time series. For this, we used hourly averaged, global horizontal irradiation measured in 2005 in de Bilt, the Netherlands. Lower irradiance levels indoors were accounted for by a simple attenuation factor of 0.05, which is similar to a constant daylight factor (DF) of 5%. When the SPM is sunbathed, available irradiance levels will be (much) larger. For this

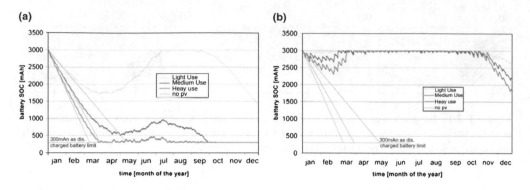

Figure E.2.2   Battery state of charge for an SPM that (a) is not and (b) is sunbathed on a daily basis between 11:30 a.m. and 12:30 p.m. (weekdays) and during the whole weekend. Gray lines denote battery state of charge for mice without PV cells

case, we assumed a constant DF of 60%, which implies a windowsill located in a south-facing room with direct sunlight access. During device use, PV power output is assumed to be zero, as the user's hand then covers the solar cell. Finally, PV-generated charge was assumed to be lowered by battery charge efficiency (90% constant), battery self-discharge (15% per month), and the efficiency of electronic circuitries (90% constant), which also provides maximum power point tracking (with 100% accuracy assumed). To deal with the many parameters, we developed a computer tool to simulate the energy flows in PV-powered consumer systems (Reich *et al.*, 2010b).

A calculated battery state of charge (SOC) for a whole year is shown in Figure E.2.2, comparing computer mice with and without incorporated PV. The battery SOC of mice without incorporated PV is depicted in both figures by gray lines. For all calculated battery SOC, three different weekly use times of 4 h (light use), 18 h (medium use), and 27 h (heavy use) were assumed. In addition, a sunbathed SPM (Figure E.2.2b) is compared to a SPM that is not sunbathed (Figure E.2.2a). Note the influence users have by sunbathing the SPM. The batteries are almost fully charged during the period March–October for assumed sunbathing patterns, in which the SPM is exposed to 60% of global horizontal outdoor irradiation daily from 11:30 a.m. to 12:30 p.m. as well as during the entire weekend. Comparing both figures also shows the difficulties encountered when dimensioning components, as different optimal battery capacities are associated with each scenario. For the sunbathing scenario chosen here, the optimal battery capacity would be 1900, 1250, and 500 mAh for heavy, medium, and light device use, respectively (i.e., the minimum battery capacity required for granting year-round product operation on 100% PV-generated charge). However, when mice are not sunbathed, battery capacity will be sufficient for light use only. It would therefore be very helpful if typical charge yields related to sunbathing activity could be defined quantitatively. Unfortunately, this is rather difficult, due to varying irradiation conditions (e.g., direct or diffuse). We therefore decided to focus only upon the question of how long users are willing to proactively place their SPM at a specific indoor location, to harvest considerable higher "daylight" (diffuse irradiance) or direct sunlight, respectively.

E.2.3   Design Criteria

Based on the exploratory studies and energy balance estimations, we defined design criteria. First, the following energetic requirements were agreed upon:

- When fully charged, the SPM must allow 5 days of work without irradiation.
- The SPM must indicate when sunbathing is required (before the battery runs flat).
- The design should stimulate the user to "sunbathe" the SPM.
- The incorporated PV and charge storage unit should provide year-round product operation, if the mouse is "sunbathed" adequately.
- The user manual must clearly state how much "sunbathing" is required.

Regarding more general product functions, it was agreed upon that the SPM must be equipped with a scroll wheel and two push buttons, guarantee a product lifetime of at least 5 years, and be compliant with the restriction of the use of certain hazardous substances in electrical and electronic equipment (RoHS, 2002). The concept should also enable the manufacturing of prototypes at a later stage. The potential concepts should be evaluated as if the SPM was to be produced in a commercial setting. To allow for (theoretical) mass production, the individual components must be commercially available. Finally, users should perceive the SPM as a quality product, not as a gadget. The focus group investigations indicated users could perceive the SPM as a (high)-quality product based solely on design aesthetics. Therefore, only a single PV cell that follows the encasing's shape should be used, to avoid bad user perception of the SPM prototype design. Figure E.2.3 shows the various components of an SPM in accordance with the defined device criteria. The solar cell is, as depicted, flat and rigid, albeit the double-bent transparent cover and the encasing of the mouse lead to a "streamlined geometry" without any corners or edges.

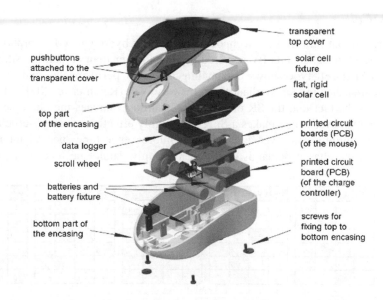

Figure E.2.3   Exploded view of the SPM (See Plate 27 for the colour figure)

E.3 Component Selection

In order to be able to build prototypes, a certain system setup must be chosen and each system component must be dimensioned. The following section addresses the selection of an adequate battery and PV type, solar cell incorporation, and charge controller options.

E.3.1 Battery Unit

First, typical battery parameters such as energy density (Wh/kg and Wh/cm^3), cycle life, charging time, cell voltage, discharge rate, operation temperature, and so on were evaluated. This narrowed the possible choices down to the following battery types: nickel–cadmium (NiCd), nickel metal hydride (NiMH), lithium-ion (Li-ion), sealed lead–acid (SLA), and rechargeable alkaline manganese (RAM).

Each technology was assessed on five aspects: cost, efficiency, design, durability, and environmental aspects. The performance of specific battery technologies was categorized as very bad ($--$), bad ($-$), good ($+$), and very good ($++$) in the SPM product case (Figure E.3.1). Categorization took into account information from handbooks as well as the particular expertise of the involved PV and industrial product designs experts. From this, only two types of batteries were considered further, the NiMH and the Li-ion, as these did not have a very negative score in any category (NiMH) or fell short only on cost and durability grounds (Li-ion). Finally, the choice was made for NiMH batteries for various reasons. NiMH batteries are available in different shapes and capacities, and voltages are multiples of 1.2 V, which allows a lot of freedom regarding system design. Furthermore, NiMH battery types can be found in electronic retailers or supermarkets, making it easy for the user to replace these secondary batteries if required. In addition, NiMH batteries with almost no self-discharge are available.

E.3.2 PV Cell

It is often argued that thin-film PV technology, especially hydrogenated amorphous silicon (a-Si:H)–based solar cells, suits PV-powered products better than crystalline silicon cells. Indeed, the a-Si:H cell type shows outstanding low-light performance, and the spectral response (SR) matches indoor irradiance spectra very well (Reich *et al.*, 2005). Especially advantageous is the fact that the SR of a-Si:H cells almost matches the photopic response curve of the human eye, which makes a-Si:H cells highly effective at energy-efficient artificial-lighting conditions. However, under typical fluorescence tube spectral intensities of ~100–150 Lux, the charge demands of the SPM simply cannot be met. Thus, the SPM

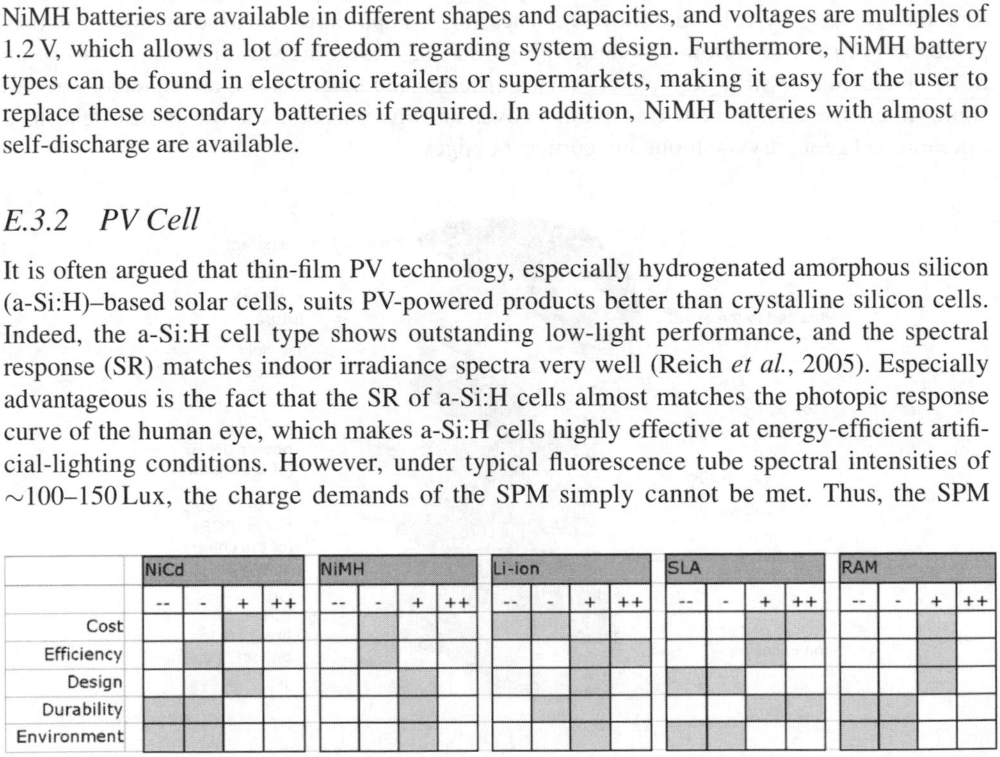

	NiCd				NiMH				Li-ion				SLA				RAM			
	--	-	+	++	--	-	+	++	--	-	+	++	--	-	+	++	--	-	+	++
Cost																				
Efficiency																				
Design																				
Durability																				
Environment																				

Figure E.3.1 Assessment of different battery types for use in the SPM requires solar energy as an irradiance source, because the energy flux of solar radiation is orders of magnitude higher than that of (energy-efficient) artificial lighting. Especially if the SPM is placed at or close to a windowsill, rather high irradiance intensities will be available for charge generation. Solar cell efficiencies at higher irradiance levels (i.e., 100–1000 W/m^2) are generally much better for mono- or multicrystalline silicon (c-Si and mc-Si) than for a-Si:H solar cells. Commercially available c-Si cells have greater than 20% efficiency at Standard Test Conditions (STC), outperforming a-Si:H cells with roughly 7–10% STC efficiency by a factor of 2–3. We thus opted for crystalline silicon–based cells.

E.3.3 Encasing

Placing the solar cell underneath a double-bent, transparent plastic cover within the SPM encasing reduces irradiance intensity, which in addition is unequally distributed across the solar cell due to shading effects. Interestingly, the product encasings' shading effect is slightly overcompensated by optical concentration, as calculated by the software tool 3D-PV (Reinders, 2007). Since the 3D-PV tool could not calculate double-bent interface characteristics, however, this does not include transparent cover transmission characteristics, which we estimated together with transmission characteristics of a two-paned window-glazing system using Fresnel equations in a separate model. Here, we found PV module tilts (i.e., the tilt of the solar cell within the SPM encasing) of 20–30° to be optimal. In the final concept, the solar cell is tilted only 10°, due to a trade-off between the SPM's ergonomics and the required space for electronics, batteries, and mechanics. This permits relatively large cells to be incorporated without sacrificing too much internal space. In order to improve light-harvesting properties, support structures should be designed for fixed and secured positioning of SPMs with about a 45° tilt angle when placed for a sunbath at a windowsill.

E.3.4 Charge Controller

The most simple charge controller would require only a single diode, preventing battery discharge over the solar cell. Overcharge protection could already be achieved by just another diode, which short-circuits the solar cell above a certain voltage threshold. Another option would be a self-regulating system design by matching maximum solar cell voltage to the maximum voltage of the storage unit. This would be desirable, owing to the sheer simplicity of the concept. However, matching battery and PV voltage is difficult, because battery voltage depends on the battery state of charge, and PV voltage depends on irradiance intensity. Voltage converters, in contrast, require rather complex electronics but also lead to greater design freedom. Theoretically, when incorporating a voltage stepping unit, all combinations for the three categories of *storage*, *module setup*, and *PV technology* become possible, as depicted in Figure E.3.2. Voltage converters apply DC/DC up- or down-conversion (of PV voltage), so they are most logically considered in combination with maximum power point tracking. With the single-cell concept already chosen, we opted for charge controller electronics that perform voltage up-conversion and maximum power point tracking.

On top of charge controlling, the electronics should provide a battery status indication, as defined in the design criteria. It was decided to use only battery voltage as the battery SOC

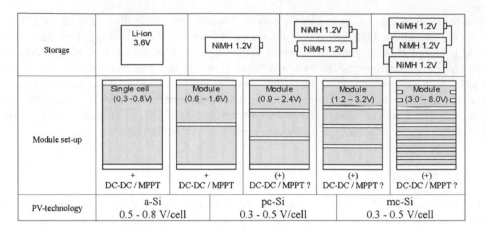

Figure E.3.2 Morphological chart of possible SPM designs concerning solar cell implementation and the possible voltage ranges of the different subcomponents

indication. In this case, the selected NiMH battery type only allows indicating battery SOC as either full or almost empty, due to the rather SOC-independent voltage potential of NiMH batteries. It would be desirable to indicate battery SOC through a bar graph composed of four or more elements. Moreover, indicating the positive influence of sunbathing may encourage users to sunbathe the mouse more regularly. However, we opted for a relatively simple but cheap and fast solution.

E.4 Final SPM Product

A total of 15 SPMs were manufactured using commercial mice electronics and rapid prototyping-based plastic encasings. Although the incorporated PV cell is flat and rigid, curved lines and double-bent surfaces dominate the design. Figure E.4.1 shows the final SPM design together with the support structure to give an impression of the design aesthetics.

E.4.1 SPM Specifications

The 15 SPM prototypes were manufactured in accordance with the different options discussed in Section E.3. A voltage converter (DC/DC voltage up-conversion) charges the battery (of NiMH type of AAA size and 800 mAh capacity) when the PV voltage is above 0.3 V. The voltage converter applies maximum power point tracking by measuring the IV curve of the incorporated c-Si or mc-Si solar cell every 10 s and adopting switching frequency based on pulse width modulation. The voltage converter is incorporated onto the PCB of the additionally incorporated charge controller, which protects the batteries from both deep discharge and overcharge. Should defined voltage thresholds be exceeded, either the solar cell or the mouse PCB is disconnected from the battery. Battery SOC indication already calls for improvement without user tests: A red LED blinks when battery voltage is below a certain

Figure E.4.1 Rendered pictures of the final SPM

voltage threshold. Nothing, however, indicates an (almost) fully charged battery. Note that we incorporated a smaller battery capacity than we consider optimal for a commercial product, to ease prototype assembly. In addition to the SPM design features discussed in Section E.3, it is possible to recharge the battery by connecting the SPM to a USB port. This allows test users to continue working with the SPM should PV-generated charge be insufficient.

E.4.2 SPM User Tests

Eighteen users tested the SPM at their office desktops in Delft, Twente, and Utrecht, the Netherlands, during October and December 2007. Six participants tested the SPM for over 7 weeks, and the other 12 for only a month. User interviews allowed in-depth understanding of individual user concerns; however, the limited number of test users allowed only qualitative conclusions to be drawn.

The general satisfaction of users with the SPM prototypes greatly diverged. At the start of the test period, most users indicated the mouse to be very big or even bulky. Nevertheless, as the test phase continued, users adjusted to product dimensions so that, once the test period came to an end, some users actually indicated a preference for a larger mouse with dimensions equal to those of the SPM. The influence of different product geometries, however, was apparently underestimated during the design process, as this issue (negatively) concerned almost half of the users at the beginning.

All users, irrespective of their willingness to sunbathe the SPM, perceived the quality of battery SOC indication as insufficient. Suggestions on possible improvements greatly diverged. The majority would be satisfied with a simple battery status indicator, for example a bar graph such as in simple mobile phones. Some desired more sophisticated solutions, especially those users who were sunbathing the SPM on a daily basis. Independently, three of these test users suggested that the SPM energy balance should be accessible via special computer software to adopt optimal sunbathing strategies.

Although sunbathing was accepted by users remarkably well, it was interpreted in a way that did not include direct sunlight exposure. Users working at indoor locations with no access to direct sunlight did not move their mice into direct sunshine, that is, to a windowsill at another room when the current weather condition was sunny. However, users agreed that placing the mouse on the windowsill at their specific workplace could be easily synchronized with daily or weekly work routines. Consequently, roughly two-thirds of all users placed their SPM on the windowsill almost every weekend, roughly one-third did so on a daily basis, and only one user refused in principle to do so. Future research should investigate more quantitatively how many users are willing to sunbathe their PV device. Particularly interesting would be an examination of how far charge-harvesting potentials related to sunbathing are affected when direct sunshine is defined as obligatory. This is important, as direct sunshine implies much higher energy density and thus shorter sunbathing durations. Some of the SPM prototypes were tested at indoor locations, in which solely diffuse solar radiation reached the windowsill. Hence, sunbathing requirements differ widely from those assumed in the preliminary assessments in this study, which suggested sunshine exposure for sunbathing activity.

Sunbathing in the preliminary assessments was assumed during lunchtime, whereas sunbathing of SPM test users occurred before 9:00 a.m., as the mice were most often placed at the windowsill at around 5:00 p.m., when people left work. Note that this would be particularly beneficial for SPM located at east- or west-facing windowsills, respectively, due to direct sunshine exposure for this case. However, time-course data for various use locations and a variety of users would be needed for representative results on whether or not sunbathing will fully provide the SPM charge demand by the incorporated solar cells. This is especially true when considering the great range of irradiance levels that prevail at different indoor locations, when distributing the SPM to the different test users. For the desktop workplaces where the SPMs were used, the average irradiance intensity of the diffuse fraction of daylight was $\sim$2.5 W/m^2 without and $\sim$4 W/m^2 with artificial lighting switched on as the average for all the test sites, with notable deviations, however, of up to -300% to $+700\%$ regarding daylight availability. The irradiance intensity of the diffuse solar fraction at windowsills was 13–20 W/m^2 on average, depending on SPM orientation (horizontal or mounted at 45°), with -1000% to $+300\%$ deviation. Illumination levels of artificial lighting were found between 450 and 700 Lux (it can be noted that all sites had fluorescence tubes as artificial-lighting type).

E.5 Conclusion

This case study presents the design process for the creation of a wireless solar-powered mouse (SPM). Fifteen SPM prototypes have been assembled, based on the concept evaluation presented in this case study. Preliminary results from prototype testing indicate the product concept to be feasible, from both technical and user perspectives. To guarantee autonomous, solar-powered operation of the SPM for medium or heavy device use also, sunbathing of the SPM is required. The conducted user tests suggest that users are willing to "sunbathe" their SPM (at least on a weekly basis, i.e., during the weekend). Future research should also lead to the development of simulation tools that model irradiance reaching a PV unit within complex product geometries. This would facilitate the design of product encasings with better light-harvesting properties, with the goal to eventually reduce sunbathing requirements during daily PV-powered product use.

E.6 Acknowledgments

We would like to thank Bas Wardenaar and Gert-Jan Langendijk from ECN for laser-scribing c-/mc-Si cells, Angele Reinders and Hugo de Wit from Twente University for rendering illumination distributions using the 3D-PV tool, Bamshad Houshiani for help during prototype assembly, and finally all members of the SYN-Energy project team – particularly Boelie Elzen for his efforts regarding user-related aspects and preparing qualitative user interviews; Menno Veefkind for his design help; Matthijs Netten, without whom prototype assembly and the qualitative user interviews would not have been possible; and Sioe Yao Kan, Jaap Jelsma, Sacha Sylvester, Wim Sinke, Wim Turkenburg, and last but not least Erik Alsema for contributions made during the various stages of the project. This work was financially supported by the NWO/SenterNovem Energy Research program.

References

Alsema, E.A., Elzen, B., Reich, N.H. *et al.* (2005) Towards an optimized design method for PV-powered consumer and professional appications – the SYN-Energy project. Proceedings of the 20th European Photovoltaic Solar Energy Conference, Barcelona, WIP, pp. 1981–1984.

Kan, S.Y., Van Beers, S., and Brezet, J. (2004) Maturely designed and sustainable PV powered products. Eurosun 2004 proceedings, Freiburg.

Percept Technology Labs (2004) Mouse battery life comparison test report http://download.microsoft.com/download/1/3/9/139a8c30-34cc-4453-a449-7a1c586a3ae5/MicrosoftMouseBatteryExecSummary.pdf (accessed April 12, 2012).

Reich, N.H., van Sark, W.G.J.H.M., Alsema, E.A. *et al.* (2005) Weak light performance and spectral response of different solar cell types. Proceedings of the 20th European Photovoltaic Solar Energy Conference, Barcelona, WIP, pp. 2120–2123.

Reich, N.H., Veefkind, M., van Sark, W.G.J.H.M. *et al.* (2009) A solar powered wireless computer mouse: Industrial design concepts. *Sol. Energy*, **83**, 202–210.

Reich, N.H., van Sark, W.G.J.H.M., Alsema, E.A. *et al.* (2010a) Crystalline silicon solar cell performance at low light intensities. *Sol. Energ. Mat. Sol. C.*, **93**, 1471–1481.

Reich, N.H., van Sark, W.G.J.H.M., Turkenburg, W.C., and Sinke, W.C. (2010b) Using CAD software to simulate PV energy yield: The case of product integrated photovoltaic operated under indoor solar irradiation. *Sol. Energy*, **84**, 1526–1537.

Reich, N.H., van Sark, W.G.J.H.M., and Turkenburg, W.C. (2011) Charge yield potential of indoor-operated solar cells incorporated into Product Integrated Photovoltaics (PIPV). *Renew. Energ.*, **36**, 642–647.

Reinders, A.H.M.E. (2007) A design method to assess the accessibility of light on PV cells in an arbitrary geometry by means of ambient occlusion. Proceedings of 22nd European Photovoltaic Solar Energy Conference, Milan, WIP, pp. 2737–2739.

RoHS (2002) Regulation of Hazardous Substances, European Directive 2002/95/EC.

Veefkind, M.J. (2004) Industrial Design and Solar Power, challenges and barriers. Proceedings of ISES World Congress, Gothenburg: ISES.

Case F

Light Urban Mobility

Satish Kumar Beella, Sacha Silvester and Han Brezet

Delft University of Technology, Faculty of Industrial Design Engineering, Landbergstraat 15, Delft, The Netherlands

F.1 Introduction

From the chariots to the bicycles and from the steam engines to the internal combustion engines (ICEs), the size of the mobility option is dependent on its energy source. The advantage of using an ICE car in comparison to walking a kilometer in simple terms is comfort, safety, and social status. The same seen in terms of spent energy is from 150 wh to 30 kwh, which is next-to-tailpipe emissions, the use of road infrastructure, and an accessible parking place. The size of a vehicle is also an important feature in the context of urban areas as it occupies statically and dynamically urban space for parking as well as for intended mobility trips over a period of time. The challenge for urban space management multiplies with the existence of various mobility modes and different vehicle speeds.

In general, the ability to stop a vehicle in time and distance more or less defined the vehicle speeds in urban areas and ultimately decided the power specifications of different modes. Economic growth and global developments have resulted in an increase in the amount of traffic on roads, and the dedicated infrastructures based on pro-car policies have increased the speeds of vehicles. As the urban population increased with time, the demand for mobility has been multiplied and resulted in increased congestion, accidents, and emissions. Preventive measures took place in terms of regulation to control urban vehicle speeds and emissions. More focus on pedestrian safety, congested areas, and the need for door-to-door travel increased the re-emergence of car-free zones and lower speeds. The urban areas challenge future mobility to decrease the local emissions, sound pollution, and accidents. The state-of-the-art urban areas are characterized by congested roads, less maneuverable spaces, and regulated emission-free zones.

The Power of Design: Product Innovation in Sustainable Energy Technologies, First Edition.
Edited by Angèle Reinders, Jan Carel Diehl and Han Brezet.
© 2013 John Wiley & Sons, Ltd. Published 2013 by John Wiley & Sons, Ltd.

The emission regulations are based on the reduction of local urban emissions and are to reduce dependence on oil resources. The implementation of alternative energy sources, which would help in the reduction of vehicle emissions, does need a change in vehicle configurations and typologies. Sources such as electricity and hydrogen do need rechargeable batteries or capacitors, and offer an advantage to limit emissions to the power or production plant. Energy sources such as petroleum gases (liquefied and compressed) and biofuels do need modifications to the existing ICE configuration and onboard fuel storage. The availability of different fuels at present is leading toward a multifuel economy (Lee, 2011). Which fuel has the proper capacity and endurance to fulfill a population's mobility needs and emission requirements is still a question to many policy makers and automobile manufacturers. Within this paradigm of increasing pressure to reduce local emissions, a demand for the implementation of alternative fuels is expected to play an important role in the coming decades.

F.2 Background

One of the solution directions in this case study is to solve current mobility problems by making use of technology developments. Emerging technologies could be applied to create an artifact or system that uses clean and quiet nonrenewable energy sources. The term *clean* here is contextual, and in most cases starts with being local emission free and with other emissions being controllable at the source such as at the production plant.

The Rechargeable battery technology is the most commonly used technology as of today, next to the ICE; it is applied in the vehicle industry to achieve the objectives discussed here. Currently, battery-powered electric vehicles (BEVs) are introduced in different fields of transportation, especially in places where air and noise pollution are to be avoided. Examples include forklift trucks, automatic guided vehicles in plants, and electric trains in mines. Recreational vehicles on golf courses and other sport arenas are well-known applications of BEVs as well. The battery technology is a proven technology with one limitation: long charging times. As battery performance improves, the performance of BEVs improves (Rand et al., 2005). The upcoming vehicle propulsion technology that addresses the problems of nonrenewable energy source consumption, emissions, and noise is the fuel cell (FC) technology. The FC technology could play an advanced role in lightweight constructions and energy storage technologies (Hwang *et al.*, 2005). Until now, in the transport section the main focus and research have been put on proton exchange membrane (PEM) FCs that uses hydrogen as fuel. With regard to the use of nonrenewable energy sources and emissions, hydrogen is a good option in comparison to other FC fuels (Johnston *et al.*, 2005; Van den Hoed, 2004).

The stimulus of modal shifts is the second solution direction that aims at improving the current mobility situation that is employed in this case study. Here the shifts to less polluting forms of mobility described by Geerlings and Peters (2002) are broadened: It is suggested to aim for shifts to mobility forms that as little as possible contribute to all of the mobility-related problems. This implies stimulating shifts away from car mobility, or at least away from the current patterns of using cars that often are single occupied during peak times and for shorter trips. The purpose here is to travel door to door in order to fulfill the mobility need. The need could be fulfilled with a customized vehicle or chain mobility, which is a combination of different modes of transport.

In addition to organized infrastructure and public transport, the solutions for the future lie in three different aspects:

• Niche and customized vehicles for urban situations and to extend or link the use of public transport.
• The existence and availability of multiple fuels for private use.
• Consumers' conscious selection of a particular mode of transport for a particular trip.

The organization of new infrastructures is also a challenge when the maturity of a particular fuel option is in question. Each type of fuel has its limitation or critical aspect to be accepted by consumers and also in order to fit within the city logistics. For BEVs, the vehicle charging and driving range are two crucial aspects for success. The establishment of comprehensive infrastructures for vehicle charging is prerequisite to the realization of electric mobility (Boulanger *et al.*, 2011; Hatton *et al.*, 2009; Silvester *et al.*, 2009). Early electric vehicle (EV) infrastructure projects should be tested in a controlled manner before they are deployed on a large scale.

F.3 Mobility and Design

The design goal is to explore new and alternative means of mobility for car travel. The noticeable aspect here is that most people do prefer to use a car in most situations, even though a car comes with many problems. The need for new mobility designs to move away from cars has a strong sustainability motivation. The combination of design and technology needs to pave the way for consumers to show the development and demonstration of alternative means of transportation, the application of new energy sources, demand management, niche mobility, and services. The objective is to fine-tune the balance between the individual transport market and public transport opportunities.

One of the main themes over recent years has been identifying options for the optimization of products and services; this includes chain mobility research, multimodality research, and consumer research. From the previous studies, several conclusions for the roadmap of urban mobility can be drawn (Silvester *et al.*, 2010; Van Timmeren *et al.*, 2010). The most important ones are:

• Mobility product development and demonstration provide more opportunities for integral testing and for improving the market worthiness of artifacts.
• Advanced technologies, alternative energy sources, and materials are applied to functionalities like lightweight commuting and intermodal chain mobility options.
• Cooperation of knowledge institutions and companies in the initial stages of product development gives the practitioners help and reflection from policy, design, market feasibility, and engineering.
• Sustainable perspective and system thinking are combined for more comprehensive solutions.
• More focus is on soft mobility designs, chain mobility options, product service systems, and sustainable urban infrastructure (Beella *et al.*, 2009).
• Alternative mobility is integrated into the urban environment (Silvester *et al.*, 2010).

The design and mobility efforts at Delft University of Technology include European cooperation and the collaboration of three technical universities (3TU) at doctoral- and graduation-level projects. The mobility theme activities that are undertaken by researchers and students, in cooperation with the external partners of the faculty of Industrial Design Engineering (IDE), are the basis of these efforts. The co-existence of design and research cycles will further help to strengthen the program and realize the roadmap.

F.4   Role and Importance of Energy

Energy is the backbone of any mobility artifact. A cyclist spends around 0.2 kWh for normal cycling, and a car driver spends 30 kWh for the same displacement. The gross weight and maximum speed of a vehicle are directly proportional to engine specifications. The important elements related to energy when urban mobility is considered are as follows:

- The usage of energy and other material resources by a mobility mode in its lifetime is the most important and deciding factor for assessing the impact and its mere existence. For example, the environmental impact of the lithium battery used in BEVs is small. The operation phase of the car, mostly recharging the batteries, is the dominant contributor to the environmental burden caused by the transport service. The contribution of the lithium within the battery has a share of about 2% of the environmental impact of the total transport service.
- The energy source or carrier will basically dictate the typology and functionality of a mobility mode. The safety and precautionary steps needed to be taken for the energy carrier for any unnecessary collisions or impact will decide the necessary active and passive safety features. The time taken to refill, swap, or recharge the carrier will decide the form and functionality of the mobility artifact. Size and weight matter for any mobile product, so any aspect of the energy source that will improve the power-to-weight ratio is a winner.
- The shape and safety requirements of energy sources decide the context in which the vehicle will operate. For example, the photovoltaic cells applied on a vehicle for energy production need to have more open flat vehicle surface, and hydrogen storage has specific safety requirements.
- Tank-to-wheel efficiency and emissions are also important, especially in the case of urban mobility, for reducing the local emissions. For example, BEVs are 30–40% greener than ICE cars when life cycle efficiency is considered. The same are almost 90% greener when only tank-to-wheel efficiencies are considered.

The maturity of the application and capacity in supply of an energy source plays an important role when pursuing a political agenda and consumer acceptance. The availability of a particular energy source in abundance for a lifetime of a vehicle system is also very important.

F.5   Light Urban Mobility

A bicycle or pedelec (*ped*al *elec*trical bicycle) is a good example of light urban mobility. The word *light* illustrates the relatively small footprint of a vehicle while it is moving and its relative low energy consumption. Three products are illustrated here to show how urban mobility can be further developed. The focus of these products is to be part of niche or customized situations and to extend or link the use of public transport.

F.5.1   Urban Mobility Concept

Commuter mobility, like all mobility, is on the rise. However, the greatest increase is among car users, and this contributes to a number of problems such as reliance on nonrenewable fuels, emissions, congestion, and safety. Stimulating modal shifts, encouraging people to use less harmful modes of transport, and making use of state-of the-art technologies are strategies employed in this case to contribute to improved mobility patterns.

The main disadvantage of public or shared transport, from the user's point of view, is the lack of flexibility due to the fixed start and end points of the journey. The urban mobility concept (UMC), the advanced portable transport solution designed within this project, intends to provide an efficient connection between the user and public or shared transport, thereby making it a more attractive mode of transportation for commuters. The UMC could also be used in combination with a car, by parking the car outside the city and using the UMC to access the city center (see Figure F.5.1).

The design research framework consisted of extensive user research and analysis of other important aspects of the design context such as market, usage environment, cultural influences, and legislation. The UMC's final design is as a light personal BEV that can be folded within seconds thanks to a new folding principle. The electric drive is integrated in the rear wheel, and the battery pack including controllers is placed in the main body frame. In the folded state, the wheels are positioned next to each other and the UMC can be rolled in a balanced manner, making it easy to handle in a busy environment. When riding the UMC, a choice can be made between sitting and standing postures. The sitting posture is intended for

Figure F.5.1   The UMC prototype and illustrations of its use (See Plate 28 for the colour figure)

somewhat longer distances, and the standing posture could be used for shorter distances, making the folding procedure even quicker. As for appearance, the UMC has a strong visual identity, making it stand out among other vehicles on the market with similar functionality.

A mock-up was built of the design to test if the folding, rolling, and riding functions could be carried out satisfactorily. It was concluded that the folding actions are natural and smooth, the rolling works adequately, and the riding (which could best be evaluated for standing posture) is fun and comfortable due to the wheel size and the combination of steering angle and fork rake.

F.5.2   MeeneemFiets

A second case we want to discuss is the MeeneemFiets. The common scene among commuters in recent years is more and more use of folding bicycles. This shows how efficient the combination of a train and folding bicycle is. The folding bicycle is also a welcome change for motorists who need to park their vehicles outside the city centers and be flexible with a bicycle.

Current folding bicycles are very convenient and well-developed products in their unfolded state. The problem starts only when they are folded. As a packet, a folding bike is too large to carry, and it is difficult to move or maneuver in congested places. This is likewise the reason why people with folding bicycles limit themselves to the entrances of train doors or corridors while traveling and are not allowed into buses. The present design is developed or suited easily for longer distances; in other words, the folding bicycle is designed for its unfolded state. A passenger using a folding bicycle normally needs to cover not more than 4 km. Consequently, design input is required more for the folded state than the unfolded state. The aim in this case is to develop an alternative for the folding bicycle with a view on the distance to be covered in relation to comfort while carrying in its folded state (Tenwolde, 2007).

The final design, called MeeneemFiets, is an elegant combination between a step and a folding bicycle (see Figure F.5.2). The MeeneemFiets is more functional to ride than a step and more comfortable to carry than a folding bicycle. Stepping on the pedal operates the MeeneemFiets, and the foot can be exchanged from time to time. The MeeneemFiets in its folded state becomes a thin packet that is easy to carry into the train or other mode of transport. The detailing of the MeeneemFiets is derived from the compact folded state and results into the desired form while unfolded. The only question that remains is how fast the consumer will get used to this unusual way of cycling and accept it as a connection to the public transport.

F.5.3   Bull

The third case, carried out at Silvestris BV, focused on the "design of a structure and package of an electric vehicle for the urban environment" (Van de Kieft, 2009). Sustainability and environmental control are currently recognized as top issues that could fit the company's targets and philosophy for an exclusive sustainable urban vehicle.

In this project a combination of methods was used. First of all, a derivative of the vision in product design method (VIP) was used in the analysis phase. With this method, a character

Figure F.5.2 The MeeneemFiets in folded and unfolded stands

description was given that, together with a user environment description, formed the base for choosing the main components, structure, theme, and package of the vehicle to be designed (Hekkert et al., 2011). The findings of the analysis, in combination with progressing insight, led to the following starting points: The vehicle will have three wheels, it will be meant for one occupant in standing position, and it will be positioned in the 25 km/h class in which the use of a helmet is not required.

In order to combine dynamic driving with good stability, a tilting mechanism was needed. In the synthesis phase, two alternatives were tested through a *design by experience method*: Four proof-of-concept models were made combining the alternatives with different setups. This led to a configuration with one larger front wheel, an in-wheel electric motor, and two smaller rear wheels, which are linked through a parallelogram. This provides a free-leaning mechanism that enables the occupant to lean while turning, in order to improve stability. The mechanism has the same dynamic characteristics as a normal bike ride.

With the outcome of the synthesis phase, a design was developed and put into practice with the final prototype. The steering was placed in the opposite direction of bike steering. This functional feature evokes the appearance of a bull and served as the inspiration for the design of other components. For this reason, the design was named *BULL* (see Figure F.5.3).

The prototype is fully functional with an electronically operated blocking mechanism applied to the parallelogram, in order to keep the vehicle upright when parked or driving at low speeds. The design aims at a target group that is willing to pay more for quality and exclusive design. However, the prototype and this project will function as a base for further development where alternative variations on this vehicle will also be taken into consideration. This development should result in a production-ready design and finally a market introduction.

Figure F.5.3 The BULL prototype and an illustration of its riding posture

F.6 Conclusions

Different design methods have been applied in all three cases in order to achieve the intended results (see Table F.6.1). In all cases, the design demonstration and evaluation play important roles in establishing the new functionality or new way of using a mobility product. Focus on the service design is part of the approach next to the existing design methods. The experience prototyping and design demonstration does play a crucial role, in translating consumer demands into marketable products to create new experiences and interactions.

The existence of multiple fuels sources for mobility is a must in the coming decades. It is necessary to be ready and prepared from design and technology for the co-existence of different energy sources depending on the power and size requirements of different functionalities. Niche and customized vehicles play an important role in fulfilling these various power needs. The gap between public transport and private transport will also be narrowed in the coming years. Para-transit and shared mobility, with strong sustainable motivation, might help to realize the implementation of alternative energy sources into mobility. Consumer's conscious selections will still hold the cue for the particular selection of a vehicle for a particular trip.

The developments in the field of energy storage and energy transfer hold the key for the future. These aspects decide the size, speed, infrastructure, and life span of mobility products. The selection and development of an energy source application need long-term vision as they involve multifaceted development and need long-term commitment from different stakeholders.

Table F.6.1 Design methods involved in the product development phase

	Design method	Features	Output
UMC	• Benchmarking	• Foldable and portable to carry in personal or public transportation	• Design demonstration
	• Morphological analysis	• Electric motor with rechargeable battery	• Part of service design
	• Conjoint analysis	• Chain mobility option	• Evaluation-ready product
	• Focus group		
	• Context mapping		
MeeneemFiets	• Trend analysis	• Foldable and portable to carry in personal or public transportation	• Design demonstration
	• Morphological analysis		• First evaluation prototype
	• Design for folding		
Bull	• Vision in product design	• Electric motor with rechargeable battery	• Market-ready product
	• Design by experience	• Vehicle tilting mechanism for easy steering	• Positive user evaluation
	• Experience prototyping		• In production

References

Beella, S. K., S. Silvester and J. C. Brezet (2009) Electric vehicles are here to stay, in *Joint actions on Climate Change* (eds. A. Remmen and H. Riisgaard), University of Aalborg, Aalborg.

Boulanger, A. G., A. C. Chu, S. Maxx and D. L. Waltz (2011) Vehicle electrification: Status and issues. *P. IEEE*, **99**(6), 1116–1138.

Geerlings, H. and Peters, G. (2002) *Mobiliteit Als Uitdaging: Een Integrale Benadering*, 010 Publishers, Rotterdam.

Hatton, C., S. K. Beella, J. C. Brezet and Y. Wijnia (2009) Charging stations for urban settings: the design of a product platform for electric vehicle infrastructure in Dutch cities. *World Electric Vehicle J.*, **3**, http://repository.tudelft.nl/view/ir/uuid%3A6e209248-7741-4743-bf7c-b59d1e1323d8/ (accessed April 12, 2012).

Hekkert, P. and van Dijk, M. (2011) *Vision in Product Design: Handbook for Innovators*, BIS Publishers, Amsterdam.

Hwang, J. J., D. Y. Wang and N. C. Shih (2005) Development of a lightweight fuel cell vehicle. *J. Power Sources*, **141**(1), 108–115.

Johnston, B., M. C. Mayo and A. Khare (2005) Hydrogen: The energy source for the 21st century. *Technovation*, **25**(6), 569–585.

Lee, S. (2011) Automobile use, fuel economy and CO2 emissions in industrialized countries: Encouraging trends through 2008? *Transport Policy*, **18**(2), 358–372.

Rand, D.A.J. and Dell, R.M. (2005) The hydrogen economy: A threat or an opportunity for lead acid batteries? *J. Power Sources*, **144**(2), 568–578.

Silvester, S., S. K. Beella and J. N. Quist (2009) Advent of electric vehicle: A research agenda, in *Joint Actions on Climate Change* (eds. A. Remmen and H. Riisgaard), University of Aalborg, Aalborg.

Silvester, S., S. K. Beella, T. A. van, P. Bauer, J. N. Quist and S. J. v. Dijk (2010) Integration of electric mobility into the built environment. Schiphol the Grounds, Delft.

Tenwolde, D. (2007) De Meeneemfiets. Master's thesis, Delft University of Technology, Delft.

Van de Kieft, J. (2009) Design of an urban electric vehicle for the high-end market. Master's thesis, Delft University of Technology, Delft.

Van den Hoed, R. (2004) Driving fuel cell vehicles: how established industries react to radical technologies. Delft University of Technology, Delft.

Van Timmeren, A., P. Bauer, S. Silvester, S. K. Beella, J. N. Quist and S. van Dijk (2010) Use of design oriented scesnarios and related tools in research by design. Design Rescarch in the Netherlands 2010, TU Eindhoven, Eindhoven.

Case G

From Participatory Design to Market Introduction of a Solar Light for the BoP Market

Jan Carel Diehl and Jeroen Verschelling
Delft University of Technology, Faculty of Industrial Design Engineering,
Landbergstraat 15, Delft, The Netherlands

G.1 Introduction

In Cambodia, 80% of the population does not have access to electricity from the public grid and rely on costly and low-grade lighting sources such as candles, kerosene lanterns, and car batteries to provide light at night. Solar energy is abundant in Cambodia (over 1900 kWh/m2 per year) (Reinders *et al.*, 2007). Solar-powered lighting has been demonstrated to be a good alternative for people in rural areas with low incomes; it has reliable high-quality light and has a lower environmental impact than traditional lighting (Ramani and Heijndermans, 2003).

In this case, we illustrate how a multidisciplinary team brought together user context research, business development, sustainability, and new technologies and innovations (see Figure G.1.1) to develop the "Moonlight."

The Moonlight was developed at Kamworks, a social enterprise in Cambodia whose mission is to provide affordable sustainable energy systems for low-income consumers in Cambodia and to locally manufacture solar products in order to develop a sustainable market for solar energy for rural electrification. In Kamworks' vision, local production is necessary so that expertise and spare parts are easily available and so that repair and service are guaranteed. Besides that, local production will also create jobs and income for young Cambodians. Cambodia has a large proportion of its population under the age

The Power of Design: Product Innovation in Sustainable Energy Technologies, First Edition.
Edited by Angèle Reinders, Jan Carel Diehl and Han Brezet.
© 2013 John Wiley & Sons, Ltd. Published 2013 by John Wiley & Sons, Ltd.

Figure G.1.1   An integral product development approach for the BoP (Diehl *et al.*, 2008)

of 20 years (60%), and employment generation for young people is an especially pressing issue.

G.2   Methods

G.2.1   Project Setup

From experiences with earlier base-of-the-pyramid (BoP) projects at Delft University of Technology (Kandachar *et al.*, 2009), it was concluded that creating products for BoP markets requires a deep understanding of the real daily needs and context of the people within it. In such a situation it is clear that designers, in this case a team of four students, should engage with the cultures directly in order to better understand local people (Rodriguez *et al.*, 2006). Because of this reason, a special emphasis of the project was put on local context research in the field. Knowing the context and observing and interacting with users in their context help designers to understand the users' latent needs and come up with new appropriate product solutions that meet their real lighting needs. As a result the most essential part of the project took place in rural areas of Cambodia during a period of 3 months, during which time designers explored and understood the local context and needs as well as developed local suitable design solutions for the lighting needs of rural Cambodians.

G.2.2   Participatory Market and Context Research

As a first step of the new product development trajectory, a thorough participatory market and context research was executed in the field. The market and context research existed of two parts: (1) observatory research and (2) participatory research.

As a first start for the designers to get familiarized with the local socioeconomic context and make some first detailed insights into the problems and needs for electrical light, observatory research was carried out (Alvarez *et al.*, 2008). The circumstances and use of current

Figure G.2.1   Observatory research in rural Cambodia

Figure G.2.2   Participatory research in rural Cambodia

lighting products were investigated in households in Kandal Province, Cambodia, to gain insight into the current use of lighting and electronic products. The design team executed observations by visiting local households and conducting interviews (see Figure G.2.1). Questions were asked by the students with the support of local translators in relation to the way and purpose of people's use of light.

During the second phase, participatory methods were used to gain a deeper insight into daily life in order to identify suitable product–market combinations as well as insight into people's needs. Focus groups sessions and day mapping (see Figure G.2.2) were used to identify and map the needs and wishes of rural Cambodian consumers for (electrical) light in their daily life.

G.2.3   Participatory Field Research: User Needs

One of the first outcomes of the participatory field research was a map of the current needed and wanted (electrical) light functions in and around the household, such as studying, managing the shop, eating, cooking, watching the animals, washing the dishes, visiting friends, and so on. Figure G.2.3 provides a more detailed overview of needed and wanted light functions. It is clear that light is needed for multiple functions.

The participatory field research led to the following main conclusion: The new light device should be completely substitute the kerosene lamp. The poor quality of kerosene light, its flammability and health hazards, and its highly volatile fuel prices are the main drivers. In addition, the new lamp should be portable: Different rooms need to be lit, but most households cannot afford more than one lamp. Furthermore, a dimmed light during the night was needed, to orient in the dark and feel safe while saving energy at the same time. The dimmed light has to last for only a few hours per night, and about 3 hours of full light are needed during the evening. In addition, the inventive character of the Cambodians and the completely

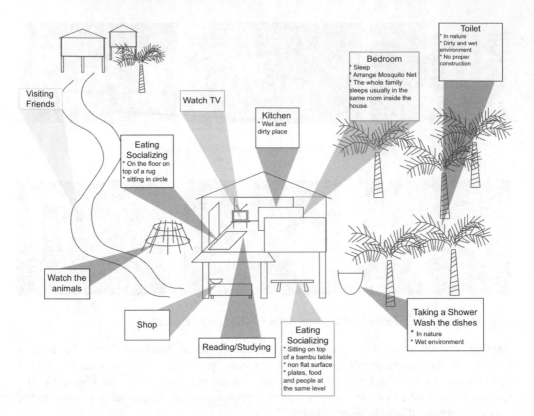

Figure G.2.3   The current as well as wanted use for light in rural Cambodia

improvised style of their houses called for a flexible product that people could use as they wish, without too many restrictions. The product should provide enough luminosity for reading. Last but not least, the lamp should be shock, water, and dirt resistant. Incorporating all these requirements would create a specific added value for rural BoP Cambodian households.

G.2.4   Technological Challenges

Parallel to the participatory field context and user needs research, the team executed technical "lab" research focusing on efficiency, reasonable purchase costs, and low total costs of ownership. For the energy supply of the system, two options were possible:

- A battery-charging system powered by photovoltaic (PV) cells with low initial costs (for the user) but higher running costs. The batteries are taken out of the products and charged by a local entrepreneur by PV cells.
- A totally independent system with higher initial costs but no running costs for the end user. The product is directly charged by the PV cells (integrated in the product or connected through a cable with an external PV panel).

The decision was taken to continue with the last option and to design an independent PV white LED–light system. To keep the product simple and the cost as low as possible, it was

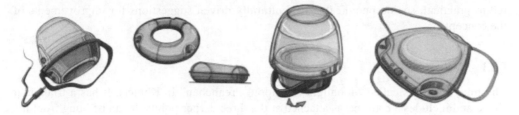

Figure G.2.4 Four concepts for new light devices for rural Cambodia

decided not to integrate the PV cell into the lamp, but to develop a separate lamp and PV panel. Simply said, a PV–LED light product consists in that case of one or more LEDs, a rechargeable battery, and a small PV module. However, from former studies it was concluded that electronics plays a crucial role in providing a series of necessary functions. The electronics have to (1) match the solar power to the battery characteristics, (2) control the charging and discharging of the battery, and (3) control the power to the LED (*driver*).

As a result of the electronics, the light device will be more energy efficient, perform better, have a longer lifetime, and be more convenient to handle. This way the Kamworks light products can provide more constant quality and reliability, higher performance, and added value to the end user compared to the cheaper PV lamps. Quality and performance are two of the most important selling points in the rural Cambodian BoP (Rijke, 2008).

G.2.5   Co-Development

Based upon the inputs of the extended participatory local-context research, the competences of Kamworks, and technological development, four concepts were developed (see Figure G.2.4). This is in line with "normal" product development in any market.

In order to develop appropriate design solutions that really fit into the BoP context of Cambodia, local stakeholders were intensively involved in the concept development, the so-called co-design. *Co-design* can be defined as a cooperative, contentious process bringing everyday people together with design professionals to find new and better ideas for daily life (Simanis and Hart, 2008).

Potential end users as well as potential sales channels like micro entrepreneurs were confronted with product ideas in an early stage (see Figure G.2.5). Direct feedback from the field

Figure G.2.5   Co-development with local stakeholders (See Plate 29 for the colour figure)

led to practical, socioeconomically and culturally driven suggestions for improvements of the concepts.

G.3 Results

The final design is called Moonlight ("Ampoul Preahchan" in Khmer). It has a triangular shape and includes a cord that is attached at the three corner points. It can be hung from the wall or ceiling, carried by hand, or hung around the neck. It has six wide-angle LEDs, which is equivalent to the light output of about 2–3 kerosene lamps. It comes with a 0.7 Wp solar panel, which can be fixed to a bamboo pole with a standard clamp. This option was chosen as several people had stated during the interviews that they were so afraid of the solar panel getting stolen they would prefer to keep the panel inside all day, leaving a window open for charging. Currently, this "antitheft" technique is used for TV antennas, so this technique is not new to the people.

The strap represents the most crucial handling feature, allowing one to wear it comfortably around the neck and easily connect the product to building constructions. The upper shell is transparent with a reflector to enhance the bright LED light, hiding at the same time inner components such as the batteries and electronics. The light has three settings: bright (for reading or work), medium (for eating or socializing), and low (for orientation or safety). When used at full power, the product delivers about 30 lm by six efficient LEDs during 4 hours. In medium dimmed mode, it produces diffuse – that is, amenity – light for 20 hours; and in fully dimmed mode it lasts 40 hours. The expected lifetime is 5 years, with battery replacement every 2 years.

G.4 Feedback from the Field

Several prototypes of the Moonlight were made at the local Kamworks workshop and tested with families in rural Cambodia at nighttime and daytime. The final user tests pointed out that the product is indeed an appropriate solution for the local context. The product was appealing to them, and the usage intuitive: People could easily understand and use the product, hanging it around their neck or placing it at the walls or ceiling of their houses (see Figure G.4.1). Most of the families of the final user test were enthusiastic and even willing to buy the prototypes on the spot.

Figure G.4.1   Field tests in rural Cambodia with working prototypes

Figure G.4.2   Moonlight production at Kamworks, Cambodia

Based upon the positive feedback from potential end users, it was decided to do a more extensive pre-production test. Ten "Moonlights" were produced that included a "data logger" to record the charging as well as using behavior. These Moonlights were used and observed during one month in 10 rural households at two different locations. In addition, tests were done with the batteries, PV panels, LEDs, and printing circuit boards in the local context to test their efficiency and duration. The outcomes of these field tests led to several inputs for improvement of the Moonlight. Based upon the positive consumer and technical feedback from the field, it was decided to proceed with production (see Figure G.4.2).

G.5 Market and Business Considerations

G.5.1 Costs

The market price of the Moonlight, including the accompanying PV panel, is around US$25. Compared to a kerosene lamp of less than US$1, it is still a relatively high initial investment for a low-income family. What counts, in the end, is the life cycle cost, also called the *cost of ownership*: After purchase, operating costs determine the costs of ownership of a certain lighting option. Costs of operation comprise replacements of spare parts and costs of energy to power light, such as electricity for grid-connected lighting and fuel for kerosene lamps. Kerosene lamps have very high costs of ownership of US$12.00 per 1000 lux-hours, mainly due to fuel consumption. In contrast, a 1 W PV-LED lamp has costs of ownership of US$0.22 per 1000 lux-hours (Gooijer *et al.*, 2008). In other words, a PV-LED-powered Moonlight has high initial costs, but in the short term these costs can be overcome and such light will save money for households.

G.5.2 Challenges with Market Implementation

Although feedback on the lanterns themselves is positive, and the return on investment is within one year, sales of the product are very challenging. In-field research revealed several main bottlenecks for the adoption of solar lanterns, such as the following:

Figure G.5.1 Business-in-a-box for rental entrepreneurs

- *Up-front investment:* Rural Cambodians often do not have the financial capacity to save money for large expenditures. Current lighting expenditure (purchasing kerosene) is done in small quantities, a few times a week. An investment of US$25 poses a barrier to purchasing a solar lantern.
- *Education:* Most rural villagers have little or no education. Basic education on the functioning of the product as well as the benefits of solar energy is needed for the rural population to adopt the (for them) new energy technology.

In order to remove the barrier of upfront investment, Kamworks is also piloting a Moonlight rental scheme. In this scheme, a village entrepreneur rents out 30–40 Moonlights to villagers at the same price as their daily average kerosene expenditure (approximately US$0.08 per day). In this way, the villager does not have to take any risk and does not have to make the upfront investment, but gets a much brighter, safer, and cleaner light than kerosene. The (carefully selected) village entrepreneur takes out a solar loan from the Micro Finance Institution and uses part of his or her revenue to pay it off. A rental business-in-a-box has all the components for a rental entrepreneur to start a business (Figure G.5.1).

G.5.3 A "Rent-to-Own" Business Model

However, a rental model, where customers pay a small weekly fee, also has multiple drawbacks: Customers have to pay a usage fee for an infinite period, making the product expensive in the long term; and the entrepreneur runs the risk that people can stop renting the lantern at any moment. Additionally, people are likely to be more careless with the lantern when they are only users and not owners. One solution directions can be found in a "rent-to-own" business model (see Figure G.5.2). In this setup, the customer rents the solar lantern from the entrepreneur for one year, and after that period, he or she will own the lantern (Elkhuizen, 2010).

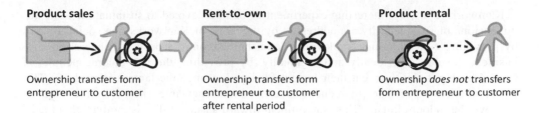

Product sales

Ownership transfers form entrepreneur to customer

Rent-to-own

Ownership transfers form entrepreneur to customer after rental period

Product rental

Ownership *does not* transfers form entrepreneur to customer

Figure G.5.2   Different business models

The advantages of this setup are:

- This eliminates an upfront investment barrier.
- Local availability and services are provided by trusted entrepreneurs who are embedded in the local community.
- This increases the customer's care with the product and loyalty to the program, as well as fair pricing compared to a long-term rental scheme.
- It supports rural livelihoods.
- It can contribute to increased evening business activity, improved study conditions, and better general health and safety for rural villagers.

G.6   Discussion

In participatory design, some methods have proved to be essential for Western designers and enterprises in unfamiliar contexts. The sequence of steps – observation, focus group selection and day mapping during the analysis phase, co-design with potential end users and salespeople during the concept development phase, small-scale field tests with prototypes, improvements, extended field tests with the zero series ($N = 20$), and final adaptation for the first series ($N = 2000$) – is quite common in product and service development all over the world. But the steps have to be carefully elaborated because of the unfamiliarity of the designer or developers with the specific sociocultural context. Involving and educating local people, like Kamworks is doing, are important requirements for establishing this context research.

The strength of Kamworks lies in the offering of sustainable high-quality solutions fitted to local needs. It is not just a product but also the services like the rental scheme, education, communication about PV technology, financing, and after sales of the lighting solutions. These services make them competitive with the low-priced imported products offered in local shops. The services make it possible to build up a long-lasting relationship with their customers.

For rural Cambodians with low household electricity needs, (clean) PV solar energy is the most economic solution. However, local expertise and spare parts are needed to provide after-sales service, and in order to develop the market in a sustainable way. Unfortunately, most solar products currently on the Cambodian market are low quality and lack after-sales support.

Kamworks is a very interesting experiment of people devoted to sustainability. All the dimensions of sustainability are addressed by the work of Kamworks: the social aspects by providing education and employment, and the economic aspects by generating new business development locally and regionally, by providing the low-income people of Cambodia a possibility to cut their budget for lighting once the lantern is paid off. The environmental aspects are to reduce the use of nonrenewable energy sources and to improve the indoor climate. This commitment toward sustainability is motivating already a considerable amount of students from all over the world to contribute to the fulfillment of Kamworks' ambitions.

Because of the upcoming importance of and interest in the needs of people living at the so-called base of the economic pyramid, design professionals and educators should invest more in research and education for "designing for the BoP." As this project illustrates, providing lighting to the people of Cambodia – that they love to use and is affordable – is not simply "designing" a product. This project is a challenging example of a transdisciplinary approach, needed for a successful development and introduction of PV-powered lighting. By using input from different design knowledge domains like sustainability, user context, technology, and business, a locally fine-tuned solution is developed.

References

Alvarez, A., Papantoniou, L. *et al.* (2008) *Affordable Lighting for Rural Cambodia*, Delft University of Technology, Delft.

Diehl, J.C., Silvester, S. *et al.* (2008) Lighting for rural Cambodia: An example of social responsible design. TMCE, Izmir, Delft University of Technology.

Elkhuizen, W. (2010) *Solar Lantern Rental Scheme in Rural Cambodia Industrial Design Engineering*. Master's thesis, Delft, Delft University of Technology.

Gooijer, H.d., Reinders, A. *et al.* (2008) Solar powered LED lighting – human factors of low cost lighting for developing countries. EU PV conference, Valencia.

Kandachar, P., Jongh, I.d. *et al.* (2009) *Designing for Emerging Markets: Design of Product and Services*, Delft University of Technology, Delft.

Ramani, K.V. and Heijndermans, E. (2003) *Energy, Poverty and Gender: A Synthesis*, World Bank, Washington, DC.

Reinders, A., Gooijer, H.d. *et al.* (2007) How participatory product design and micro entrepreneurship favor the dissemination of photovoltaic systems in Cambodia. International Photovoltaic Science and Engineering Conference, Fukuoka, Japan.

Rijke, K. (2008) *Strategy and design of a solar shop in Cambodia. Industrial Design Engineering*. Master's thesis, Delft University of Technlogy, Delft.

Rodriguez, J., Diehl, J.C. *et al.* (2006) Design toolbox for contextualizing users in emerging markets. IEA2006, Maastricht, Elsevier.

Simanis, E. and Hart, S. (2008) *BoP Protocol: Towards Next Generation BoP Strategy*. World Bank, Washington, DC.

Index

The Power of Design: Product Innovation in Sustainable Energy Technologies, First Edition.
Edited by Angèle Reinders, Jan Carel Diehl and Han Brezet.
© 2013 John Wiley & Sons, Ltd. Published 2013 by John Wiley & Sons, Ltd.